Lecture Notes in Physics

Volume 894

Founding Editors

W. Beiglböck
J. Ehlers
K. Hepp
H. Weidenmüller

Editorial Board

B.-G. Englert, Singapore, Singapore
P. Hänggi, Augsburg, Germany
W. Hillebrandt, Garching, Germany
M. Hjorth-Jensen, Oslo, Norway
R.A.L. Jones, Sheffield, UK
M. Lewenstein, Barcelona, Spain
H. von Löhneysen, Karlsruhe, Germany
M.S. Longair, Cambridge, UK
J.-M. Raimond, Paris, France
A. Rubio, Donostia, San Sebastian, Spain
M. Salmhofer, Heidelberg, Germany
S. Theisen, Potsdam, Germany
D. Vollhardt, Augsburg, Germany
J.D. Wells, Geneva, Switzerland

The Lecture Notes in Physics

The series Lecture Notes in Physics (LNP), founded in 1969, reports new developments in physics research and teaching-quickly and informally, but with a high quality and the explicit aim to summarize and communicate current knowledge in an accessible way. Books published in this series are conceived as bridging material between advanced graduate textbooks and the forefront of research and to serve three purposes:

- to be a compact and modern up-to-date source of reference on a well-defined topic
- to serve as an accessible introduction to the field to postgraduate students and nonspecialist researchers from related areas
- to be a source of advanced teaching material for specialized seminars, courses and schools

Both monographs and multi-author volumes will be considered for publication. Edited volumes should, however, consist of a very limited number of contributions only. Proceedings will not be considered for LNP.

Volumes published in LNP are disseminated both in print and in electronic formats, the electronic archive being available at springerlink.com. The series content is indexed, abstracted and referenced by many abstracting and information services, bibliographic networks, subscription agencies, library networks, and consortia.

Proposals should be sent to a member of the Editorial Board, or directly to the managing editor at Springer:

Christian Caron
Springer Heidelberg
Physics Editorial Department I
Tiergartenstrasse 17
69121 Heidelberg/Germany
christian.caron@springer.com

More information about this series at
http://www.springer.com/series/5304

Ilya Feranchuk • Alexey Ivanov • Van-Hoang Le •
Alexander Ulyanenkov

Non-perturbative Description of Quantum Systems

Springer

Ilya Feranchuk
Physics Department
Belarusian State University
Minsk
Belarus

Alexey Ivanov
Physics Department
Belarusian National Technical University
Minsk
Belarus

Van-Hoang Le
Department of Physics
Ho Chi Minh City University of Pedagogy
Ho Chi Minh City
Vietnam

Alexander Ulyanenkov
Atomicus GmbH
Karlsruhe
Germany

ISSN 0075-8450 ISSN 1616-6361 (electronic)
Lecture Notes in Physics
ISBN 978-3-319-13005-7 ISBN 978-3-319-13006-4 (eBook)
DOI 10.1007/978-3-319-13006-4
Springer Cham Heidelberg New York Dordrecht London

Library of Congress Control Number: 2014958312

Printed on acid-free paper

Springer is part of Springer Science+Business Media (www.springer.com)

Dedicated to the memory of
Professor Lev Komarov

Preface

There are many excellent books and monographs dedicated to the fundamentals of quantum mechanics. The content of these books is mainly focused on what has to be done to describe a physical system in a quantum-mechanical way, namely to solve the Schrödinger equation for a wave function or statistical operator. This book is an attempt to discuss how to describe a real quantum mechanical system, and we use a variety of examples from the broad spectrum of physical problems in support of this attempt.

In contrast to equations of condensed matter physics or hydrodynamics, the direct numerical solution of the Schrödinger equation, using a finite-element approximation for differential operators, is practically ineffective due to the large number of variables. Therefore, the most widely used approach for the solution is to approximate the physical system by some models, which have an exact solution. Then construct the perturbation series for the perturbation operator, which is defined by the difference between the Hamiltonian of real system and the Hamiltonian of the model. The applicability of this approach depends on the existence of small parameters for the perturbation operator.

Modern theoretical physics, however, deals with physical systems, the majority of which do not permit us to define a small parameter bearing a physical sense and mathematically fitting to the formal method of perturbation theory. Even after being constructed, the perturbation series often have an asymptotic nature and zero convergence radius. Moreover, there are many practical physical problems where the parameters of Hamiltonian vary within a broad range of values, and the perturbation operator is not sorted out. These systems require a non-perturbative approach for the development of the quantum theory without a physical small parameter.

There are several methods capable of theoretically investigating quantum systems without introducing a small parameter. The direct numerical integration of the Schrödinger equation by using expansion over an artificial parameter, a step of finite-difference approximation derivatives, is a frequently used technique to avoid implementation of a small parameter. The use of lattice models of quantum field theory is another example of non-perturbative methods. The lattice constant is a non-physical parameter which determines the accuracy of the calculation for functional integrals when the integration is replaced by summation. The drawbacks of numerical methods are the exponential growth of calculations and instability of the algorithm with increased degrees of freedom for the investigated physical system.

The physical non-perturbative methods include, for example, the variational principle, the Hartree–Fock method, a method of approximating Hamiltonians, the density functional theory, and others. In the framework of these methods, the estimate for the ground state energy can be obtained; however, the calculation of successive approximations with required accuracy and the calculus of the spectrum of the system provoke essential difficulties.

This monograph introduces and demonstrates the application of the original non-perturbative method for the description of quantum systems, which overcomes most of the above-mentioned obstacles typical for non-perturbative techniques. This method, named the operator method (OM), was first introduced and published in numerous works by authors in the early 1980s. For the construction of zeroth approximation of the operator method and further successive approximations there is no necessity to solve differential and integral equations because all calculations are reduced to algebraic calculus with matrix elements of the operators. A very prominent feature of the operator method is an ability to obtain zeroth approximation for the eigenfunctions and eigenvalues of the Hamiltonian, which are uniformly suitable in the entire range of physical parameters and quantum numbers of the system. The successive approximations of operator method deliver the converging sequence, which enables the calculation of physical characteristics with any required accuracy.

The book presents both the mathematical fundamentals of the operator method and its practical applications for real quantum systems, including ones with an infinite number of the degrees of freedom. A further generalization of the method is given for the quantum statistics, which implements the additional physical parameter, temperature. In comparison with the description of pure quantum states, the calculation of observables in statistics using the density matrix includes an additional complicated procedure: the summation over all states of physical system. For this purpose, the standard procedure of the operator method is supplemented by a non-perturbative algorithm for the summation over quantum states. This combined technique calculates the thermodynamic characteristics of physical systems in a wide diapason of temperatures and remains effective for the systems with multiple degrees of freedom.

The authors dedicate this book to the memory of Professor Lev Ivanovich Komarov, who was one of the pioneers of OM development and contributed essential research on operator methods. The major part of the studies described in this monograph was conducted by the scientists and alumnus of the Department of Theoretical Physics and Astrophysics of the Belarusian State University (Minsk, Belarus). We are thankful to all of our colleagues for their studies published in the recent decades together with the authors of this book: I.K. Dmitrieva, G.I. Plindov, L.I. Gursky, I.V. Nichipor, T.S. Romanova, A.L. Tolstik, S.I. Fisher,

I. Tsvetkov, P.A. Khomyakov, V.V. Triguk, A.V. Leonov, X.H. Ly, Z.A. Chan, N.T. Vu, A.T. Le, and N.T. Hoang-Do. We also appreciate C. Klein for the help with stylistic corrections.

The target audience for this monograph is researchers, post-graduate students, and students studying theoretical and mathematical physics.

Minsk, Belarus — Ilya Feranchuk
Minsk, Belarus — Alexey Ivanov
Ho Chi Minh City, Vietnam — Van-Hoang Le
Karlsruhe, Germany — Alex Ulyanenkov
July 2014

List of Acronyms

AM	Approximate method
BWPT	Brillouin-Wigner perturbation theory
CAO	Coupled anharmonic oscillators
CE	Cumulant expansion
CQAO	Coupled quantum anharmonic oscillators
CGF	Coulomb Green's function
DFT	Density functional theory
ECM	Effective charge model
EF	Eigenfunction
EV	Eigenvalue
HF	Hartree–Fock
LDA	Local density approximation
MTF	Model Thomas-Fermi
PT	Perturbation theory
PF	Partition function
OM	Operator method
QAO	Quantum anharmonic oscillator
RPT	Regular perturbation theory
RSPT	Rayleigh-Schrödinger perturbation theory
RWA	Rotating wave approximation
SCA	Strong coupling approximation
SE	Schrödinger equation
TLS	Two level system
TPT	Thermodynamical perturbation theory
UAA	Uniformly available approximation
WKB	Wentzel-Kramers-Brillouin
VM	Variational method

Contents

1 Capabilities of Approximate Methods in Quantum Theory 1
1.1 Effectiveness Criteria for Approximate Methods 2
1.2 Perturbation Theory for Solution of Stationary Schrödinger Equation 6
1.3 Non-perturbative Methods for Stationary Schrödinger Equation 13
References 24

2 Basics of the Operator Method 27
2.1 The Zeroth Approximation Choice 28
2.2 Iteration Scheme for Calculation of the Successive Approximations 38
2.3 Calculation Accuracy of the Wave Function 45
2.4 Iterative Solution for Inverse Problem 48
2.5 Non-perturbative Approach in the Theory of Classical Nonlinear Oscillations 51
2.6 Why Do the OM Successive Approximations Converge? 59
2.7 Calculation of Energy and Level Width of Quasi-Stationary States 65
2.8 States with a Broken Symmetry (Integrals of Motion) 72
References 79

3 Applications of OM for One-Dimensional Systems 81
3.1 Anharmonic Oscillator with High Anharmonicity 82
3.2 Anharmonic Oscillator with Non-symmetric Potential 88
3.3 Morse Potential 92
3.4 Solution of the Mathieu Equation 95
3.5 Quasienergies and Wave Functions of the Two-Level System in a Classical Monochromatic Field 103
3.6 More Applications of Operator Method 109
3.7 Operator Method for Uniformly Suitable Approximation of Integrals and Sums 117
References 127

4 Operator Method for Quantum Statistics 129
4.1 General Algorithm for Calculation of PF 129
4.2 Statistics of Non-interacting Systems with One-Dimensional Energy Spectrum 133
4.3 Coupled Quantum Anharmonic Oscillators (CQAO) 160
4.4 Density Matrix 166
4.5 Calculation of Physical Characteristics 176
References 184

5 Quantum Systems with Several Degrees of Freedom 187
5.1 Analytical Approximation for the Energy Levels of CQAO 188
5.2 Comparison with Known Analytical and Numerical Results 195
5.3 Regular Perturbation Theory for Two-Electron Atoms 203
5.4 Energies of the Excited States 208
References 213

6 Two-Dimensional Exciton in Magnetic Field with Arbitrary Strength 215
6.1 The Schrödinger Equation through the Levi-Civita Transformation 216
6.2 Solving the Schrödinger Equation by the Operator Method 218
6.3 Exact Numerical Solutions 224
6.4 Schrödinger Equation with Asymptotic Components 229
6.5 Highly Accurate Analytical Solutions 233
6.6 OM Application to Complex Two-Dimensional Atomic Systems 241
References 249

7 Atoms in the External Electromagnetic Fields 251
7.1 Hydrogen-Like Atom and Harmonic Oscillator 252
7.2 Analytical Estimate for Rydberg States of a Hydrogen Atom in an Electric Field 259
7.3 Iterative Calculation of Energy for Quasi-Stationary States 264
7.4 Operator Method for Hydrogen Atom in Magnetic Field 266
7.5 Two Level System in a Single-Mode Quantum Field 271
References 285

8 Many-Electron Atoms 287
8.1 Oscillator Model of Atom 289
8.2 Continuous Oscillator Model in the Limit $Z \gg 1$ 294
8.3 Coulomb Based Atomic Model 303
8.4 Effective Charges Model for Many-Electron Atom 323
References 329

9 Systems with Infinite Number of Degrees of Freedom 331
9.1 OM for Strong Coupling Polaron 332
9.2 One-Dimensional Polaron 340
9.3 UAA for Three-Dimensional Polaron Energy 349
9.4 Particle-Field Interaction Model with a Divergent Perturbation Theory 355
References 357

Index 359

Chapter 1
Capabilities of Approximate Methods in Quantum Theory

The majority of physical phenomena in condensed matter, atomic and molecular systems is defined by electromagnetic interactions and governed by quantum mechanics laws. The systems possess an entirely defined Hamiltonian and the physical properties are described by the corresponding solutions of Schrödinger equation. The quantum description has an universal character, which assumes the wave functions of complex systems are the solutions of the linear equations, which have similar mathematical structure for the physical systems with essentially different physical properties. The mathematics plays a special role in a quantum mechanics [1], and any new method for the solution of Schrödinger equation results in essential progress in the description of numerous physical systems.

For the most of the problems in a quantum theory of many-body systems, the numerical solution of Schrödinger equation using a finite-element approximation for differential operators is ineffective, even by using modern powerful computers, because of the complexity of algorithms and large volume of processed information. The use of the functional integrals for quantum theory [2] faces a similar difficulty. Therefore, the development of new methods for approximate description of quantum systems plays a crucial role for both analytical investigations and design of the algorithms for numerical calculation of physical properties of the objects.

In the Chap. 1, we discuss the general criteria for the effectiveness of the approximate methods (AM), and make an assessment of most frequently used AMs in quantum theory. In this monograph, we apply exclusively the operator (Schrödinger's) formulation of quantum mechanics, and do not discuss the approximate methods for quantum theory in the form of Feynman's functional path integrals [2]. The analytical and approximate methods in the latter form have been widely presented in numerous monographs (for example, [3] and references therein). However, it is worth to mention that the non-perturbative investigations in both forms of quantum mechanics are pretty similar.

I. Feranchuk et al., *Non-perturbative Description of Quantum Systems*,
Lecture Notes in Physics 894, DOI 10.1007/978-3-319-13006-4_1

1.1 Effectiveness Criteria for Approximate Methods

The criteria for mathematical methods to be effective listed below have rather subjective nature, however, they are nevertheless important for the assessment of general effectiveness. To our opinion, they are:

- **universality**,

meaning the representation of the calculus scheme in a form, which is not related to the specific features of the physical problem. However, there is no universal method exists which delivers exact solution of Schrödinger equation for the system with an arbitrary Hamiltonian. Instead of, the method should include an iteration algorithm, which establishes the procedure for the successive approximations. Therefore, the following criteria have to be added:

- **high accuracy of the zeroth approximation**,

which has to describe correctly the most principal properties of the physical system, and

- **uniform convergence of successive approximations**, for calculation of the solution with any required accuracy. For practical applications, the following criterion is important:
- **simple enough algorithm for calculation of zeroth and successive approximations**, to be applicable for the systems with a large number of the degrees of freedom.

The detailed assessment of widely used AMs based on the above-mentioned criteria will be given in the following sections. The perturbation theory (PT) and its modifications (for example, [4, 5] and citations therein) seems to possess a highest universality in the sense of criteria listed above. The PT series can be constructed by the extraction of the operator $\hat{H}_0$ from the entire Hamiltonian $\hat{H}$, which has the known spectrum of eigenvalues (EV) and eigenfunctions (EF), and the operator of perturbation $\hat{V}$, both satisfying the condition $\hat{H} = \hat{H}_0 + \lambda\hat{V}$. The dimensionless parameter $\lambda \ll 1$ defines the characteristic value of the perturbation amplitude in relation with the distance between the energetic levels of the non-perturbed system with Hamiltonian $\hat{H}_0$. Following a simple receipt, formally *applicable to arbitrary quantum system*, the series over the parameter λ for both EV and EF can be constructed. However, for the majority of physical systems, the series over λ, obtained on the basis of canonic form of PT, have an asymptotic nature. These series [5] are divergent and *do not permit to find a solution of Schrödinger equation* by a simple summation of the terms of series. Therefore, the quantitative description of the physical systems using PT is only possible in narrow diapason of small values of the parameter λ and for low energies of excitation.

All aforesaid is also true for the limit of strong coupling ($\lambda \gg 1$), where a small parameter is λ^{-1} [6], and for the quasi-classic approximation (with Planck constant $\hbar$ as a small parameter), and for various modifications of adiabatic expansion, where a small parameter is placed in a front of the operator of kinetic energy [7]. The asymptotic character of the expansions is manifested in the fact, that two-sided approximations being constructed for the same system, for instance, using parameters λ and λ^{-1}, result in a different functional dependence of EV on Hamiltonian parameters, which does not permit the continuous transition from one series to another. Thus, using the PT in real applications, the important property of *universality is lost*, because of the focus is switched from the mathematical procedure for construction of successive approximations to the deep understanding of the physics of processes, taking into account the qualitative characteristics of physical systems in operator $\hat{H}_0$. This behavior is well described by the statement after L.Landau "There is no physical theory without a small parameter".

The modern theoretical physics deals with the problems, the majority of which do not allow to select a small parameter having a certain physical sense. On the contrary, there are many problems exist where all parameters of Hamiltonian are varied in a wide range. Therefore, the development of the methods, which are able to build *a theory of quantum system without a small parameter*, or so called non-perturbative methods, is an actual trend of modern theoretical physics. For certainty, we mean here the avoidance of small parameters in the Hamiltonian of the system. At the same time, this does not exclude the use of supplementary or artificially introduced parameters in non-perturbative methods, which govern the accuracy of the numerical calculations.

There are also many methods exist, which investigate quantum systems without introduction of a physical small parameter (see, for example, [8–11] and citations therein). In a first turn, the direct numerical integration of Schrödinger equation using expansion over the artificially introduced parameter, for instance, the step of the finite-element approximation of derivatives, is an example of non-perturbative method. Another example is a lattice model of the quantum field theory, where the non-physical parameter is a lattice constant, which defines the accuracy of calculation of functional integrals for the transition from integration to summation [3]. The drawbacks of direct numerical methods are the exponential growth of the calculation volume and the loss of the algorithm stability when the dimension of the system s increases, for example, at $s \geq 3$ their effectiveness is already essentially low.

There are more examples of non-perturbative methods having a physical nature: variational principle, Hartree–Fock method for multi-electron atoms, approximating Hamiltonian method, etc. In the most cases, these methods deliver an approximate estimate for the energy of ground state and *are not capable to calculate the entire energy spectrum of the system with a required precision.*

In the majority of applications, the divergence of PT series is not related to the real properties of the physical system, but just points to necessary rearrangement of the received expansion over the Hamiltonian parameter to provide the analytical extension of EV and EF outside of the convergence region of the initial series. This task is partly solved by the summation methods for asymptotic series [12] and various modifications of Pade-approximation [13]. However, all these methods are not really universal in the sense mentioned above, and cannot be generalized for the systems with the multiple degrees of freedom.

For quantitative characterization of the effectiveness and the accuracy of non-perturbative methods, the supplementary definitions have been introduced [14]. Let us assume that some characteristics of the quantum system is described by the function $F_n(\lambda)$, which depends on the quantum number n (quantum number set) of the state and on the parameter λ (parameter set), defining the perturbation amplitude in the system. We assume further that the non-perturbative method delivers both zeroth $F_n^{(0)}(\lambda)$ and successive $F_n^{(s)}(\lambda)$, $s = 0, 1, 2, \ldots$ approximations for the function in question. Then these functions yield a *uniformly available approximation (UAA)* for the physical system, provided the following conditions are fulfilled for the entire range of parameter λ and for all quantum numbers n:

$$\left| \frac{F_n^{(s)}(\lambda) - F_n(\lambda)}{F_n(\lambda)} \right| \leq \xi^{(s)}, \tag{1.1}$$

where each parameter $\xi^{(s)} < 1$ is independent on n and λ and defines the accuracy of the approximation. The condition of the *convergence of non-perturbative method* corresponds to decreasing sequence of parameters $\xi^{(s)}$:

$$\lim_{s \to \infty} F_n^{(s)}(\lambda) = F_n(\lambda). \tag{1.2}$$

The asymptotic series obtained from PT are obviously not satisfying the conditions (1.1) and (1.2), because of they approximate the function in question in a narrow range of n, λ and are not convergent. At the same time, the two-side asymptotic expansions, corresponding to limits $\lambda \ll 1$ and $\lambda \gg 1$, allow to control the conditions of UAA for various non-perturbative methods [14]. In contrast to the asymptotic expansions, the UAA approximates the value with high relative accuracy $\xi^{(s)}$ *in the entire range of the variation of physical parameters and quantum numbers of states*. Thus, the effectiveness of the non-perturbative method can be quantitatively estimated from the infinitesimality of $\xi^{(s)}$ in primary approximations and the speed of their decay with the increase of s.

In this short introductory chapter dedicated to non-perturbative methods for quantum systems, we illuminate the techniques, which are to some extent related to the approach taking a central place in this monograph: operator method (OM). First of all, one of this techniques is a non-perturbative method utilizing the self-similar approximation for calculation of the dependence of system characteristics on the parameters of Hamiltonian. The basic ideas of this approach have been published for the first time in [15], and later it has been successfully applied to numerous problems of quantum mechanics and field theory (see, for instance, the review [16] and citations therein).

The non-perturbative method for calculation of functional integrals for the systems with a non-linear operation (see, for example, [17] and citations therein) has been used in the theory of strong interaction. The scheme of non-linear scaling of coupling constant in quantum chromodynamics makes possible the calculation of the processes in the range of energy variables, where the standard form of PT fails. The application area for this approach can be essentially extended, if the method is generalized for the theory of condensed matter and quantum mechanics of multiple-particle systems.

The method presented in this monograph is a kind of universal algorithm for rearrangement of PT series and it satisfies all the requirements for non-perturbative techniques listed above. This method is named an operator method in the sense that both for zeroth approximation and for high-order approximations there is no necessity to solve differential or integral equations: all calculations are reduced to the algebraic calculus with matrix elements of operators. In the book, we illustrate the application of OM to the description of various quantum systems, including the systems with infinite number of the degrees of freedom. The results demonstrate that already zeroth approximation gives a uniformly available approximation for EV and EF of Schrödinger equation at arbitrary values of Hamiltonian parameters, and the successive approximations converge to the exact solution for all values of quantum numbers of the system and for the entire range of parameters. The algebraic nature of calculations simplifies the development of the algorithms to obtain the higher approximations for the systems with a large number of the degrees of freedom, and to find the eigenvalues and eigenfunctions with a high accuracy.

A special attention is paid to the generalization of operator method for the quantum statistics. In this case, the physical system obtains one additional parameter: a temperature. In comparison with the description of pure quantum states, the calculation of the observed values in statistics involves a complex procedure of summation over all states of the system. Thus, the operator method for calculation of EV and EF is supplemented by the algorithm for the summation over the quantum states. This approach permits to calculate thermodynamical characteristics of physical systems in a wide range of temperature and is effective for the systems with the multiple degrees of freedom.

Prior the start of the detailed description of operator method in Chap. 2, we remind in current chapter the basic relationships from other recognized methods used for approximate evaluation of quantum systems. Their ability and limitations will be demonstrated by applying them to the model systems: the quantum anharmonic oscillator (QAO) and the coupled anharmonic oscillator (CAO), which both are widely used for approval of approximate methods in quantum [18, 19] as well as in classic [20] theories. The analysis of these problems shed the light on the difficulties of approximate methods, which are successfully overcome by the operator method.

1.2 Perturbation Theory for Solution of Stationary Schrödinger Equation

In 1939 Paul Dirac [21] introduced the notations and terminology, which are perfectly fitting the representation of perturbation theory in algebraic form. The state of the quantum system in arbitrary representation [22] is given by ket-vector $|\Psi_n\rangle$, where index n defines the set of quantum numbers corresponding to this state. In general case, these numbers may have both discrete (discrete spectrum) and continuous (continuous spectrum) values. The bra-vector $\langle\Psi_n|$ corresponds to Hermitian conjugate vector of the state. The wave function is determined by the projections of the state vector onto the axes of Hilbert space [22], which corresponds to this representation. For example, for wave functions in coordinate representation:

$$\Psi_n(x) \equiv \langle x|\Psi_n\rangle;\ \Psi_n^*(x) \equiv \langle\Psi_n|x\rangle;\ \int dx\Psi_m^*(x)\Psi_n(x) \equiv \langle\Psi_m|\Psi_n\rangle. \quad (1.3)$$

In these notations, the stationary Schrödinger equation for eigenvalues $E_n(\lambda)$, complete Hamiltonian and normalized eigenfunction $|\Psi_n\rangle$ take the following form:

$$\hat{H}|\Psi_n\rangle = E_n(\lambda)|\Psi_n\rangle, \quad \hat{H} = \hat{H}_0 + \lambda\hat{V};\ \langle\Psi_n|\Psi_n\rangle = 1. \quad (1.4)$$

For the unperturbed system:

$$\hat{H}_0|\psi_n\rangle = \epsilon_n|\psi_n\rangle, \quad \langle\psi_n|\psi_n\rangle = 1. \quad (1.5)$$

where $\hat{H}_0$ is a Hamiltonian of zeroth approximation, for which a complete set of eigenvalues ϵ_n and eigenfunctions $|\psi_n\rangle$ of the solutions of Eq. (1.5) is known. We assume that the ratio of the matrix elements of perturbation operator $\lambda\hat{V}$ to the difference between energy levels of non-perturbed system has an amplitude determined by the dimensionless physical parameter λ.
The state vector $|\Psi_n\rangle$ can be expanded into full set $|\psi_m\rangle$:

$$|\Psi_n\rangle = \sum_{m=0}^{\infty} C_{nm}|\psi_m\rangle, \quad (1.6)$$

and after substitution it into (1.5), we find an exact system of equations for the coefficients C_{nm} [4]:

$$(E_n - \epsilon_k)C_{nk} = \lambda\sum_m V_{km}C_{nm}; \quad V_{km} = \langle\psi_k|\hat{V}|\psi_m\rangle, \quad (1.7)$$

which include matrix elements V_{km} of the perturbation operator.

Formally the perturbation theory corresponds to the expansion of values in question into series over the λ:

$$C_{nk} = C_{nk}^{(0)} + \lambda C_{nk}^{(1)} + \lambda^2 C_{nk}^{(2)} + \ldots;$$
$$E_n(\lambda) = E_n^{(0)} + \lambda E_n^{(1)} + \lambda^2 E_n^{(2)} + \ldots \tag{1.8}$$

There are two ways to obtain this expansion: the algorithm of Rayleigh-Schrödinger for perturbation theory (RSPT) [23, 24] and the method of Brillouin-Wigner for perturbation theory (BWPT) [25, 26]. They differ each from the other by the normalization of the state vector. In case of RSPT, the corrections for the coefficient $C_{nn}^{(l)}$ in each approximation order are selected from the condition for normalization $\langle \Psi_n | \Psi_n \rangle = 1$ to be satisfied with the accuracy up to the terms of the order λ^l. Assuming all the eigenvalues of Hamiltonian $\hat{H}_0$ are non-degenerate ($\epsilon_m \neq \epsilon_n$ for all $m \neq n$), we obtain [4]:

$$E_n(\lambda) = \epsilon_n + \lambda V_{nn} - \lambda^2 \sum_{m \neq n} \frac{|V_{mn}|^2}{\epsilon_m - \epsilon_n} + \ldots,$$

$$|\Psi_n\rangle = |\psi_n\rangle - \lambda \sum_{m \neq n} \frac{V_{mn}}{\epsilon_m - \epsilon_n} |\psi_m\rangle + \lambda^2 \Bigg[\sum_{m \neq n} \sum_{k \neq n} \frac{V_{mk}}{(\epsilon_m - \epsilon_n)} \frac{V_{kn}}{(\epsilon_k - \epsilon_n)} |\psi_m\rangle -$$
$$- \sum_{m \neq n} \frac{V_{nn} V_{mn}}{(\epsilon_m - \epsilon_n)^2} |\psi_m\rangle - \sum_{m \neq n} \frac{|V_{mn}|^2}{2(\epsilon_m - \epsilon_n)^2} |\psi_n\rangle \Bigg] + \ldots \tag{1.9}$$

In the case of degeneration of the states of unperturbed Hamiltonian, each energy level ϵ_n has a corresponding set of the distinguished vectors:

$$|\psi_n 1\rangle; |\psi_n 2\rangle; \ldots |\psi_n s_n\rangle, \tag{1.10}$$

where s_n is a multiplicity of degeneration. In this case, to obtain the RSPT series, a new set of states for zeroth approximation has to be constructed [4]:

$$|\psi_n^\sigma\rangle = \sum_{s=1}^{s_n} c_s^\sigma |\psi_n s\rangle; \; \sigma = 1, 2, \ldots s_n. \tag{1.11}$$

Here the coefficients c_s^σ and new eigenvalues ϵ_n^σ are calculated from the solution of the system of linear equations:

$$(\epsilon_n^\sigma - \epsilon_n) c_s^\sigma = \lambda \sum_{s'} V_{ss'}^{(n)} c_{s'}^\sigma; \; V_{ss'}^{(n)} = \langle \psi_n s' | \hat{V} | \psi_n s \rangle; \; \langle \psi_n^{\sigma'} | \psi_n^\sigma \rangle = \delta_{ss'}. \tag{1.12}$$

This additional operation does not influence the convergence and the general form of the series RSPT, if the set of indices n, σ is included in the definition of quantum number n.

If the BWPT form of the expansion over the parameter λ is used, the following fact is exploited: due to the linearity of Schrödinger equation (1.4) the solutions are defined with the accuracy of constant, and therefore the normalization vector can be found by applying the alternative normalization condition:

$$\hat{H}|\tilde{\Psi}_n\rangle = E_n(\lambda)|\tilde{\Psi}_n\rangle,$$
$$\langle\psi_n|\tilde{\Psi}_n\rangle = 1;\ |\Psi_n\rangle = A|\tilde{\Psi}_n\rangle;\ A^2 = [\langle\tilde{\Psi}_n|\tilde{\Psi}_n\rangle]^{-1}. \tag{1.13}$$

As a result, the system of equations for coefficients of the expansion of state vector is found:

$$|\tilde{\Psi}_n\rangle = |\psi_n\rangle + \sum_{k\neq n}\tilde{C}_{nk}|\psi_k\rangle,$$
$$E_n = \epsilon_n + \lambda V_{nn} + \lambda\sum_{k\neq n}\tilde{C}_{nk}V_{nk};$$
$$\tilde{C}_{nk} = \frac{\lambda}{E_n-\epsilon_k}\left[V_{kn} + \sum_{m\neq n}\tilde{C}_{nm}V_{mk}\right];\ k\neq n. \tag{1.14}$$

The successive approximations of the BWPT series are found by iterating the last equation in the expression (1.14):

$$\tilde{C}^{(0)}_{nk} = 0;\ \tilde{C}^{(1)}_{nk} = \frac{\lambda V_{kn}}{E_n-\epsilon_k};$$
$$\tilde{C}^{(2)}_{nk} = \frac{\lambda}{E_n-\epsilon_k}\left[V_{kn} + \lambda\sum_{m\neq n}\frac{V_{nm}V_{mk}}{E_n-\epsilon_m}\right];\ldots, \tag{1.15}$$

and the energy levels for each approximation are found as the solutions of transcend equation. For example, in the second order BWPT this equation has a form:

$$E_n = \epsilon_n + \lambda V_{nn} + \lambda^2\sum_{k\neq n}\frac{V_{nk}V_{kn}}{E_n-\epsilon_k}. \tag{1.16}$$

The expansion (1.14) is generalized for the case of degenerate states using the substitution of $|\psi_n\rangle, \epsilon_n$ by the states $|\psi^{\sigma}_n\rangle, \epsilon^{\sigma}_n$ in zeroth approximation. Thus, the formulas (1.9) and (1.14) define the *universal* method for calculation of eigenvalues and eigenfunctions as a series over the ratio of matrix elements of perturbation operator to the distance between the energy levels of the zeroth approximation:

$$\xi_n = \sum_{k \neq n} \frac{\lambda V_{kn}}{\epsilon_n - \epsilon_k}. \tag{1.17}$$

However, these series allow to calculate eigenvalues and eigenfunctions within a limited range of the parameters of perturbation operator even for simple physical systems. As an example, we consider here one-dimensional QAO, which is often used for approbation of the solution of Schroödinger equation or as a basic model in the field theory with non-quadratic Hamiltonian [18]. The problem is reduced to the solution of the following equation (further we use the units system with Planck constant $\hbar$ and particle mass m both equal unity) [19]:

$$\begin{aligned} \hat{H}|\Psi_n\rangle &= E_n(\lambda)|\Psi_n\rangle, \\ \hat{H} &= \frac{1}{2}(\hat{p}^2 + x^2) + \mu x^2 + \lambda x^4; \quad \hat{p} = -i\frac{d}{dx}. \end{aligned} \tag{1.18}$$

We start with the consideration of the case with $\lambda = 0$, which gives a good illustration of several obstacles in the canonic perturbation theory [27]. The perturbation operator is chosen as:

$$\hat{H}_0 = \frac{1}{2}(\hat{p}^2 + x^2); \; \hat{V} = \mu x^2; \; \mu > -\frac{1}{2}. \tag{1.19}$$

Here we use the algebraic calculations in the particle number representation [4], which is based on the canonic transformation to the creation and annihilation operators for unperturbed oscillator:

$$\begin{aligned} \hat{x} &= \frac{1}{\sqrt{2}}\left[\hat{a} + \hat{a}^+\right], \\ \hat{p} &= i\frac{1}{\sqrt{2}}\left[\hat{a}^+ - \hat{a}\right], \end{aligned} \tag{1.20}$$

The operators of annihilation $\hat{a}$ and creation $\hat{a}^+$ satisfy to the permutation relation:

$$\left[\hat{a}\,\hat{a}^+\right] = 1. \tag{1.21}$$

In this representation the operators $\hat{H}_0$ and $\hat{V}$ have the following form:

$$\hat{H}_0 = \frac{1}{2}\left[1 + 2\hat{a}^+\hat{a}\right], \tag{1.22}$$

$$\hat{V} = \frac{\mu}{2}\left[1 + 2\hat{a}^+\hat{a} + \left(\hat{a}^+\right)^2 + (\hat{a})^2\right]. \tag{1.23}$$

The eigenfunctions for unperturbed oscillator in this representation coincide with the eigenvectors of the operator of the excitation numbers $\hat{n}$:

$$\begin{aligned}
|n\rangle &= \frac{1}{\sqrt{n!}}\left[\hat{a}^{+}\right]^{n}|0\rangle\,, \quad n = 0, 1, 2, 3, \ldots, \\
\hat{n} &= \hat{a}^{+}\hat{a}, \quad \hat{n}\,|n\rangle = n\,|n\rangle\,, \\
\hat{a}\,|n\rangle &= \sqrt{n}\,|n-1\rangle\,, \quad \hat{a}^{+}\,|n\rangle = \sqrt{n+1}\,|n+1\rangle\,,
\end{aligned} \tag{1.24}$$

and the ground state of unperturbed system follows from the expression:

$$\hat{a}\,|0\rangle = 0. \tag{1.25}$$

Thus, the zeroth approximation for eigenvalues and eigenfunctions in the Eq. (1.19) is given by:

$$E_n^{(0)} = \frac{1}{2}(2n+1), \qquad \left|\Psi_n^{(0)}\right\rangle = |n\rangle\,. \tag{1.26}$$

Using the algebra (1.21) for the creation and annihilation operators, the first terms of the series over the operator $\hat{V}$ can be found from the formula (1.9):

$$E_n^{(1)} = \langle n|\,\hat{V}\,|n\rangle = \frac{\mu}{2}(2n+1), \tag{1.27}$$

$$E_n^{(2)} = -\frac{\left|\langle n+2|\,\hat{V}\,|n\rangle\right|^2}{E_{n+2}^{(0)} - E_n^{(0)}} = -\frac{\mu^2}{8}(n+1)(n+2), \tag{1.28}$$

and in a similar way for further terms. As follows from the expression (1.27), even for the ground state with $n = 0$ the series of the perturbation theory converges only in the domain $|\mu| < \frac{1}{2}$. At the same time, the operator (1.19) is evidently a Hamiltonian of the harmonic oscillator with frequency:

$$\omega(\mu) = \sqrt{1+2\mu}, \tag{1.29}$$

with known set of eigenfunctions and exact spectrum of eigenvalues:

$$E_n = \omega(\mu)\left(n+\frac{1}{2}\right) = \left(n+\frac{1}{2}\right)\sqrt{1+2\mu}. \tag{1.30}$$

The eigenvalues (1.30), considered as the functions of parameter μ, have a singularity at $\mu = -\frac{1}{2}$, because of at $\mu < -\frac{1}{2}$ the Hamiltonian (1.19) does not possess a discrete spectrum. In general, the convergence radius of the power series

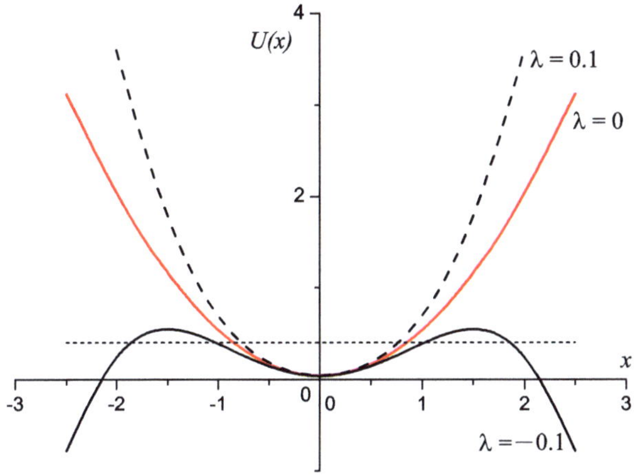

Fig. 1.1 The potential energy of anharmonic oscillator for different values of λ

is determined by the distance to the nearest singular point in complex plane μ [5], that limits the convergence of PT series to the values of perturbation parameter in interval $|\mu| < \frac{1}{2}$. However, from (1.29) follows that all the values of the parameter are permissible in the interval $\left(-\frac{1}{2};\infty\right)$. Thus, even in this simple situation the calculation of EV spectrum for the entire range of Hamiltonian parameter requires an essential reconstruction of PT series. For the operator (1.18) with $\mu = 0$, $\lambda \neq 0$ the situation becomes even more complicated:

$$\hat{H}_0 = \frac{1}{2}(\hat{p}^2 + x^2);\ \hat{V} = \lambda x^4. \tag{1.31}$$

The potential energy $U(x) = x^2/2 + \lambda x^4$ at various λ for the oscillator is shown in Fig. 1.1. At the values $\lambda < 0$, the motion of particle in this potential becomes infinite due to the subbarrier tunneling. That means the system has not discrete spectrum at infinitesimal negative λ and the functions $E_n(\lambda)$ have a singularity at $\lambda = 0$. In this case, the power series over this parameter have a zeroth convergence radius and are not converging at any values of λ. Such series are called asymptotic, and they can be used with a limited number of terms and only for estimate of eigenvalues at small λ [5]. The described above properties of PT series are confirmed when the formulas (1.9) are used for calculations in the particle number representation for operators (1.31):

$$\hat{H}_0 = \frac{1}{2}\left[1 + 2\hat{a}^+\hat{a}\right],\quad \hat{V} = \frac{\lambda}{16}\left[\hat{a}^+ + a\right]^4. \tag{1.32}$$

Using the Eq. (1.24) for matrix elements of the creation and annihilation operators, the following expressions are found for the first terms of the RSPT series [4]:

$$E_n(\lambda) = \left(n + \frac{1}{2}\right) + \frac{3}{4}\lambda[2n^2 + 2n + 1] -$$
$$-\frac{\lambda^2}{8}[34n^3 + 51n^2 + 59n + 21] +$$
$$+\frac{\lambda^3}{16}[375n^4 + 750n^3 + 1416n^2 + 1041n + 333] + \ldots \tag{1.33}$$

For the ground state of the system ($n = 0$), the calculations are easier, and we show here more terms of the series [19]:

$$E_0(\lambda) = \sum_{s=0}^{\infty} \lambda^s A_0^s = \frac{1}{2} + \frac{3}{4}\lambda - \frac{21}{8}\lambda^2 + \frac{333}{16}\lambda^3 - \frac{30885}{128}\lambda^4 + \frac{916731}{256}\lambda^5 - \ldots \tag{1.34}$$

In the formula (1.33) the effective expansion parameter is λn, and for high excitation levels the RSPT series can only be used for very small λ. The graph on the Fig. 1.2, calculated on the basis of expression (1.34) for the function

$$E_0^{(l)}(\lambda) = \sum_{s=0}^{l} \lambda^s A_0^s,$$

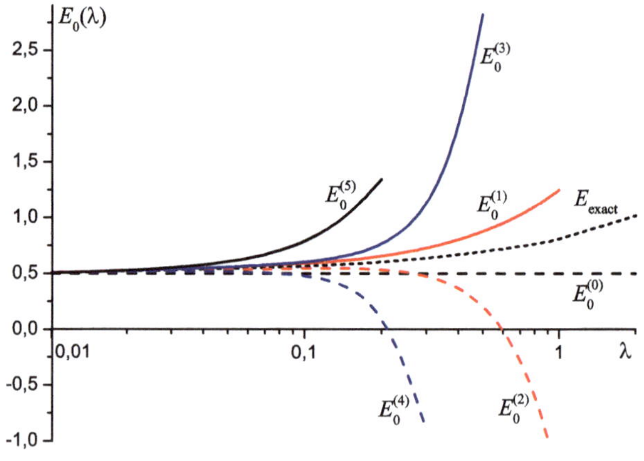

Fig. 1.2 The estimate for the energy of ground state of QAO for different numbers of terms $E_0^{(l)}$ in (1.34). The accurate values for $E_0(\lambda)$ are obtained in [28] from numerical solution of Schrödinger equation

illustrates the asymptotic character of the series with zeroth convergence radius: the larger the number of the involved terms in the expansion, the narrower is the applicability range of the series.

The divergence of RSPT series for the eigenvalues follows from its representation as power series, for example, for ground state:

$$E_0(\lambda) = \sum_{s=0}^{\infty} \lambda^s A_0^{(s)}. \tag{1.35}$$

The general behavior of the coefficients of the series at $s \gg 1$ was found in [18]:

$$A_0^{(s)} \approx (-1)^{s+1} \left(\frac{6}{\pi^3}\right)^{1/2} \Gamma\left(s + \frac{1}{2}\right) 3^s, \tag{1.36}$$

and using the general expression for the convergence radius R of the power series [5], we obtain:

$$R = \lim_{s\to\infty} \left| \frac{A_0^{(s)}}{A_0^{(s+1)}} \right| \approx \lim_{s\to\infty} \frac{1}{3(s+1/2)} = 0. \tag{1.37}$$

Summarizing all written above, the demonstrated in this section examples show that even for relatively simple physical systems the canonic perturbation theory is not able to find an uniformly available approximation for the solution of Schrödinger equation.

1.3 Non-perturbative Methods for Stationary Schrödinger Equation

The other than canonic perturbation theory methods are usually formulated not universally and relate to the specific properties of quantum system. To illustrate the principle idea of these methods, we again use the quantum anharmonic oscillator problem. One of the effective non-perturbative method is the *strong coupling approximation* (SCA), which is conveniently formulated in the coordinate representation of Schrödinger equation:

$$-\frac{1}{2}\frac{d^2\Psi_n(x)}{dx^2} + \frac{1}{2}x^2\Psi_n(x) + \lambda x^4\Psi_n(x) = E_n(\lambda)\Psi_n(x). \tag{1.38}$$

The basic idea of SCA in the range $\lambda \gg 1$ consists of the transformation of independent variable in such a way that new small parameter is introduced in the equation, which depends on $\lambda^{-1} \ll 1$. By re-scaling the variable in the Eq. (1.35) as:

$$x = \xi y;\ \Psi_n(y\xi) = \Phi_n(y);$$
$$-\frac{1}{2\xi^2}\frac{d^2\Phi_n(y)}{dy^2} + \frac{1}{2}\xi^2 y^2\Phi_n(y) + \lambda\xi^4 y^4\Phi_n(y) = E_n(\lambda)\Phi_n(y), \quad (1.39)$$

and choosing the parameter ξ under the condition that the coefficients at highest derivative in the equation and the largest term in the limit $\lambda \gg 1$ coincide:

$$\xi^{-2} = \lambda\xi^4;\ \xi = \lambda^{-1/6};$$
$$-\frac{1}{2}\frac{d^2\Phi_n(y)}{dy^2} + \frac{1}{2\lambda^{2/3}}y^2\Phi_n(y) + y^4\Phi_n(y) = \epsilon_n(\lambda)\Phi_n(y);$$
$$E_n(\lambda) = \lambda^{1/3}\epsilon_n(\lambda), \quad (1.40)$$

we reduce the problem to the form, which can be treated by the perturbation theory with the effective parameter $\lambda^{-2/3} \ll 1$ and the equation for zeroth approximation, which is independent on λ:

$$-\frac{1}{2}\frac{d^2\phi_n(y)}{dy^2} + y^4\phi_n(y) = \epsilon_n^{(0)}\phi_n(y). \quad (1.41)$$

In strong coupling approximation, contrary to the case $\lambda \ll 1$, the equation for zeroth approximation does not have an analytical solution for eigenvalues and eigenfunctions, and therefore the dependence on λ is manifested as a series:

$$E_n(\lambda) = \lambda^{1/3}\sum_{s=0}^{\infty} B_n^s\lambda^{-2s/3}, \quad (1.42)$$

and for the coefficients B_n^s the numerical calculation of EV and EF is required from the differential equation (1.41) and matrix elements of the perturbation operator $\frac{1}{2\lambda^{2/3}}y^2$ (see, for example, [19]). In the next chapters we compare the analytical results after OM with the numerical calculations, and here we show some few results for the coefficients B_0^s to demonstrate the asymptotic character of SCA series, illustrated in Fig. 1.3, where the following functions are presented:

$$\tilde{E}_0^{(l)}(\lambda) = \lambda^{1/3}\sum_{s=0}^{l} B_n^s\lambda^{-2s/3};$$
$$B_0^0 \approx 0.6680;\ B_0^1 \approx 0.1437;\ B_0^2 \approx -0.0088. \quad (1.43)$$

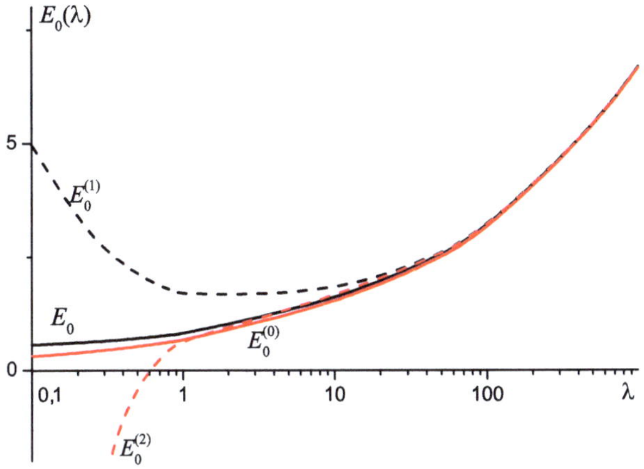

Fig. 1.3 The estimate for the energy of ground state of QAO for different numbers of terms $\tilde{E}_0^{(l)}$ in (1.43) for the case of SCA

As has been previously discussed in the Sect. 1.2, the applicability area of the perturbation theory is determined both by parameter λ and small values of a quantum number n. In the range $n \gg 1$, the *quasi-classic approximation* of Wentzel-Kramers-Brillouin (WKB) [4] can be used for calculation of eigenvalues. For the Eq. (1.38), this method is reduced to the Bohr-Sommerfeld quantization procedure:

$$\int_{x_n}^{x_n'} [2E_n - x^2 - 2\lambda x^4]^{1/2} dx = \pi \left(n + \frac{1}{2}\right) + O(1/n), \tag{1.44}$$

where stationary points x_n, x_n' are defined as real solutions of the equation:

$$2E_n - x^2 - 2\lambda x^4 = 0.$$

In general case of arbitrary λ, the use of WKB is ineffective because of requires the time-consuming calculation of the elliptic integrals [29]. However, in the limit $\lambda \gg 1$ these integrals can be approximated by using the expansion over the parameter $\lambda^{-1/3}$ and the series similar to (1.42) is obtained with approximate analytical expressions for coefficients [19]:

$$B_n^0 \approx 3^{4/3}\pi^2 \left[\Gamma\left(\frac{1}{4}\right)\right]^{-8/3} \left(n + \frac{1}{2}\right)^{4/3};$$
$$B_n^1 \approx 3^{4/3} 4\pi^3 \left[\Gamma\left(\frac{1}{4}\right)\right]^{-16/3} \left(n + \frac{1}{2}\right)^{2/3}; \; B_n^2 \approx -\frac{1}{32} + 6\pi^4 \left[\Gamma\left(\frac{1}{4}\right)\right]^{-8}. \tag{1.45}$$

Another asymptotic expression used further for the analysis of the effectiveness of methods is obtained from the numerical solution of the Eq. (1.38) by using the power series [5]. To execute this procedure, the asymptotic behavior of wave function at $|x| \gg 1$ has to be studied, where the wave function is determined from the solution of the equation:

$$-\frac{1}{2}\frac{d^2\Psi_n(x)}{dx^2} + \lambda x^4 \Psi_n(x) \approx 0;$$
$$\Psi_n(x) \sim e^{-|x|^3\sqrt{2\lambda}} + O\left(\frac{1}{|x|}\right). \tag{1.46}$$

Using the design of the wave function in the form:

$$\Psi_n(x) = e^{-|x|^3\sqrt{2\lambda}} \sum_{k=0}^{\infty} a_{nk} x^k,$$

the coefficients a_{nk} satisfy to the system of recurrent equations, which can be investigated analytically in the limiting case:

$$\lambda \to 0;\ n \to \infty;\ \lambda n = \beta < 1,$$

and for eigenvalues the following asymptotic expansion is obtained [19]:

$$E_n = (n + \frac{1}{2})\left[1 + \frac{3}{2}\beta - \frac{17}{16}\frac{4 + 9\beta}{(1 + 3\beta)^2}\beta^2 + \dots\right]. \tag{1.47}$$

Summarizing the above presented approaches, all the approximate methods described in this chapter are able to estimate the functions $E_n(\lambda)$, each method in a particular domain of the parameters n, λ. However, neither algorithm provides the uniformly available approximation in the entire range of the parameters. Another essential drawbacks of all methods are the divergence of the successive approximations and cumbersome form of wave functions for zeroth approximations.

The *variational method* (VM) is used often for the evaluation of ground state energy for arbitrary amplitude of perturbation. This method is based on the fact that Schrödinger equation for the wave function $\Psi_0(\{\zeta\})$, which depends on the set of variables $\{\zeta\}$ parameterizing the physical system, corresponds to the extremum of the following functional [5]:

$$I_0[\Psi] = \int d\{\zeta\}\Psi_0^*(\{\zeta\})[\hat{H} - E_0]\Psi_0(\{\zeta\});$$
$$\frac{\delta I}{\delta \Psi_0^*} = 0;\ \int d\{\zeta\}\Psi_0^*(\{\zeta\})\Psi_0(\{\zeta\}) = 1. \tag{1.48}$$

The principle idea of VM consists of the replacement of the exact solution of Schrödinger equation by the function, which is modeled by analytical expression (trial function) containing the set of variational parameters $\omega_i, i = 1, 2, \ldots s$:

$$\Psi_0(\{\zeta\}) \approx \psi_0(\{\zeta\}, (\{\omega_i\}); \int d\{\zeta\}\psi_0^*(\{\zeta\})\psi_0(\{\zeta\}) = 1. \tag{1.49}$$

By substituting (1.49) into (1.48), we find the approximate value of the energy, which depends on these parameters:

$$E_0 \rightarrow \epsilon_0(\{\omega_i\}) = \int d\{\zeta\}\psi_0^*(\{\zeta\})\hat{H}\psi_0(\{\zeta\}),$$

and the best approximation for the energy on the selected class of functions is obtained on the basis of the choice of variational parameters from the minimum condition for $\epsilon_0(\{\omega_i\})$:

$$\frac{\partial \epsilon_0(\{\omega_i\})}{\partial \omega_i} = 0; \; i = 1, 2, \ldots s; \; \Rightarrow \{\omega_i^{(0)}\};$$
$$E_0 \approx \epsilon_0(\{\omega_i^{(0)}\}). \tag{1.50}$$

In general case, the function $\epsilon_0(\{\omega_i\})$ is non-linear. Therefore, in this formulation of VM it is difficult to design the regular procedure for improvement of the accuracy of zeroth approximation as well as to evaluate the convergence of the resulted estimate for the large number of parameters. Moreover, the use of VM for excited states becomes complicated due to the accounting of additional orthogonality of wave function to the wave functions of all lower states, which have to be implemented in the variation of the functional.

$$I_n[\Psi] = \int d\{\zeta\}\Psi_n^*(\{\zeta\})[\hat{H} - E_n]\Psi_n(\{\zeta\}); \int d\{\zeta\}\Psi_n^*(\{\zeta\})\Psi_n(\{\zeta\}) = 1;$$
$$\int d\{\zeta\}\Psi_m^*(\{\zeta\})\Psi_n(\{\zeta\}) = 0, \; m = 0, 1, \ldots (n-1). \tag{1.51}$$

To some extent, these problems are eliminated when using the variational principle based on the Ritz-Bubnov-Galerkin method [30]. In this method, the trial function consists of limited number of terms of the series (1.6) in the expansion of the state vector:

$$|\Psi^{(N)}\rangle = \sum_{m=0}^{N} C_m|\psi_m\rangle. \tag{1.52}$$

The set of the coefficients C_m represents the variational parameters in this case. By substituting expression (1.52) into functional (1.48), and calculating the variational derivatives, the system of N linear uniform equations for these coefficients is found and the determinant Δ_N is written as:

$$(E - \epsilon_k)C_k = \lambda \sum_{m=0}^{N} V_{km} C_m;$$
$$\Delta_N(E, \lambda) = \| (E - \epsilon_k)\delta_{km} - \lambda V_{km} \| . \tag{1.53}$$

The solution of the equation $\Delta_N(E, \lambda) = 0$ delivers $(N + 1)$ eigenvalues $E_n^N(\lambda); n = 0, \dots N$ and corresponding eigenvectors $|\Psi_n^{(N)}\rangle$. This method is one of the techniques to numerically solve the Schrödinger equation, which has been used in [28] for calculation of eigenvalues and eigenfunctions for QAO. However, this method is not able to investigate analytically the qualitative behavior of quantum system for various ranges of physical parameters, which is a key subject for the presented in this monograph operator method. There are also substantial difficulties of the application of Ritz-Bubnov-Galerkin method for the physical systems with a large number of the degrees of freedom because of the cumbersome matrix elements and the increase of the dimension of the determinant in (1.53). To illustrate the non-perturbative nature of VM, we consider here the functional (1.48), which corresponds to the Eq. (1.38) for the ground state of anharmonic oscillator:

$$I_0[\Psi] = \int_{-\infty}^{\infty} dx \{ |\frac{d\Psi_0(x)}{dx}|^2 + \Psi_0^*(x)[x^2 + 2\lambda x^4 - E_0]\Psi_0(x)\};$$
$$\int_{-\infty}^{\infty} dx |\Psi_0(x)|^2 = 1. \tag{1.54}$$

The trial function is chosen in compliance with the wave function of harmonic oscillator, but with arbitrary frequency ω [4]:

$$\Psi_0(x) \approx \tilde{\Psi}_0(x) = C \exp\left[-\frac{\omega x^2}{2}\right], \tag{1.55}$$

and the normalizing constant C and the frequency ω are treated as variational parameters. The calculation of integrals in (1.54) with the function (1.55) results in:

$$I_0[\Psi] \approx \tilde{I}_0[\omega, C] = C^2 \sqrt{\frac{\pi}{\omega}} \left[\frac{1}{2}(\omega + \frac{1}{\omega}) + \frac{3\lambda}{\omega^2} - \tilde{E}_0 \right]; \; C^2 \sqrt{\frac{\pi}{\omega}} = 1. \tag{1.56}$$

The calculation of the derivatives over C from the functional gives to the equation for energy, and the value C is found from the normalization condition:

$$\tilde{E}_0 = \frac{1}{2}\left(\omega + \frac{1}{\omega}\right) + \frac{3\lambda}{\omega^2}; \quad C = \left(\frac{\omega}{\pi}\right)^{1/4}, \tag{1.57}$$

and the derivative over ω determines the dependence of the variational parameter on the coupling constant λ:

$$\omega^3 - \omega - 6\lambda = 0. \tag{1.58}$$

Thus, the Eqs. (1.57) and (1.58) define the parametric form of the function $\tilde{E}_0(\lambda)$, which can be used for the estimate of the eigenvalue $E_0(\lambda)$ in the entire range of the parameter λ. The analytical solution of these equations at $\lambda \ll 1$ and $\lambda \gg 1$ results in the following expansions:

$$\tilde{E}_0(\lambda) \approx \frac{1}{2} + \frac{3}{4}\lambda - \frac{9}{4}\lambda^2 + \ldots, \ \lambda \ll 1;$$

$$\tilde{E}_0(\lambda) \approx \lambda^{1/3}\left[\left(\frac{3}{4}\right)^{4/3} + \frac{1}{4(36\lambda)^{2/3}} - \frac{1}{144\lambda^{4/3}} + \ldots\right], \ \lambda \gg 1. \tag{1.59}$$

These expansions are in a good agreement with the asymptotic series RSPT (1.34) and SCA (1.43), respectively, but contrary to them are convergent for all λ. Thereby the variational method delivers an uniformly available approximation for the eigenvalues in a zeroth approximation. It is worth to notice that in VM the wave function of physical system is modeled and not the Hamiltonian, as in the case of perturbation theory. Being used in the systems with several degrees of freedom, VM faces some specific issues, which can be illustrated by a simple model of two coupled harmonic oscillators [31]. The dimensionless form of Hamiltonian for this system is written as [32]:

$$\hat{H} = \frac{1}{2}\hat{p}_x^2 + \frac{1}{2M}\hat{p}_y^2 + \frac{1}{2}x^2 + \frac{1}{2}y^2 + \lambda\, xy, \tag{1.60}$$

where M is the ratio of the oscillator masses and λ is the interaction parameter. In spite of its simplicity, this Hamiltonian is often used for the approbation of various approximate methods in many-particle quantum theory [31]. The classical trajectories of the system are described by rather complicated Lissajous figures which demonstrates the essential dependence of the quantum levels on the interaction and mass parameters. The exact eigenvalues of the Hamiltonian are found straightforward:

$$E_{nm} = \nu_1\left(n + \frac{1}{2}\right) + \nu_2\left(m + \frac{1}{2}\right), \tag{1.61}$$

where $\nu_{1,2}$ are defined by the expression:

$$\nu_{1,2}^2 = \frac{1}{2M}(1 + M \pm \sqrt{(1-M)^2 + 4\lambda^2 M}). \tag{1.62}$$

The formula (1.62) shows that the system energy has singularities if it is considered as an analytical function in the complex plane of the parameters λ and M. Thus, the series in terms of powers of these parameters have the finite convergence radii. This is the mathematical reason of the restrictions for various approximate methods, as has been discussed earlier for one-dimensional system.

The analogous restrictions for the convergence of the series appear when the interactions between the oscillators are considered by some approximate method. Let us consider the results of VM in one-particle approximation used for the Hamiltonian (1.60). The wave functions of the system in the zeroth order are chosen as the product of one-particle functions, i.e. Hartree approximation is applied (the symmetrization of the function in the case of $M = 1$ is not essential for our discussion):

$$\Psi_{VM}(x,y) = \varphi(x)\chi(y); \int_{-\infty}^{\infty} dx|\varphi(x)|^2 = \int_{-\infty}^{\infty} dy|\chi(y)|^2 = 1. \quad (1.63)$$

The system of the approximate equations for one-particle functions follows from the exact variational principle (1.48):

$$\left\{\frac{1}{2}\hat{p}_x^2 + \frac{1}{2}x^2 + \lambda\, x y_{mm} - \epsilon_n\right\} \varphi_n = 0;$$

$$\left\{\frac{1}{2M}\hat{p}_y^2 + \frac{1}{2}y^2 + \lambda\, y x_{nn} - \epsilon_m\right\} \chi_m = 0. \quad (1.64)$$

Both equations correspond to the uncoupled harmonic oscillators with displaced equilibrium positions defined as:

$$\bar{x} = y_{mm} = \int_{-\infty}^{\infty} dy\chi_m(y)^* y\chi_m(y);\ \bar{y} = x_{nn} = \int_{-\infty}^{\infty} dx\varphi_n(x)^* x\varphi_n(x). \quad (1.65)$$

The energy spectrum of the system in this approximation is:

$$E_{nm}^{VM} = \left(n + \frac{1}{2}\right) + \frac{1}{\sqrt{M}}\left(m + \frac{1}{2}\right) - \frac{1}{2}\lambda^2(x_{nn}^2 + y_{mm}^2) - \lambda x_{nn} y_{mm}. \quad (1.66)$$

This expression is actually the power series of the parameter λ, and taking into account the conditions of self-consistency for the values $\bar{x}, \bar{y}$:

$$\bar{x} = -\lambda\bar{y};\ \bar{y} = -\lambda\bar{x},$$

one finds $\bar{x} = \bar{y} = 0$ for arbitrary value of λ.

Thus, the zeroth-order approximation of E_{nm}^{VM} differs essentially from the corresponding exact value. Certainly, the consequent corrections take into account the particle correlations, but in any case VM fails to describe the energy levels over the entire range of parameters λ and M. The so-called adiabatic approximation ($M \gg 1$) being applied to the Schrödinger equation with the Hamiltonian (1.60) leads to analogous problem. A similar calculation has also been considered in [31] to illustrate the general method for calculation of high-order corrections for the adiabatic approximation. In the adiabatic zeroth-order approximation the operator $\hat{p}_y^2$ should be neglected and the "adiabatic" terms $\epsilon_n(y)$ are defined by the energy levels of that part of the Hamiltonian which depends on the "quick" variable x:

$$\epsilon_n(y) = n + \frac{1}{2} - \frac{1}{2}\lambda^2 y^2.$$

These values play the role of the potential energy in the Schrödinger equation for the "slow" oscillator y in the next order of the approximation. In the result, the energy spectrum of the system, taking into account two orders of the series in the parameter $1/\sqrt{M}$, has the form:

$$E_{nm}^{AA} = n + \frac{1}{2} + \sqrt{\frac{1-\lambda^2}{M}}\left(m + \frac{1}{2}\right). \tag{1.67}$$

Comparing this expression with formula (1.62), it becomes evident that the adiabatic (Born-Oppenheimer) [33] approximation also does not lead to the uniformly suitable estimation for the energy levels, even for such a simple system. Certainly, the same restrictions of the considered methods appear for more complicated model when the anharmonicity of the oscillators is included. In any case, the singularities of the energy, considered as an analytical function of the Hamiltonian parameters, define the finite radii of convergence for the power series in these parameters and do not allow to find an uniformly suitable approximation.

As we have shown above, while working with the real physical systems, the perturbation methods applied for the solution of the Schrödinger equation have led to the divergent asymptotic series for the energy corrections. However in some cases it is possible to develop the methods of regular summation of such series. We will mention here one of these methods—the Borel's summation [34], which is very effective for calculation of the divergent or very slowly convergent series. This method was developed in the end of the nineteenth century by Emile Borel and then generalized by Gösta Mittag-Leffler [35].

First of all, we recall some main ideas of the method, starting with the investigation of power series:

$$f(z) = \sum_{k=0}^{+\infty} c_k z^k. \tag{1.68}$$

The series (1.68) could be divergent at a certain area of value z, or very slowly convergent. Therefore, instead of (1.68) we consider another series called the Borel transform of (1.68):

$$F(z) = \sum_{k=0}^{+\infty} \frac{1}{k!} c_k z^k. \tag{1.69}$$

The Borel transform (1.69) is obviously converging faster than (1.68), and assuming it converges to the analytical function near the point $z = 0$, the analytical continuation can be performed along the positive real axis. The following integral is defined:

$$f(z)(Borel) = \int_0^{+\infty} e^{-t} F(zt) dt, \tag{1.70}$$

which is called the Borel summation of (1.68). The function (1.70) can converge to a certain value even in the region where the series (1.68) is divergent.

This method is very effective for the calculation of divergent series, which depends on single variable. The method was generalized and adjusted to different problems, for example, in [36–40]. Here we generalize the Borel summation (1.69) by applying the Pade approximant [38,39] of the order m/n to obtain the result with a better convergence:

$$P(z) = \frac{a_0 + a_1 z + \ldots + a_m z^m}{1 + b_1 z + \ldots + b_n z^n}, \tag{1.71}$$

where the coefficients are defined by the following equations:

$$\begin{aligned} P(0) = F(0),\ P'(0) = F'(0),\ P''(0) = F''(0), \ldots \\ \ldots P^{(m+n)}(0) = F^{(m+n)}(0). \end{aligned} \tag{1.72}$$

These equations reflect the fact that the Taylor expansion of $P(z)$ at zero point has the first $m+n$ terms coinciding with the first $m+n$ terms of $F(z)$. Using the notation for Pade approximant as $[m/n]_F(z)$, the Borel summation (1.70) is rewritten as:

$$f^{[m/n]}(z)(Borel - Pade) = \int_0^{+\infty} e^{-t} [m/n]_F(zt) dt. \tag{1.73}$$

The method of Borel summation with the Pade approximant can be applied to the problem of the anharmonic oscillator. In quantum mechanics, the use of the perturbation theory delivers the energy in the form of a power series of the perturbation parameter:

$$E(\lambda) = E^{(0)} + \sum_{k=1}^{+\infty} C_k \lambda^k. \tag{1.74}$$

Here we consider a specific case of the anharmonic oscillator with the perturbation term λx^4. Using the Rayleigh-Schrödinger scheme, the energy of the ground state is written in the form:

$$\begin{aligned} E_0(\lambda) = 1 &+ 0.214286 \times \left(\frac{7\lambda}{2}\right) - 0.107143 \times \left(\frac{7\lambda}{2}\right)^2 \\ &+ 0.121356 \times \left(\frac{7\lambda}{2}\right)^3 - 0.200990 \times \left(\frac{7\lambda}{2}\right)^4 \\ &+ 0.426130 \times \left(\frac{7\lambda}{2}\right)^5 - 1.087689 \times \left(\frac{7\lambda}{2}\right)^6 + O[\lambda^7]. \end{aligned} \tag{1.75}$$

The series (1.75) has a zero radius of convergence [40], which means even for a very small parameter λ, the high-order corrections for the energy do not result in a right value. For large values of λ, the terms of high order in (1.75) grow very fast, and the series (1.75) quickly becomes divergent. However, applying the Borel-Pade technique described above in the formulae (1.68–1.73), the correct values of energy $E_0(\lambda)$ can be obtained from (1.75) in the wide range of the parameter λ.

For the illustration of this technique, we reproduce in the Table 1.1 the energy values with the different orders of Pade approximation [38]. The results show the convergence to the known values of the energy for anharmonic oscillator. In the work [39], the Borel-Pade method of regular summation has been applied to obtain the asymptotic energy values for large parameter $\lambda \gg 1$: $E_0(\lambda) \sim \lambda^{1/3}$.

Table 1.1 Energy of ground state $E_0^{[m/n]}(\lambda)$ for different orders of Pade approximant

$m/n = m$	$\lambda = 0.1$	$\lambda = 0.2$	$\lambda = 1.0$
1	1.063829787234	1.111111111111	1.272727272727
2	1.065217852490	1.117540578275	1.348289096707
3	1.065280680051	1.118183011861	1.373799864956
4	1.065285049128	1.118272722955	1.383756497228
5	1.065285455239	1.118288405206	1.388075603389
6	1.065285502030	1.118291631128	1.390103754651
7	1.065285508357	1.118292382860	1.391116612108
8	1.065285509335	1.118292576357	1.391648018148
9	1.065285509503	1.118292630404	1.391938365335
10	1.065285509535	1.118292646573	1.392102495074
11	1.065285509541	1.118292651703	1.392198009942
12	1.065285509543	1.118292653416	1.392255010021

The table demonstrates the fast convergence of the method, however, the generalization of this technique for calculation of the energies of high excited states, and especially for multi-dimensional case, encounters substantial difficulties. Moreover, the partial summation of the series (1.74) for energy does not define the wave function and other physical characteristics of the system.

References

1. E. Wigner, *Symmetries and Reflections. Scientific Essays* (Indiana University Press, Bloomington, 1967)
2. R.P. Feynman, A.R. Hibbs, *Quantum Mechanics and Path Integrals* (McGraw-Hill, New York, 1965)
3. H. Kleinert, *Path Integrals in Quantum Mechanics, Statistics, Polymer Physics, and Financial Markets*, 5th edn. (World Scientific, Singapore, 2009)
4. L.D. Landau, E.M. Lifshitz, *Quantum Mechanics* (Fizmatgiz, Moscow, 2004)
5. P.M. Morse, H. Feshbach, *Methods of Theoretical Physics* (McGraw-Hill, New York, 1953)
6. N.N. Bogolubov, *Selected Works*, vol. 2 (in russian) (Navukova Dumka, Kiev, 1972), p. 499
7. B.-G. Englert, *Semiclassical Theory of Atoms* (Springer, Berlin, 1988)
8. F. Strocchi, *An Introduction to the Non-Perturbative Foundations of Quantum Field Theory* (Oxford University Press, Oxford, 2013)
9. T. Borne, G. Lochak, H. Stumpf, *Nonperturbative Quantum Field Theory and the Structure of Matter* (Springer, Heidelberg, 2001)
10. J. Fröhlich, *Non-Perturbative Quantum Field Theory. Mathematical Aspects and Applications* (World Scientific, Singapore, 1992)
11. F.M. Fernandez, E.A. Castro, *Algebraic Methods in Quantum Chemistry and Physics* (CRC Press, Boca Raton, 1996)
12. G.N. Hardy, *Divergent Series* (Clarendon Press, Oxford, 1963)
13. L. Wuytack, *Pade Approximations and its Applications* (Springer, New York, 1979)
14. I.D. Feranchuk, A.A. Ivanov, J. Phys. A Math. Gen. **37**, 9841 (2004)
15. V.I. Yukalov, Theor. Math. Phys. **28**, 652 (1976)
16. V.I. Yukalov, E.P. Yukalova, Chaos Solitons Fractals **14**, 839 (2002)
17. A.N. Sissakian, I.L. Solovtsov, Phys. Element. Part. Atom. Nucl. (in russian) **30**, 1057 (1999)
18. C.M. Bender, T.T. Wu, Phys. Rev **184**, 1231 (1969)
19. F.T. Hioe, D. MacMillen, E.W. Montroll, Phys. Rep. **335**, 307 (1978)
20. A.H. Nayfeh, Perturbation Methods (Wiley, New York, 1973)
21. P.A.M. Dirac, *The Principles of Quantum Mechanics* (Oxford University Press, Oxford, 1981.)
22. A. Messiah, *Quantum Mechanics* (North-Holland, Amsterdam, 1976)
23. J.W.S. Rayleigh, *Theory of Sound I*, 2nd edn. (Macmillan, London, 1894)
24. E. Schrödinger, Annalen der Physik **80**, 437 (1926)
25. L. Brillouin, Le Journal de Physique et le Radium Series VII **3**, 373 (1932)
26. E.P. Wigner, Math. und Natur. Anzeiger der Ungarischen Akademie der Wissenschaften **L3**, 475 (1935)
27. F.M. Fernandez, E.A. Castro, Phys. Lett. A **88**, 5083 (1982)
28. F.T. Hioe, E.W. Montroll, J. Math. Phys **16**, 1945 (1975)
29. I.S. Gradshtein, I.M. Ryzhik, *Tables of Integrals, Sums, Series and Products* (FIZMATGIZ, Moscow, 1963)
30. S.G. Mikhlin, *Variational Methods in Mathematical Physics* (Pergamon Press, Oxford, 1964)
31. F.M. Fernandez, Phys. Rev. A **50**, 2953 (1994)
32. I.D. Feranchuk, A.L. Tolstik, J. Phys. A Math. Gen. **32**, 2115 (1999)
33. M. Born, J.R. Oppenheimer, Ann. Phys. **84**, 457 (1927)

34. E. Borel, Ann. Sci. École Norm. Sup. **16**, 9 (1899)
35. G.H. Hardy, *Divergent Series* (Clarendon Press, Oxford, 1949)
36. O. Costin, *Asymptotics and Borel Summability* (CRC Press, Boca Raton, 2009)
37. J.J. Loeffel, A. Martin, B. Simon, A.S. Wightman, Phys. Lett. B **30**, 656 (1969)
38. S. Graffi, V. Grecchi, B. Simon, Phys. Lett. B **32**, 631 (1970)
39. S. Graffi, V. Grecchi, G. Turchetti, Il Nuovo Cimento B **4**, 313 (1971)
40. C.M. Bender, T.T. Wu, Phys. Rev. Lett. **27**, 461 (1971)

Chapter 2
Basics of the Operator Method

The majority of approximate methods for the solution of Schrödinger equation (SE) demonstrated in Chap. 1 is not sufficiently universal and theirs applications in the case of system with many degrees of freedom are bound up with serious difficulties. In the following chapters we consider the method which is proposed as an universal procedure for transformation of the perturbation series and its further applications for various quantum systems. We call this technique *operator method* (OM) in the sense that all calculations are reduced to the algebraic manipulations with the operator matrix elements without solving any differential or integral equations, in zeroth-order approximation as well as in calculating the successive approximations. The results prove that the OM zeroth-order approximation provides an uniformly fitted estimation of the SE eigenvalues and eigenfunctions in the entire range of the Hamiltonian parameters. As a result, the OM sequential approximations converge to the exact solution in the entire range of the system parameters and quantum numbers. The algebraic nature of the method also allows to construct the effective procedure for the calculation of the higher approximations and to determine the systems of the eigenvalues and the eigenfunctions with a high accuracy. We will demonstrate that the OM successive approximations converge as the geometric progression with the denominator defined by the ratio of non-diagonal to diagonal matrix elements of the Hamiltonian. This value depends on the choice of the initial basis set of quantum states but is less then unity under all possible conditions.

The operator method had been initially introduced in the papers [1–3]. Some ideas, which we used in our concept of the OM, were described earlier in the papers [4–10] and the development of the technique has been contributed by Fernandes, Meson and Castro [11–16] and other authors [17–20]. The similar results were obtained later by many authors (for example, [21–24]) and there are many publications where OM applications for the specific systems were reported (for example, [25–43]).

Any new method for the solutions of the SE describing the real quantum systems has to be testified upon the nontrivial model problems. In the case of nonrelativistic

I. Feranchuk et al., *Non-perturbative Description of Quantum Systems*,
Lecture Notes in Physics 894, DOI 10.1007/978-3-319-13006-4_2

quantum mechanics, the role of such test is performed by the problem of one-dimension system corresponding to QAO (1.18). The abstract quantum system with the Hamiltonian (1.18) allows to characterize the real physical systems successfully. This physical system has important applications in atomic, molecular and solid state physics, and can be considered as one-dimensional field quantum theory modeling typical features of the four-dimension quantum field systems. Due to these reasons, practically all known analytical methods have being approved for calculation of the QAO eigenvalue and eigenfunction. In the present paper we also use the QAO problem in order to introduce main ideas of the operator method for the solution of SE, to discuss its possibilities in comparison with other methods, to consider the calculation procedure of the successive approximations to the exact eigenvalues and eigenfunctions and the convergence of this procedure [44].

2.1 The Zeroth Approximation Choice

We start here with the simple example (1.19) which has been already considered within the framework of PT: the eigenvalues and eigenfunctions of the operator

$$\hat{H} = \frac{1}{2}(\hat{p}^2 + \hat{x}^2) + \mu\hat{x}^2, \tag{2.1}$$

have to be found, where $\mu > -\frac{1}{2}$.

The second quantization representation in a general form is used for this purpose, which is a further extension of (1.19):

$$\hat{x} = \frac{1}{\sqrt{2\omega}}\left[a(\omega) + a^+(\omega)\right], \quad \hat{p} = i\sqrt{\frac{\omega}{2}}\left[a^+(\omega) - a(\omega)\right], \tag{2.2}$$

where ω is an arbitrary positive real value. The operators of annihilation $a(\omega)$ and creation $a^+(\omega)$ satisfy the standard permutable relations:

$$[a(\omega), a^+(\omega)] = 1. \tag{2.3}$$

By substituting (2.2) into (2.1) and reducing the operator $\hat{H}$ to the normal form (all operators of creation are to the left side of annihilation operators) by the permutable relations (2.3) we obtain:

$$\begin{aligned} \hat{H} = \tfrac{\omega}{4}&\left[1 + 2a^+(\omega)a(\omega) - \left(a^+(\omega)\right)^2 - \left(a(\omega)\right)^2\right] + \\ &+ \tfrac{1+2\mu}{4\omega}\left[1 + 2a^+(\omega)a(\omega) + \left(a^+(\omega)\right)^2 + \left(a(\omega)\right)^2\right]. \end{aligned} \tag{2.4}$$

Operator of the excitations number $\hat{n}(\omega) = a^+(\omega)a(\omega)$ can be defined with the eigenvectors, which are the following normalized state vectors:

$$|n(\omega)\rangle = \frac{1}{\sqrt{n!}}\left[a^+(\omega)\right]^n |0(\omega)\rangle, \qquad n = 0, 1, 2, 3\ldots$$
$$a^+(\omega)a(\omega_0)|n(\omega)\rangle = n|n(\omega)\rangle. \tag{2.5}$$

The state of vacuum of the excitations is then defined by the equation:

$$a(\omega)|0(\omega)\rangle = 0. \tag{2.6}$$

In fact, this procedure is the generalization of the full set of the state vectors as a further extension of (1.25). In the case of PT, the Hamiltonian was considered in the analogous form but at $\omega = 1$:

$$\hat{H} = \hat{H}_0 + \hat{V}, \tag{2.7}$$

$$\hat{H}_0 = \tfrac{1}{2}[1 + 2a^+(1)a(1)],$$
$$\hat{V} = \tfrac{\mu}{2}\left[1 + 2a^+(1)a(1) + (a^+(1))^2 + (a(1))^2\right]. \tag{2.8}$$

The PT series for (2.7) has been shown to converge only in the range of $|\mu| < 1/2$, whereas this parameter may vary in the range of $(-\frac{1}{2}, \infty)$. Thus, the analytical continuation of the PT series is necessary to find the spectrum in the closed form and to build a new PT series converging at all possible values of the parameter μ. That can be done by choosing other type of zeroth approximation for Hamiltonian, namely to include the part of the operator $\hat{H}$ (2.4) which commutes with the excitation number operator $\hat{n}(\omega) = a^+(\omega)a(\omega)$ at arbitrary ω into $\hat{H}_0$, which means a new splitting of the operator $\hat{H} = \hat{H}_0 + \hat{H}_1$ has to be used:

$$\hat{H}_0 = \frac{1}{4\omega}(\omega^2 + 1 + 2\mu)[1 + 2a^+(\omega)a(\omega)]; \tag{2.9}$$

$$\hat{H}_1 = -\frac{1}{4\omega}(\omega^2 - 1 - 2\mu)[(a^+(\omega))^2 + a^2(\omega)]. \tag{2.10}$$

Thus in zeroth approximation of the operator (2.4) the eigenvalues have the following form:

$$E_n^{(0)}(\omega) = \frac{1}{4\omega}(\omega^2 + 1 + 2\mu)(1 + 2n); \qquad n = 0, 1, 2\ldots \tag{2.11}$$

In order to find the parameter, used in the powers series of the eigenvalue expansion of the operator (2.4) in PT with a new perturbation Hamiltonian (2.10), one has to use (2.9) and (2.10) in the standard PT series, however, with different perturbation operator, and this leads to the following series:

$$E_n = (1+2n)\frac{1}{4\omega}(\omega^2+1+2\mu)\left[1-\frac{1}{2}\left(\frac{\omega^2-1-2\mu}{\omega^2+1+2\mu}\right)^2+\ldots\right]. \tag{2.12}$$

In this case the perturbation parameter is

$$\epsilon = \left(\frac{\omega^2-1-2\mu}{\omega^2+1+2\mu}\right)^2,$$

which takes the values $\epsilon < 1$ at any real positive ω and $\mu > -\frac{1}{2}$, and therefore the new PT series converges in this range. Thus, the above mentioned choice of the zeroth approximation Hamiltonian allows one to construct an analytical continuation of the converging PT series in order to calculate Hamilton operator eigenvalues and eigenfunctions at all possible values of the parameter μ [14].

We would like to draw the attention to the fact that the parameter ω in (2.9)–(2.11) has not yet been fixed. This gives us an additional degree of freedom allowing the acceleration of the convergence of the rebuilt PT series. Since the exact eigenvalues of the total hermitian Hamiltonian don't depend on the choice of the particular representation of the full wave function set (the parameter ω in our case), the following condition has to be satisfied for exact energies:

$$\frac{\partial E_n}{\partial \omega} = 0. \tag{2.13}$$

To make further PT calculations more convenient, the parameter ω is chosen to satisfy the zeroth approximation (2.11) to the condition (2.13) or

$$\frac{\partial}{\partial \omega} E_n^{(0)}(\omega) = 0. \tag{2.14}$$

Using (2.11) in the considered example one can find:

$$\frac{1}{4}\left(1-\frac{1+2\mu}{\omega^2}\right)(2n+1) = 0; \quad \omega = \omega_0 = \sqrt{1+2\mu};$$

$$E_n^{(0)}(\omega_0) = \left(n+\frac{1}{2}\right)\sqrt{1+2\mu}. \tag{2.15}$$

Hence both the described procedure of the construction of Hamiltonian for zeroth approximation and the condition (2.14) lead to the exact solution for the eigenvalues and eigenfunctions of the operator (2.1) and the perturbation Hamiltonian (2.10) becomes equal to zero at $\omega = \omega_0$. This example proposed for the first time in [14] illustrates well the ideas of the zeroth approximation of the OM. Furthermore, we testify the suitability of this method in solving the non-trivial problem before making any generalization, for example, considering the QAO with Hamiltonian:

$$\hat{H} = \frac{1}{2}(\hat{p}^2 + \hat{x}^2) + \lambda\hat{x}^4, \qquad \lambda \geq 0. \tag{2.16}$$

As shown in the Sect. 2.1, the operator (2.16) has no discrete spectrum at $\lambda < 0$, which means the PT series with perturbation operator $\lambda\hat{x}^4$ has zero radius of convergence in a complex plane of the parameter λ. To construct the Hamiltonian (2.16) zeroth approximation in the above mentioned way, we substitute the canonical transformation (2.2) in (2.16) and reduce the operator $\hat{H}$ to the normal form:

$$\hat{H} = \hat{H}_0 + \hat{H}_1,$$

where

$$\hat{H}_0 = \frac{1}{4\omega}(\omega^2 + 1)(1 + 2a^+a) + \frac{3\lambda}{4\omega^2}\left[1 + 2a^+a + 2(a^+a)^2\right]; \tag{2.17}$$

$$\begin{aligned} \hat{H}_1 = &-\tfrac{1}{4\omega}(\omega^2 - 1)\left[(a^+)^2 + a^2\right] + \\ &+\tfrac{\lambda}{4\omega^2}\left[2(a^+)^2(2a^+a + 3)) + 2(2a^+a + 3)a^2 + (a^+)^4 + a^4\right]. \end{aligned} \tag{2.18}$$

We omit here the argument ω of the annihilation and creation operators and assume that it should be clear from the text to which values of ω the operators correspond. The eigenvectors of the OM zeroth approximation for the Hamilton operator (2.17) are the vectors (2.5) and the corresponding eigenvalues are:

$$E_n^{(0)}(\omega, \lambda) = \frac{1}{4\omega}(\omega^2 + 1)(1 + 2n) + \frac{3\lambda}{4\omega^2}(1 + 2n + 2n^2); \qquad n = 0, 1, 2 \ldots \tag{2.19}$$

According to the condition (2.13), we choose the parameters ω from the equation at each particular value of the eigenvalue of the operator (2.17):

$$\frac{\partial}{\partial\omega}E_n^{(0)}(\omega, \lambda) = 0, \tag{2.20}$$

which has the following form for the definite n:

$$\omega_n^3 - \omega_n - 6\lambda\frac{1 + 2n + 2n^2}{1 + 2n} = 0. \tag{2.21}$$

Contrary to the case of the harmonic oscillator and usual form of PT, the parameter $\omega = \omega_n$ appears to be different for different states in the problem of QAO. Due to this fact the dependence of the energy levels on the quantum number n is sufficiently complex even in zeroth approximation. Combining (2.19) and (2.21), we find the zeroth approximation of the QAO energy levels:

$$E_n^{(0)}(\lambda) = \frac{1}{4}\left(3\omega_n + \frac{1}{\omega_n}\right)\left(n + \frac{1}{2}\right), \tag{2.22}$$

where ω_n is the solution for the algebraic equation (2.21). It seems unusual that a parameter in the Hamiltonian depends on the energy level, however, we emphasize again that the parameter ω is not the real parameter of the system, but has been introduced in the Hamiltonian artificially in order to rebuild the perturbation operator. The condition (2.20) reconstructs the independence of the eigenvalues on this artificial parameter in zeroth approximation. Actually, the Eqs. (2.20) and (2.22) define the energy of the n-th level as a function of λ in a parametric form.

The condition (2.20) has a formal resemblance with the variational principle, however, the coincidence occurs only for the energies of the lowest ($n = 0$) and the first excited ($n = 1$) states. The difference between the conditions (2.20) and the variational evaluation at $n \geq 2$ is determined by the fact that in the given approach the eigenvectors of the operator (2.17) are not orthogonal to the vectors with the lower numbers. Indeed, the vectors $|n(\omega_n)\rangle$ and $|n(\omega_m)\rangle$ are not orthogonal because of according to the Eq. (2.21) they correspond to the wave functions of the harmonic oscillators with different frequencies. At the same time, according to (2.5), every vector $|n(\omega_n)\rangle$ provides an accurate symmetry and corresponding number of branch points of the wave function in coordinate representation. This is sufficient to obtain an appropriate approximation of $E_n{}^{(0)}(\lambda)$ to the exact values of $E_n(\lambda)$ (see below). However, we should use only the full orthonormal set of the state vectors in order to compute higher approximations for the energy level corresponding to the quantum number n. The diagrams in Fig. 2.1 illustrate the relation between the states used in the canonical PT and in the OM. Evidently, the main difference between these approaches is in the fact that the different full sets of wave functions are used to compute the various eigenvalues. The set of the functions for definite state does not coincide with those introduced to consider other states.

The method for the choice of the parameter ω introduced in the Hamiltonian by transformation (2.4) is closely connected with the transformations introduced by other authors in order to exceed the bounds of PT in computing QAO energy levels. First of all, it is the Principle of Minimum Sensitivity reported by Caswell [5]. The coordinate scale transformation was introduced by Dmitrieva and Plindov [7] as a method for the approximate calculation of the QAO levels. The results in these papers are similar to ours in the case of the lowest energy levels in the zeroth approximation. However, the approaches considered by these authors were

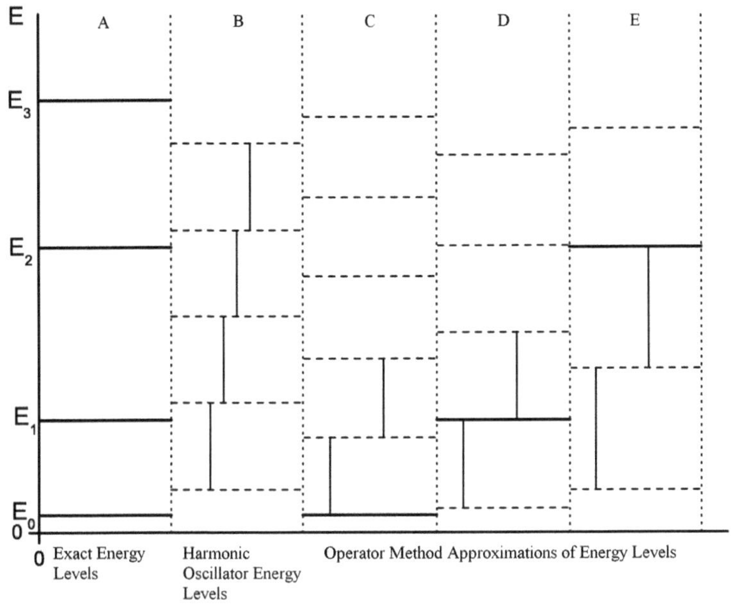

Fig. 2.1 The scheme of the full sets of states used in the standard PT series and the OM zeroth approximation. *A*—exact levels of QAO; *B*—harmonic oscillator levels for PT; *C*, *D* and *E* are the different full states used for the ground, 1st and 2nd levels correspondingly in the OM zeroth approximation

not presented in the universal form because they were based on the peculiarities of the QAO system. Besides, the regular method of calculation of successive approximations to exact eigenvalues was not proposed in these papers, too.

The delta-expansion method has been suggested as a non-perturbation approach in the quantum field theory (see, for example, the paper [45]). This method is similar to the OM with respect to several details: for example, an artificial parameter was introduced in order to rebuild the perturbation operator and to optimize the convergence of the PT series. However, alternatively to this method, the OM contains the additional idea how to include all diagonal elements of the perturbation operator to the Hamiltonian of zeroth approximation. This condition has a principal significance for the convergence of the successive approximations (see below Sect. 2.4).

The canonical transformation, being more common then (2.2), was considered in [46]. This transformation corresponds to the complex values of the parameter ω and defines a three-parametric group of transformation of the operators a and a^+. The investigation has shown that the choice of real ω provided the correct evaluation of the energy of the QAO stationary states.

To assess the effectiveness of the methods, let us compare the eigenvalue in zeroth approximation defined by the Eqs. (2.21) and (2.22) with the known exact results. In the weak coupling limit ($\lambda \to 0$), the following expansion (2.21) is valid:

$$\omega_n = 1 + 3\lambda\frac{1+2n+2n^2}{1+2n} - \frac{3}{2}\left(3\lambda\frac{1+2n+2n^2}{1+2n}\right)^2 + \\ +4\left(3\lambda\frac{1+2n+2n^2}{1+2n}\right)^3 + O(\lambda^4), \quad (2.23)$$

which being substituted in the formula (2.22) results in:

$$E_n^{(0)}(\lambda) = n + \frac{1}{2} + \frac{3}{4}\lambda(1+2n+2n^2) - \frac{9}{4}\lambda^2\frac{(1+2n+2n^2)^2}{1+2n} + \\ +\frac{27}{2}\lambda^3\frac{(1+2n+2n^2)^3}{(1+2n)^2} + O(\lambda^4). \quad (2.24)$$

This expression can be compared with the Rayleigh-Shrödinger canonic PT, applied to this problem (1.33):

$$E_n(\lambda) = n + \frac{1}{2} + \frac{3}{4}\lambda(1+2n+2n^2) - \frac{1}{8}\lambda^2(21+59n+51n^2+34n^3) + \\ +\frac{1}{16}\lambda^3(333+1041n+1416n^2+750n^3+375n^4) + O(\lambda^4). \quad (2.25)$$

Table 2.1 compares the numerical values for several coefficients of both expansions presented in the following forms:

$$E_n^{(0)}(\lambda) = \sum_{k=0}^{\infty} a_{nk}^{(OM)}\lambda^k; \qquad E_n(\lambda) = \sum_{k=0}^{\infty} a_{nk}^{(PT)}\lambda^k.$$

Table 2.1 Coefficients of the weak coupling in PT and OM series for the QAO

	k			
$a_{nk}^{(OM)}$ ($a_{nk}^{(PT)}$)	0	1	2	3
$n = 0$	0.5	0.75	−2.25	13.5
	(0.5)	(0.75)	(−2.63)	(20.8)
$n = 1$	1.5	3.75	−18.75	187.5
	(1.5)	(3.75)	(−20.63)	(244.7)
$n = 2$	2.5	9.75	−76.05	1186.38
	(2.5)	(9.75)	(−76.88)	(1254.94)
$n = 3$	3.5	18.75	−200.893	4304.8469
	(3.5)	(18.75)	(−196.875)	(4176.5625)
$n = 10$	10.5	165.75	−5232.964	330424.3
	(10.5)	(165.75)	(−4963.875)	(290771.4)

Evidently the coefficients do not fully coincide because PT leads to the asymptotically divergent series, whereas the OM series converges. In the opposite limit, when the anharmonicity parameter $\lambda \to \infty$, the functional dependence on λ in asymptotic expansion of the operator (2.16) is defined by the formula (1.42), which results in:

$$E_n(\lambda) = \lambda^{1/3} \sum_{k=0}^{\infty} b_{nk}^{(T)} (\lambda)^{-\frac{2}{3}k} \tag{2.26}$$

at $\lambda \gg 1$.

In contrast with PT series, the coefficients of asymptotic expansion (2.26) are derived numerically from the solutions of the complex differential equations. On the other hand, the solution of our algebraic equation (2.21) has the following expansion at $\lambda \gg 1$:

$$\omega_n = \lambda^{1/3}\left[6^{1/3}\left(\frac{1+2n+2n^2}{1+2n}\right)^{1/3} + \frac{1}{3\cdot 6^{1/3}}\left(\frac{1+2n}{1+2n+2n^2}\right)^{1/3}\frac{1}{\lambda^{2/3}} + O\left(\frac{1}{\lambda^2}\right)\right], \tag{2.27}$$

which results in the formula:

$$E_n(\lambda) = \lambda^{1/3}\left[\frac{3^{4/3}}{2^{8/3}}\left(1+2n+2n^2\right)^{1/3}(1+2n)^{2/3} + \right. \\ \left. + \frac{1}{4\cdot 6^{1/3}}\frac{(1+2n)^{4/3}}{(1+2n+2n^2)^{1/3}}\frac{1}{\lambda^{2/3}} - \frac{1}{144}\frac{(2n+1)^2}{(1+2n+2n^2)}\frac{1}{\lambda^{4/3}}\right] + O\left(\frac{1}{\lambda^2}\right) \tag{2.28}$$

after the substitution in (2.22). As follows from the comparison of (2.28) and (2.26), the OM zeroth approximation delivers the correct functional dependence of the parameter λ on the eigenvalues in the range of strong anharmonicity. The numerical values of the coefficients b_{nk} in formulas (2.26) and (2.28) are presented in the Table 2.2.

Table 2.2 Coefficients of strong coupling asymptotic (A) and OM series for QAO

	k		
$b_{nk}^{(OM)}$ $(b_{nk}^{(T)})$	0	1	2
$n=0$	0.68142	0.13758	−0.0069
	(0.66799)	(0.14367)	(−0.0088)
$n=1$	2.42374	0.34812	−0.0125
	(2.39364)	(0.35780)	(−0.0140)
$n=2$	4.68526	0.50027	−0.0133
	(4.69680)	(0.49397)	(−0.0125)
$n=3$	7.29111	0.63005	−0.0136
	(7.33573)	(0.61826)	(−0.0116)

Thus, comparing the formulas (2.24) and (2.28) with the formulas (2.25) and (2.26) we conclude that the function $E_n{}^{(0)}(\lambda)$ correctly reproduces a functional structure of the asymptotic series in corresponding limits. In contrast to PT series or strong coupling limit, when it is necessary to take into account higher approximations to compute addends containing new powers of λ, the function $E_n^{(0)}(\lambda)$ initially contains all necessary powers of coupling constant.

Therefore, we expect that $E_n{}^{(0)}(\lambda)$ provides a suitable approximation to the exact eigenvalues $E_n(\lambda)$ in the range of intermediate λ. Indeed, comparing the energy values defined by the formulas (2.21) and (2.22) with the numerical calculation results [47], we see that the function $E_n^{(0)}(\lambda)$ provides a valid approximation of the QAO energy levels at any n and λ (Table 2.3), the accuracy of the results is about 2–3 %. This statement is illustrated by Fig. 2.2, which shows the functions $E_n^{(0)}(\lambda)$ and $E_n(\lambda)$ and analogous functions found by PT and strong coupling approximation.

Table 2.3 Comparison of some numerical (N) and OM zeroth approximation results for QAO

	λ			
$E_n^{(N)}$ $(E_n^{(OM)})$	0.1	1	10	100
$n = 0$	0.560307	0.812500	1.53125	3.19244
	(0.559146)	(0.803771)	(1.50497)	(3.13138)
$n = 10$	17.26588	32.66349	68.17094	145.8383
	(17.35190)	(32.93326)	(68.03695)	(147.2270)
$n = 40$	94.84034	192.7883	409.8935	880.546
	(95.56017)	(194.6022)	(413.9383)	(889.325)

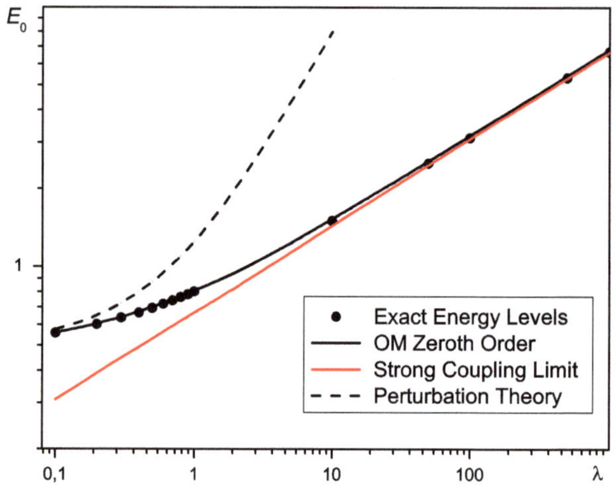

Fig. 2.2 Comparison of the function $E_0(\lambda)$ calculated in different approximations: exact values, OM, PT and strong coupling (SC) series

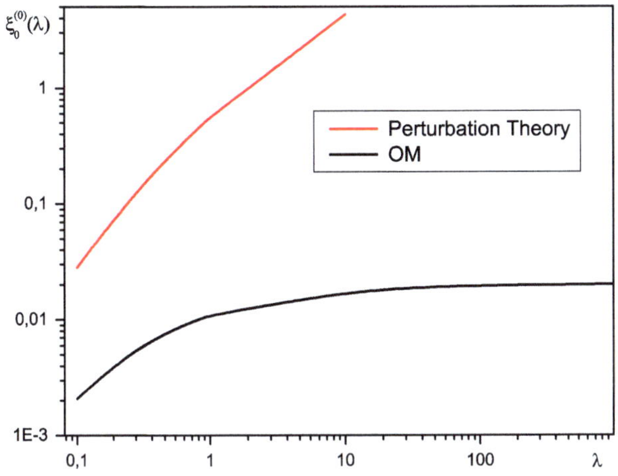

Fig. 2.3 Characteristic parameter $\xi_n^{(0)}(\lambda)$ which defines the accuracy of the OM zeroth approximation and PT for various λ and $n = 0$

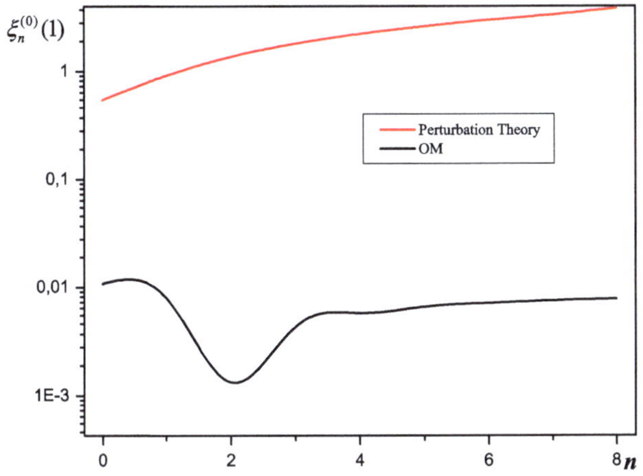

Fig. 2.4 Characteristic parameter $\xi_n^{(0)}(\lambda)$ which defines the accuracy of the OM zeroth approximation and PT for various n and $\lambda = 1$

Figure 2.3 shows the dependence of the characteristic parameter $\xi^{(0)}$, which defines the accuracy of the OM zeroth approximation for the coupling constant λ and quantum number n (Fig. 2.4). Thus, the OM zeroth approximation satisfies to the UAA condition, according to the definition in (1.1).

2.2 Iteration Scheme for Calculation of the Successive Approximations

The goodness of the function $E_n^{(0)}(\lambda)$ for approximation of the exact eigenvalue $E_n(\lambda)$ means that the operator $\hat{H}_0$ (2.17), separated from $\hat{H}$ according to the OM prescription, takes into account all the peculiarities of the initial Hamiltonian. Therefore, we may expect that the use of the operator $\hat{H}_1$ (2.18) for the solution of SE will not result in qualitative changes of the energy spectrum and makes it possible to construct the algorithm for calculation of the eigenvalues and eigenfunctions with any required accuracy at all n and λ. Thus, we consider here SE for eigenvector $|\psi_n\rangle$, corresponding to the energy E_n:

$$(\hat{H}_0 + \hat{H}_1)|\psi_n\rangle = E_n|\psi_n\rangle, \tag{2.29}$$

where operators $\hat{H}_0$ and $\hat{H}_1$ are determined by formulas (2.17) and (2.18). The vector $|\psi_n\rangle$ can be presented in terms of the eigenfunctions of operator $\hat{H}_0$:

$$|\psi_n\rangle = |n(\omega_n)\rangle + \sum_{k\neq n} C_{nk}|k(\omega_n)\rangle \tag{2.30}$$

with the coefficients C_{nk} to be defined by the Eq. (2.29). Note that all the state vectors $|k(\omega_n)\rangle$ correspond to the same parameter ω_n values at fixed n and hence, satisfy the ordinary conditions of orthogonality and completeness. The state vector (2.30) should be normalized by the following condition:

$$\langle n(\omega_n)|\psi_n\rangle = 1, \tag{2.31}$$

which essentially simplifies the further derivations. The similar condition is used for the solution of the Eq. (2.29) by means of the BWPT scheme (1.13). By substituting the expansion (2.30) in the Eq. (2.29) and calculating the projection on bra-vector $\langle n|$ by taking into account (2.31), the following expression is obtained:

$$E_n = H_{nn} + \sum_{k\neq n} C_{nk} H_{nk}. \tag{2.32}$$

We omit the index ω in the subsequent formulas:

$$(\hat{H}_0)_{nn} \equiv H_{nn}; \quad (\hat{H}_1)_{nk} \equiv H_{nk}, \quad k \neq n.$$

The equations for the coefficients C_{nm} are found by projecting the Eq. (2.29) onto different vectors $|m\rangle$ of the full set of states:

$$C_{nm} = -\Big[H_{mm} - E_n\Big]^{-1}\Big[H_{nm} + \sum_{k\neq n} H_{nk}C_{km}\Big]; \quad m \neq n. \tag{2.33}$$

By using the expressions for the matrix elements of the creation and annihilation operators ($\delta_{k,n}$ is a Kronecker symbol):

$$\langle k|a|n\rangle = \sqrt{n}\delta_{k,n-1}; \quad \langle k|a^+|n\rangle = \sqrt{n+1}\delta_{k,n+1}, \tag{2.34}$$

the expressions for the diagonal matrix elements H_{nn} and non-zero transition matrix elements H_{nk} in the case of QAO can be found:

$$H_{nn} = \frac{1}{4\omega}(\omega^2+1)(1+2n) + \frac{3\lambda}{4\omega^2}\left(1+2n+2n^2\right); \tag{2.35}$$

$$\begin{aligned} H_{n,n+2} &= H_{n+2,n} = \tfrac{1}{4}\sqrt{(n+1)(n+2)}\left[\tfrac{1-\omega^2}{\omega} + \tfrac{2\lambda}{\omega^2}(2n+3)\right]; \\ H_{n,n+4} &= H_{n+4,n} = \tfrac{\lambda}{4\omega^2}\sqrt{\tfrac{(n+4)!}{n!}}. \end{aligned} \tag{2.36}$$

The system of the nonlinear algebraic equations (2.32), (2.33) is an exact one and is completely equivalent to the initial SE (2.29). As mentioned above, the dependence of the coefficients C_{nk} on the parameter ω is determined by the fact that the form of the wave function depends on the choice of the specific representation, defined by ω_n in our case. In accordance with Fig. 2.5, these representations are different for various quantum numbers n. In contrast to SE in a coordinate representation, where the numerical solution requires a finite-element approximation for the derivatives, in the present case the problem reduces to the algebraic calculations only with the matrix elements of the Hamiltonian. On the

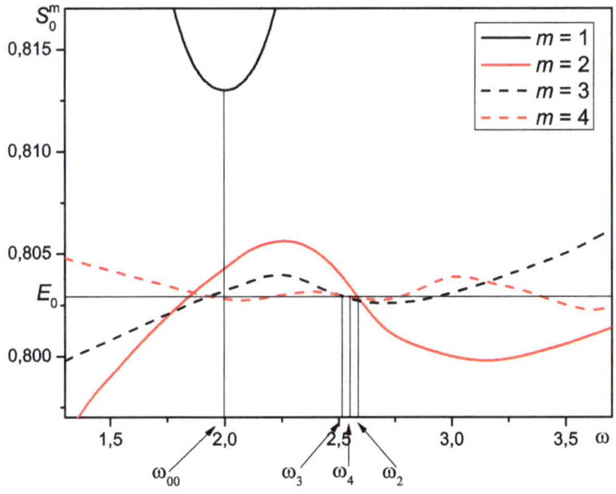

Fig. 2.5 The dependence of the OM successive approximations on the parameter ω

basis of the results, described in Sect. 2.1 for the case of QAO, there is exist a uniformly suitable zeroth approximation:

$$\omega = \omega_n; \quad E_n^{(0)} = H_{nn}(\omega_n); \quad C_{nk}^{(0)} = 0; \quad |\psi_n^{(0)}\rangle = |n(\omega_n)\rangle. \tag{2.37}$$

Based on the statements above, the non-diagonal matrix elements H_{nk} are supposed to make a negligible contribution into the energy of the system within the range of $\omega \sim \omega_n$ and this fact permits to develop a scheme of successive approximations for E_n. There are several models exists for this scheme. First of all, the calculation procedure [3,48] is chosen based on the solution of Eqs. (2.32), (2.33) by ordinary form of RSPT (1.9). By adding to the operator $\hat{H}_1$ a formal small parameter β ($\hat{H}_1 \to \beta\hat{H}_1$), the unknown quantities E_n and C_{nk} can be presented in the form of the following series:

$$E_n = H_{nn} + \sum_{s=1}^{\infty} \beta^s \epsilon_n^{(s)}; \quad C_{nk} = \sum_{s=1}^{\infty} \beta^s C_{nk}^{(s)}; \quad k \neq n. \tag{2.38}$$

The parameter β is equal to unity in the formulas (2.38) while performing all specific calculations. By substituting the expansion (2.38) in the Eqs. (2.32), (2.33) and comparing the coefficients at similar powers of β, we find the recurrence formulas connecting the values ${\epsilon_n}^{(s)}$ and ${C_{kn}}^{(s)}$ with the corrections of the lower order

$$\begin{aligned} \epsilon_n^{(1)} &= 0; \\ \epsilon_n^{(s)} &= \textstyle\sum_{k\neq n} H_{nk} C_{kn}^{(s-1)}, \quad s = 2, 3, \ldots \\ C_{mn}^{(s)} &= \left[H_{nn} - H_{mm}\right]^{-1} \left\{ \textstyle\sum_{k\neq m,n} H_{mk} C_{kn}^{(s-1)} - \sum_{t=1}^{s-1} C_{mn}^{(t)} \epsilon_n^{(s-t)} \right\}. \end{aligned} \tag{2.39}$$

Generally speaking, the parameter ω in Eqs. (2.39) remains arbitrary and it is not obvious that the choice of $\omega = \omega_n$ on the basis of condition (2.14), which is optimal in the zeroth approximation, remains the best in the consequent approximations. Moreover, the paper [11] shows that in the case of QAO, the recurrence relations (2.39) diverge at $\omega = \omega_n$ starting from some $s \geq s_0$, and the value $|{\epsilon_n}^{(s)}|$ stops to decrease at $s \geq s_0$. As shown below, the choice of different values of ω results in a reconstruction of the coefficients of the series (2.39), similarly to the partial summation of some sequences of ordinary PT terms. Therefore the sequence (2.39) converges and the rate of the convergence depends on the value ω. The parameter ${\omega_{n0}}^{(s)}$ depends on the number of iterations s, and here we discuss the procedure for calculation of the optimal value for ${\omega_{n0}}^{(s)}$ in the Eq. (2.39) for the case of QAO, as reported in [30]. We introduce the following partial sums:

$$S_n^{(t)}(\omega) = H_{nn}(\omega) + \sum_{s=2}^{t} \epsilon_n^{(s)}(\omega), \tag{2.40}$$

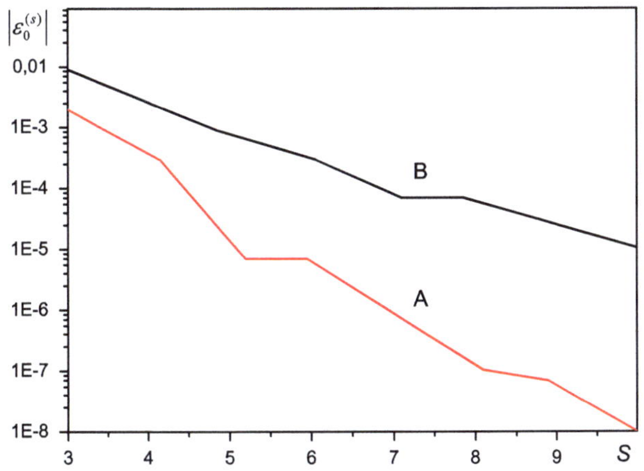

Fig. 2.6 The convergence of the OM series in cases of fixed (B) and optimized (A) parameter ω

which become equal to the exact eigenvalues E_n at $t \to \infty$ and hence should not depend on the parameter ω. This dependence disappears for QAO, and Fig. 2.5 shows several functions ${S_n}^{(t)}(\omega)$ at different n and t, computed by Eqs. (2.39) and (2.40). The algebraic nature of this calculation requires a minor computing time, which is a big advantage of the method. Computational time for the functions ${S_n}^{(t)}(\omega)$ at $t \le 10$ is practically defined by the rate of the results readout and is much less than the time of the direct numerical solution of the SE in coordinate representation by using finite-difference approximation for derivatives [49], or by other known methods [47], assuming the same accuracy of results. The analysis of the curves reproduced in Fig. 2.6 shows the convergence of the function ${S_n}^{(t)}$ to the value E_n: with t increasing, the difference $|{S_n}^{(t)} - E|$ becomes an oscillating function of ω. The oscillation magnitude, however, quickly tends to zero, and therefore the locations of the extrema of the function ${S_n}^{(t)}$ change essentially at different t and the condition (2.14) for the selection of ω becomes variational in the higher approximations. At the same time, there are such values of ω_{n0} at which ${S_n}^{(t)}$ coincides with its limiting value E_n, and this statement is valid at $t \ge 2$, if $n = 0; 1$.

The numerical algorithm for calculation of the optimal value of ω_{n0} follows from the solution of the equation:

$$
\begin{aligned}
&H_{nn}(\omega_{n0}) + \epsilon^{(2)}(\omega_{n0}) = E_n, \qquad n = 0; 1., \\
&H_{nn}(\omega_{n0}) = E_n, \qquad n \ge 2.
\end{aligned} \tag{2.41}
$$

The exact value E_n is unknown and in fact ω_{n0} is calculated as a limit of

$$\omega_{n0} = \lim_{t \to \infty} \omega_{n0}^{(t)}; \qquad \omega_{n0} > \omega_n,$$

with the parameter ${\omega_{n0}}^{(t)}$ being determined by one of the equations:

$$\sum_{s=3}^{t} \epsilon_n^{(s)}(\omega_{n0}^{(t)}) = 0; \qquad n = 0; 1,$$

$$\sum_{s=1}^{t} \epsilon_n^{(s)}(\omega_{n0}^{(t)}) = 0; \qquad n \geq 2. \tag{2.42}$$

for every eigenvalue after the calculation of t-th correction. Moreover, it is necessary to select the solution, which is the closest to the value ω_n of (2.21) to be optimal in zeroth approximation. The condition (2.42) simplifies the recurrent formulas (2.39), which take the following form:

$$C_{nk}^{(s)} = (H_{kk} - H_{nn})^{-1} \sum_{m \neq k} H_{nm} C_{mk}^{(s-1)}. \tag{2.43}$$

Figure 2.6 illustrates the convergence of the successive approximations for ω_{n0} and eigenvalues E_n by formulas (2.39)–(2.42) for QAO. The results of the numerical calculations show that for the optimal choice of the parameter ω, the successive corrections ${\epsilon_n}^{(s)}$ decrease with increasing s as the terms of the geometric progression with denominator $q \sim \frac{1}{8}$, when the relations (2.39) are used. At fixed ω [11] the convergence rate proves to be essentially less ($q \simeq \frac{1}{3}$). The comparison of exact values for QAO with the results based on the operator method with parameter optimization procedure is shown in Table 2.4.

There are another ways for adjustment of the parameter ω for calculation of high order corrections, which changes the convergence rate [11,50]. Such a change of ω is necessary for calculation of the energy and the coefficients C_{nk} in the form of series (2.38) because of the divergence of these series for fixed ω. This is an important procedure, even despite for the operator method the growth of addend takes place in the approximations of essentially higher orders than for the conventional PT. In real calculations, the optimization of ω depending on the approximation order s requires a repeated application of the recurrent formulas (2.39) at different values of ω that leads to the essential increase of the calculation

Table 2.4 Results of the optimal choice of the parameter ω

λ	E_0	$H_{00} + \epsilon_0^{(2)}(\omega_{00}^{(8)})$	$H_{00} + \epsilon_0^{(2)}(\omega_{00}^{(9)})$	$\omega_{00}^{(9)}$
1	0.80377065	0.80377079	0.80377071	2.5170107
50	2.49970877	2.49971045	2.49970957	8.7704769

λ	E_2	$H_{22} + \epsilon_2^{(2)}(\omega_{20}^{(8)})$	$H_{22} + \epsilon_2^{(2)}(\omega_{20}^{(9)})$	$\omega_{20}^{(9)}$
1	5.17929169	5.17929328	5.17929202	3.0118979
50	17.4369921	17.4370082	17.4369958	10.712937

time. Besides, the optimization algorithm is kind of a trial-and-error character in a certain sense and its application for the systems more complex than QAO may lead to time-consuming calculations. Therefore, we introduce here another scheme for the solution of the initial equations (2.32) and (2.39), which to our opinion is more effective, especially for description of the physical systems with several degrees of freedom. The present approach is based on the simple iterations in nonlinear equation system (2.32) and (2.33) and practically is a modification of operator $\hat{H}_1$ in BWPT. In this case, the calculation of the higher approximation is reduced to computing of the sequences of ${E_n}^{(s)}$ and ${C_{nk}}^{(s)}$, and not to the calculation of the corrections for zeroth order values, as it is in (2.38):

$$E_n = \lim_{s\to\infty} E_n^{(s)}; \qquad C_{nk} = \lim_{s\to\infty} C_{nk}^{(s)}. \tag{2.44}$$

Two neighboring terms of the sequences (2.44) satisfy the algebraic recurrences, which appear in the iteration scheme for the exact equations (2.32), (2.33):

$$E_n^{(s)} = H_{nn}(\omega) + \sum_{k\neq n} C_{nk}^{(s-1)} H_{kn}(\omega);$$

$$C_{nk}^{(s)} = \left[E_n^{(s-1)} - H_{kk}\right]^{-1}\left\{H_{nk}(\omega) + \sum_{m\neq k,n} H_{nm}(\omega) C_{mn}^{(s-1)}\right\}, \tag{2.45}$$

$$E_n^{(0)} = H_{nn}; \qquad C_{nk}^{(0)} = 0. \tag{2.46}$$

The volume of the calculations required to compute high order approximations by formulas (2.45) is essentially less than in using a conventional BWPT, where a complex nonlinear equation for ${E_n}^{(s)}$ should be solved to calculate the energy in s^{th} approximation [51]:

$$E_n^{(s)} = H_{nn} + \sum_{k\neq n} H_{kn}\left[E_n^{(s)} - H_{kk}\right]^{-1}\left\{H_{nk} + \sum_{m\neq k,n} H_{nm} C_{mn}^{(s-1)}\right\}.$$

At the same time, the calculation of both energy and wave functions of given state with any required precision by the iteration scheme (2.45) with the initial condition (2.46) requires a single calculation of sufficient amount of terms for recurrent sequence (2.45), which proves to converge evenly in the whole range of the parameter ω satisfying the condition:

$$\omega > \omega_n, \tag{2.47}$$

where ω_n is obtained as a solution of the Eq. (2.23). The results of calculation for QAO are demonstrated in Table 2.5, which shows that the rate of convergence strongly depends on the choice of the specific value ω. The best accuracy has been reached in [14] on the basis of these formulas for calculation of QAO energy levels.

We hope the reader, that had the patience to reach formula (2.47), has already suspected that SE solution for QAO is not an aim in itself. Therefore, in spite of the simplicity of the iteration scheme (2.45), we improve it in such a way that it is no longer connected to the specific properties of QAO. The summation over all the states $|k\rangle$ has to be performed, for which operator $\hat{H}$ provides nonzero transition probability, to calculate every iteration in (2.45) at certain n. In the case of QAO, the operator $\hat{H}$ has only 4 non-zero matrix elements at fixed n, which cuts down the volume of calculations for this specific problem. However, in the common case of non-polynomial Hamiltonian, the number of matrix elements H_{nk} will be infinite and the problem of cutting off the summation over k in (2.45) arises in real numerical calculations.

Zeroth approximation of OM $E_n{}^{(0)} = H_{nn}$ correctly describes qualitative features of the energy spectrum of the physical system, when the transitions onto the states different from $|n\rangle$ are not considered. The magnitude of the corrections to the zeroth approximation, determined by the influence of other states $|k\rangle, k \neq n$, are expected to decrease with the growth of the difference $l = |n-k\rangle$, and this fact might be settled by using the following representation:

$$H_{nk} = \sum_{l=1}^{\infty} \left\{ V_{nl}^{(+)} \delta_{k,n+l} + V_{nl}^{(-)} \delta_{k,n-l} \right\}; \quad k \neq n, \tag{2.48}$$

where

$$V_{nl}^{(\pm)} = H_{n,n\pm l}.$$

We suppose here that each term of series (2.48) has to be taken into account, starting from those elements of the sequences (2.44) for which iteration number s coincides with the "transition length" $l = |n-k|$. Then the recurrent relations (2.45) result in the following expressions:

$$E_n^{(s)} = H_{nn} + \sum_{l=1}^{s-1} \left\{ C_{n,n+l}^{(s-l)} V_{nl}^{(+)} + C_{n,n-l}^{(s-l)} V_{nl}^{(-)} \right\}; \tag{2.49}$$

$$C_{nk}^{(s)} = \frac{H_{nk}}{E_n^{(s-1)} - H_{kk}} + \sum_{l=1}^{s-1} \left\{ \frac{V_{nl}^{(+)}}{E_n^{(s-l)} - H_{kk}} C_{n+l,k}^{(s-l)} + \frac{V_{nl}^{(-)}}{E_n^{(s-l)} - H_{kk}} C_{n-l,k}^{(s-l)} \right\},$$

with the initial conditions (2.48) remaining unchanged. To calculate the approximations, the summation has to be carried out in the finite limits due to the use of formulas (2.49), which determine the OM calculation procedure for the solution of SE with any required accuracy and for the discrete spectrum states.

Table 2.5 Convergence of the OM iteration scheme (2.45)

	$(n;\lambda)$	
$E_n^{(s)} - E_n^{(s-1)}$	(10;10)	(0;1)
$s = 4$	$-0.1212077 \cdot 10^{-5}$	$-0.911329466 \cdot 10^{-8}$
$s = 8$	$-0.10778 \cdot 10^{-9}$	$0.3108703 \cdot 10^{-10}$
$s = 12$	$0.737 \cdot 10^{-11}$	$0.4712 \cdot 10^{-11}$
$s = 16$	$0.57 \cdot 10^{-13}$	$0.172 \cdot 10^{-14}$
$E_n^{(exact)}$ $(s \geq 25)$	68.80369551829225	0.803770651234274

The matrix elements for QAO (2.36) with $l = 2$ and $l = 4$ can be divided into two classes, based on the expansion (2.48). The comparison of the calculations by formulas (2.49) with the similar ones based on the full iteration scheme (2.45) shows that successive inclusion of matrix elements does not lead to tangible changes in convergence rate of OM, however, it makes this convergence more smooth. The convergence of the OM iteration scheme (2.45) is shown in Table 2.5.

2.3 Calculation Accuracy of the Wave Function

The results of the previous sections demonstrate the ability of OM to find the QAO eigenvalue spectrum with any required accuracy. However, for any method dedicated to the solution of SE, the precision of the eigenfunction calculation is even more important test because of the useability of eigenfunctions for various applications. The are different ways to assess the accuracy of the wave functions, and first of all we consider matrix elements of different operators by using zeroth approximation of OM and iteration scheme (2.49) for further approximations.

The case appears to be simple when we calculate the diagonal matrix elements since they are reduced to the well-known matrix elements of the transition between the harmonic oscillator states. For instance, the formulas for calculation of the value ${x_{nn}}^2$ in zeroth approximation $({x_{nn}}^2)_0$ and after applying s iterations $({x_{nn}}^2)_s$ have the following form:

$$
\begin{aligned}
(x_{nn}^2)_0 &= \langle n(\omega_n)|\tfrac{1}{2\omega_n}\Big(a^{+2} + a^2 + 2a^+a + 1\Big)|n(\omega_n)\rangle = \tfrac{2n+1}{2\omega_n};\\
(x_{nn}^2)_s &= \langle\psi_n^{(s)}|\tfrac{1}{2\omega}\Big(a^{+2} + a^2 + 2a^+a + 1\Big)|\psi_n^{(s)}\rangle \cdot \langle\psi_n^{(s)}|\psi_n^{(s)}\rangle^{-1};\\
|\psi_n^{(s)}\rangle &= |n(\omega)\rangle + \sum_{k\neq n} C_{nk}^{(s)}|k(\omega)\rangle;
\end{aligned} \tag{2.50}
$$

$$\langle\psi_n^{(s)}|\psi_n^{(s)}\rangle \equiv A_n^{(s)} = 1 + \sum_{k\neq n}\left[C_{nk}^{(s)}\right]^2; \tag{2.51}$$

$$(x_{nn}^2)_s = \frac{1}{2\omega A_n^{(s)}}\Bigg\{2n+1+\sum_{k\neq n}(2k+1)\left[C_{nk}^{(s)}\right]^2 + \sqrt{n(n-1)}C_{n,n-2}^{(s)}+$$

$$\sqrt{(n+1)(n+2)}C_{n,n+2}^{(s)} + \sum_{k\neq n}\Big[\sqrt{k(k-1)}C_{nk}^{(s)}C_{n,k-2}^{(s)}+$$

$$\sqrt{(k+1)(k+2)}C_{nk}^{(s)}C_{n,k+2}^{(s)}\Big]\Bigg\},$$

In formula (2.50), the parameter ω_n is taken as a solution of the Eq. (2.19) and in the Eq. (2.51) the parameter ω takes any value for which the iterations (2.49) converge. The results listed in Table 2.6 show that for calculation of diagonal matrix elements, the OM has the same characteristics as for calculation of eigenvalues: zeroth order produces evenly valid approximation at every λ. The rate of the convergence to the exact value x_{nn}^2 for the matrix element $(x_{nn}^2)_s$ is of the same order as for $E_n^{(s)}$.

The calculation of non-diagonal matrix elements requires an additional consideration. In order to be fully confident we take into account the values $x_{n,n+k}$ which are used to calculate dipole transition probability, and for this purpose the following values have to be calculated:

$$(x_{n,n+k})_0 = \langle n,\omega_n|\hat{x}|n+k,\omega_{n+k}\rangle, \tag{2.52}$$

in which different frequencies ω_n and ω_{n+k} correspond to the initial and the finite states, respectively. The operator $\hat{x}$ can be represented in terms of the operators a_ω and a_ω^+, related to arbitrary frequency. To remove this uncertainty and to preserve the algebraic nature of calculations we construct the operator $\hat{R}(\omega,\omega')$ which transforms the state $|n,\omega\rangle$ into the state $|n,\omega'\rangle$, corresponding to the frequency ω' [3]. As a basis of this construction for $\hat{R}(\omega,\omega')$ we use the invariance of coordinate $\hat{x}$ and momentum $\hat{p}$ operators under the following transformations:

$$\hat{x} = \frac{1}{\sqrt{2\omega}}(a_\omega^+ + a_\omega) = \frac{1}{\sqrt{2\omega'}}(a_{\omega'}^+ + a_{\omega'});$$

$$\hat{p} = i\sqrt{\frac{\omega}{2}}(a_\omega^+ - a_\omega) = i\sqrt{\frac{\omega'}{2}}(a_{\omega'}^+ - a_{\omega'}); \tag{2.53}$$

that is

$$a_{\omega'}^{+} = \left(\sqrt{\frac{\omega'}{\omega}} + \sqrt{\frac{\omega}{\omega'}}\right)a_{\omega}^{+} + \left(\sqrt{\frac{\omega'}{\omega}} - \sqrt{\frac{\omega}{\omega'}}\right)a_{\omega}$$

$$a_{\omega'} = \left(\sqrt{\frac{\omega'}{\omega}} + \sqrt{\frac{\omega}{\omega'}}\right)a_{\omega} + \left(\sqrt{\frac{\omega'}{\omega}} - \sqrt{\frac{\omega}{\omega'}}\right)a_{\omega}^{+} \qquad (2.53a)$$

The explicit form of the operator $\hat{R}(\omega, \omega')$ realizing the transformation (2.53*a*) can be found as:

$$\hat{R}(\omega', \omega) = \exp\left\{\tfrac{1}{4}\left(a_{\omega}^{+2} - a_{\omega}^{2}\right)\ln\tfrac{\omega'}{\omega}\right\};$$

$$a_{\omega'} = \hat{R}(\omega', \omega)a_{\omega}\hat{R}^{-1}(\omega', \omega); \qquad |n, \omega'\rangle = \hat{R}(\omega', \omega)|n, \omega\rangle, \qquad (2.54)$$

by using simple identities arising from the algebra of creation and annihilation operators. Since there are only quadratic combinations of the operators a_{ω} and a_{ω}^{+} in the exponent index of $\hat{R}(\omega, \omega')$, the latter can be reduced to the normal form [52]:

$$\hat{R}(\omega', \omega) = \exp\left(\varphi_1 a_{\omega}^{+2}\right)\exp\left[\varphi_2\left(a_{\omega}^{+}a_{\omega} + \frac{1}{2}\right)\right]\exp(-\varphi_1 a_{\omega}^{2}), \qquad (2.55)$$

where

$$\varphi_1 = \frac{\omega - \omega'}{2(\omega + \omega')}; \qquad \varphi_2 = \ln\frac{2\sqrt{\omega\omega'}}{\omega + \omega'}.$$

By using (2.55) in formula (2.52), we find the following expression for the considered matrix element in the zeroth approximation:

$$(x_{n,n+k})_0 = \left\langle n, \omega_n \left| \frac{a_{\omega_n}^{+} + a_{\omega_n}}{\sqrt{2\omega_n}} \hat{R}(\omega_{n+k}, \omega_n) \right| n + k, \omega_n \right\rangle =$$

$$\frac{1}{\sqrt{2\omega_n}}\left\{\sqrt{n}\langle n-1|\hat{R}(\omega_{n+k}, \omega)|n+k\rangle + \sqrt{n+1}\langle n+1|\hat{R}(\omega_{n+k}, \omega)|n+k\rangle\right\}. \qquad (2.56)$$

For example,

$$(x_{01})_0 = \frac{1}{\sqrt{2\omega_0}}\left[\frac{2\sqrt{\omega_0\omega_1}}{\omega_0 + \omega_1}\right]^{3/2}.$$

Table 2.6 Calculation results for QAO matrix elements and inverse problem ($\hat{H} = \frac{1}{2}(\hat{p}^2+\hat{x}^2)+\alpha\hat{x}^3+\lambda\hat{x}^4$ [30])

α	0.0	1.0	1.0
λ	1.0	1.0	0.6
$x_{01}^{(0)}$	0.501263	0.528346	0.594462
$x_{01}^{(s\geq 8)}$	0.506600	0.527097	0.595176
$x_{03}^{(0)}$	0.023587	0.029017	0.298617
$x_{03}^{(s\geq 8)}$	0.022483	0.027096	0.031825

$n_2 \to n_1$ ($\alpha = 0$)	$1 \to 0$	$3 \to 0$	$2 \to 1$
$\Omega_{n_2 n_1}^{[47]}$	1.934221	8.775756	5.025446
λ	1	2	10
$\lambda^{(0)}$	0.998562	1.999673	9.994531
$\lambda^{(10)}$	0.999998	1.999998	9.999985
$E_{n_1}^{(10)} = E_{n_1}^{[47]}$	0.803771	0.954568	5.321609

At the same time, the formula for $(x_{n,n+k})_s$ is similar to the expression (2.51) after s iterations, because of the recurrent relations (2.49) converge in a wide range of frequencies ω, and the coefficients C_{nk} of the initial and the finite states can be computed at the same values of the parameter ω:

$$(x_{n,n+k})_s = \frac{1}{\sqrt{2\omega A_n^{(s)} A_{n+k}^{(s)}}} \Bigg\{ \sqrt{n+1}\delta_{k,1} + \sqrt{n}\delta_{k,-1} + \\ + \sum_{j\neq n} C_{nj}\Big[\sqrt{j}C_{n+k,j-1} + \sqrt{j+1}C_{n+k,j+1}\Big] + \\ + \sqrt{n}C_{n+k,n-1} + \sqrt{n+1}C_{n+k,n+1}\Bigg\}. \tag{2.57}$$

Here we have taken into account the normalizing condition $\langle n|\psi_n^{(0)}\rangle = 1$ for the wave functions. The results of the calculations by formulas (2.56) and (2.57) are also presented in Table 2.6.

2.4 Iterative Solution for Inverse Problem

The possibility of the conversion of OM with respect to the Hamiltonian parameters is a good test for the iteration scheme (2.49). This operation is required for the description of real physical systems, when the potential parameters have to be found for given frequencies of transitions between energy levels (inverse problem of spectroscopy [53]). The problem can be also reduced to the recurrent formulas for calculation of the parameter λ, analogous to (2.49). The QAO Hamiltonian is represented in the following form:

$$\hat{H} = (\hat{H}_0 + \hat{H}_1) + \lambda(\hat{V}_0 + \hat{V}_1), \tag{2.58}$$

where $\hat{H}_0 + \hat{H}_1$ is the Hamiltonian of the harmonic oscillator and $\hat{V}_0 + \hat{V}_1$ is the anharmonic part of the potential, divided into diagonal and non-diagonal parts according to recipe of the OM:

$$\begin{aligned} &\hat{H}_0 = \tfrac{1}{4}\Big(\omega + \tfrac{1}{\omega}\Big)(2\hat{n} + 1); \quad \hat{H}_1 = \tfrac{1}{4}\Big(\tfrac{1}{\omega} - \omega\Big)(a^{+^2} + a^2); \quad \hat{n} = a^+ a; \\ &\hat{V}_0 = \tfrac{3}{4\omega^2}(1 + 2\hat{n} + 2\hat{n}^2); \\ &\hat{V}_1 = \tfrac{1}{4\omega^2}[6(a^2 + a^{+^2}) + a^4 + a^{+^4} + 4(\hat{n}a^2 + a^{+^2}\hat{n})]. \end{aligned} \tag{2.59}$$

We assume here that the transition frequencies are known:

$$\Omega_{mn} = E_m - E_n.$$

Then using the expansions:

$$|\psi_m\rangle = |m\rangle + \sum_{l \neq m} C_{ml}|l\rangle; \quad |\psi_n\rangle = |n\rangle + \sum_{k \neq n} C_{nk}|k\rangle,$$

the equations for energy levels E_m and E_n analogous to (2.32) are obtained:

$$E_m = H_{mm} + \lambda V_{mm} + \sum_{l \neq m} C_{ml}(H_{lm} + \lambda V_{lm}); \tag{2.60}$$

$$C_{ml} = -\Big[H_{ll} + \lambda V_{ll} - E_m\Big]^{-1}\Big\{H_{ml} + \lambda V_{ml} + \sum_{i \neq m,l} C_{mi}(H_{il} + \lambda V_{li})\Big\};$$

$$E_n = H_{nn} + \lambda V_{nn} + \sum_{k \neq n} C_{nk}(H_{kn} + \lambda V_{kn}); \tag{2.61}$$

$$C_{nk} = -\Big[H_{kk} + \lambda V_{kk} - E_n\Big]^{-1}\Big\{H_{nk} + \lambda V_{nk} + \sum_{j \neq n,k} C_{nj}(H_{jk} + \lambda V_{jk})\Big\};$$

The solution of these equations in zeroth approximation over the operators $\hat{H}_1$ and $\hat{V}_1$ permits to express the anharmonicity constant λ and the absolute value of the energy E_n in terms of the transition frequency:

$$C_{nk}^{(0)} = C_{ml}^{(0)} = 0;$$
$$\lambda^{(0)} = \frac{\Omega_{mn} - (H_{mm} - H_{nn})}{V_m^{(0)} - V_l^{(0)}}; \quad E_n^{(0)} = H_{nn} + \lambda^{(0)} V_{nn}. \tag{2.62}$$

$$\frac{\partial \lambda^{(0)}}{\partial \omega} = 0; \quad \omega = \frac{2\Omega_{mn}}{m-n}. \tag{2.63}$$

Thus, these expression can be used as the initial condition for the recurrent formulas for iteration of exact equations (2.60) and (2.61).

$$\lambda^{(s)} = \left[V_{mm} - V_{nn}\right]^{-1} \left\{ \Omega_{mn} - \sum_{l \neq m} C_{ml}^{(s-1)} \left(H_{lm} + \lambda^{(s-1)} V_{lm} \right) + \right.$$
$$\left. + \sum_{k \neq n} C_{nk}^{(s-1)} \left(H_{kn} + \lambda^{(s-1)} V_{kn} \right) - \left(H_{mm} - H_{nn} \right) \right\};$$
$$E_n^{(s)} = H_{nn} + \lambda^{(s-1)} V_{nn} + \sum_{k \neq n} C_{nk}^{(s-1)} \left(H_{kn} + \lambda^{(s-1)} V_{kn} \right); \tag{2.64}$$
$$C_{nk}^{(s)} = -\left[H_{kk} + \lambda^{(s-1)} V_{kk} - E_n^{(s-1)} \right]^{-1} \left\{ H_{nk} + \lambda^{(s-1)} V_{nk} + \right.$$
$$\left. + \sum_{j \neq n,k} C_{nj}^{(s-1)} \left(H_{jk} + \lambda^{(s-1)} V_{jk} \right) \right\}.$$

Similarly to the Eq. (2.45), the exact values of the quantities to be found are defined as the limits of the corresponding sequences:

$$\lambda = \lim_{s \to \infty} \lambda^{(s)}; \qquad E_n = \lim_{s \to \infty} E_n^{(s)}.$$

By analogy with (2.49), the iteration scheme (2.64) can be generalized for non-polynomial interaction potentials, used in the modeling of the interatomic potentials [53]. The comparison of the exact values λ with the ones calculated from (2.64) and shown in Table 2.6, demonstrates the efficiency of OM in the solution of the inverse problem.

2.5 Non-perturbative Approach in the Theory of Classical Nonlinear Oscillations

The considered above examples involve the calculation of the wave functions, which correspond to small quantum numbers and have a small number of nodes. Another test of the quality of wave functions found by OM, can be done for the systems with quickly oscillating wave functions of highly excited states. The characteristics of these systems are important in studying the properties of the quantum mechanical systems in a classical limit, which corresponds to the Plank constant $\hbar \to 0$. For these systems, we find the approximate analytical solutions for the SE for arbitrary quantum number n:

$$(\hat{H} - E_n)\psi_n(x) = 0; \tag{2.65}$$

with Hamiltonian

$$\hat{H} = \frac{1}{2}\hat{p}^2 + v(\hat{x}); \quad \hat{p} = -i\hbar\frac{d}{dx}, \tag{2.66}$$

where $v(x)$ is an arbitrary potential. With these solutions, the time dependence of the average value of the coordinate operator $\hat{x}$ at arbitrary initial state can be determined. Using the limit $\hbar \to 0$ in the obtained expression (for this, $n \to \infty$ so that the system energy should still remain a finite value), the classical law of motion $x(t)$ can be found, which is the approximate solution to the Newton's equation:

$$\frac{d^2x(t)}{dt^2} = -\frac{\partial v(x)}{\partial x}. \tag{2.67}$$

For the systems with Hamilton function (2.66), the mentioned scheme of the limit transition might be considered as a uniformly fitted method in the theory of nonlinear oscillations of classic Hamilton systems [3], which has a wider field of application in comparison with known asymptotic methods of averaging (see, e.g., [54]). This method is able to find the oscillation characteristics for any physical parameters. As an example, we consider the Eq. (2.65), assuming that a part of the solutions for this equation relates to the discrete spectrum. This assumption corresponds to the existence of the periodic solutions in the Eq. (2.65) within the classical limit. In addition, for simplicity sake we suppose function $v(x)$ to be even function of its argument. Then carrying out the canonic transformation:

$$\hat{x} = \sqrt{\tfrac{\hbar}{2\omega}}(a^+ + a); \quad \hat{p} = i\sqrt{\tfrac{\hbar\omega}{2}}(a^+ - a);$$
$$aa^+ - a^+a = 1, \tag{2.68}$$

the Hamiltonian (2.66) is reduced to the following form:

$$\begin{aligned}\hat{H} &= \tfrac{1}{4}\hbar\omega(1 + 2a^+a - a^{+2} - a^2) + \\ &+ \tfrac{1}{2\pi}\int\limits_{-\infty}^{\infty} dy\, v(y) \int\limits_{-\infty}^{\infty} dk\, e^{-iky} e^{-\frac{\hbar k^2}{4\omega}} e^{ik\sqrt{\frac{\hbar}{2\omega}}a^+} e^{ik\sqrt{\frac{\hbar}{2\omega}}a}.\end{aligned} \tag{2.69}$$

According to the recipe of the OM for the Eq. (2.65) in zeroth approximation, we extract the part of Hamiltonian (2.69) which commutates with the excitation number operator $\hat{n} = a^+a$:

$$\hat{H} = \hat{H}_0 + \hat{H}_1, \tag{2.70}$$

where

$$\hat{H}_0 = \frac{\hbar\omega}{4}(1 + 2a^+a) + \tag{2.71}$$

$$+\frac{1}{2\pi}\int\limits_{-\infty}^{\infty} dy\, v(y) \int\limits_{-\infty}^{\infty} dk\, e^{-\frac{\hbar k^2}{4\omega}} \sum_{s=0}^{\infty} \frac{(-1)^s}{(s!)^2}\left(\frac{\hbar}{2\omega}\right)^s k^{2s} (a^+)^s a^s;$$

$$\hat{H}_1 = -\frac{\hbar\omega}{4}(a^{+2} + a^2) + \tag{2.72}$$

$$+\frac{1}{2\pi}\int\limits_{-\infty}^{\infty} dy\, v(y) \int\limits_{-\infty}^{\infty} dk\, e^{-iky} e^{-\frac{\hbar k^2}{4\omega}} \left(e^{ik\sqrt{\frac{\hbar}{2\omega}}a^+} e^{ik\sqrt{\frac{\hbar}{2\omega}}a} - \right.$$

$$\left. - \sum_{s=0}^{\infty} \frac{(-1)^s}{(s!)^2}\left(\frac{\hbar}{2\omega}\right)^s k^{2s} (a^+)^s a^s \right).$$

In accordance with the definition, the eigenvectors of the operator $\hat{H}_0$ are the following vectors:

$$|n\rangle = \frac{1}{\sqrt{n!}}(a^+)^n|0\rangle; \quad n = 0, 1, 2, \ldots; \quad a|0\rangle = 0, \tag{2.73}$$

belonging to the eigenvalues:

$$E_n^{(0)} = \frac{\hbar\omega}{4}(1 + 2n) + \frac{1}{2\pi}\int\limits_{\infty}^{\infty} dy\, v(y) \int\limits_{-\infty}^{\infty} dk\, e^{-iky} e^{-\frac{\hbar k^2}{4}} \sum_{s=0}^{n} \frac{(-1)^s}{(s!)^2}\frac{n!}{(n-s)!}\left(\frac{\hbar k^2}{2\omega}\right)^s. \tag{2.74}$$

As a next step, the transition to the classical limits in the expression (2.76) for eigenvalues is carried out in the following way:

$$\hbar \to 0, \quad n \to \infty, \quad \hbar n \to \beta = const, \tag{2.75}$$

and then taking into account the fact that for $n \to \infty$:

$$\frac{n!}{(n-s)!} \to n^s,$$

we find the energy of the classical motion in the field of potential $v(x)$ using the following formula:

$$\lim_{\substack{\hbar\to 0\\ n\to\infty}} E_n^{(0)} = E = \frac{1}{2}\omega\beta + \frac{1}{2\pi}\int_{-\infty}^{\infty} dy\, v(y) \int_{-\infty}^{\infty} dk\, e^{-iky} \sum_{s=0}^{\infty} \frac{(-1)^s}{(s!)^2}\left(\frac{1}{2}k\sqrt{\frac{2\beta}{\omega}}\right)^{2s}, \tag{2.76}$$

which brings together the energy E with the parameter β. Let us note that the value β might be also considered with the respect to the arbitrary choice of the parameter ω. Taking notice of that, the sum in the right part of the Eq. (2.76) is a Bessel function $J_0\left(k\sqrt{\frac{2\beta}{\omega}}\right)$ and then using the integral [55]:

$$\int_0^{\infty} dk \cos ky J_0\left(k\sqrt{\frac{2\beta}{\omega}}\right) = \begin{cases} \sqrt{\frac{2\beta}{\omega} - y^2}, & \frac{2\beta}{\omega} - y^2 > 0; \\ 0, & \frac{2\beta}{\omega} - y^2 < 0, \end{cases} \tag{2.77}$$

we finally find the following formula:

$$E = \frac{1}{2}\omega\beta + \frac{2}{\pi}\int_0^{\sqrt{2\beta\omega^{-1}}} dy \frac{v(y)}{\sqrt{2\beta\omega^{-1} - y^2}}. \tag{2.78}$$

Expression (2.78) is derived from the formulas (2.66) for Hamiltonian eigenvalue, obtained in the OM zeroth approximation. The accounting of the corrections caused by the perturbation operator $\hat{H}_1$ in the eigenvalues leads to the change of the Eq. (2.78). However, the Eq. (2.78) remains unchanged with respect to these corrections. This situation happens when the optimal value of the parameter ω is chosen from the nullification condition for the second order correction to zeroth approximation for eigenvalues over H_1 (the OM first order correction is equal to zero by definition). The application of this condition after the transition to the classic limit results in the following equation, which determines the parameter ω (we omit calculation details, which are analogous to the transition from (2.74) to (2.78)):

$$-\frac{\omega^2\beta}{8\nu}-\frac{1}{8}\omega^2\beta^2A+\frac{\omega}{8\pi\nu}\int\limits_0^{\zeta}dy\frac{yv'(y)}{\sqrt{\zeta^2-y^2}}+\frac{2\omega\beta A}{\pi}\int\limits_0^{\zeta}dy\frac{v(y)}{\sqrt{\zeta^2-y^2}}\Big(2\frac{y^2}{\zeta^2}-1\Big)-$$

$$-\frac{1}{\pi\beta\nu}\int\limits_0^{\zeta}dy\frac{y\,v'(y)\,v(y)}{\sqrt{\zeta^2-y^2}}+\frac{2}{\pi^2\beta\nu}\int\limits_0^{\zeta}dy\frac{y\,v'(y)}{\sqrt{\zeta^2-y^2}}\int\limits_0^{\zeta}dz\frac{v(z)}{\sqrt{\zeta^2-z^2}}+$$

$$+\frac{4A}{\pi^2}\Bigg(\int\limits_0^{\zeta}dy\frac{v(y)}{\sqrt{\zeta^2-y^2}}\Bigg)^2-\frac{2A}{\pi}\int\limits_0^{\zeta}dy\frac{v^2(y)}{\sqrt{\zeta^2-y^2}}=0, \qquad (2.79)$$

where to shorten the expression we introduced the notations:

$$\zeta=\sqrt{\frac{2\beta}{\omega}};\qquad \nu=\frac{1}{2}\omega+\frac{1}{\pi\beta}\int\limits_0^{\zeta}dy\frac{y\,v'(y)}{\sqrt{\zeta^2-y^2}};\qquad v'(y)=\frac{dv}{dy};$$

$$A=\frac{\zeta^2}{4\pi\nu^2\beta^2}\int\limits_0^{\zeta}dy\frac{v''(y)}{\sqrt{\zeta^2-y^2}}\Big(2\frac{y^2}{\zeta^2}-1\Big).$$

Thus, the dependence of the classic motion energy E on the parameters ω and β is defined by the algebraic equations (2.78) and (2.79), without necessity to define the form of the potential function $v(x)$. Using the condition of the independence of eigenvalue E on the parameter ω without optimal choice of this parameter, we obtain even more simple equation instead of (2.79):

$$\frac{\partial E}{\partial\omega}=\frac{1}{2}\beta-\frac{4}{\pi\omega}\Bigg\{1+\zeta^2\int\limits_0^{\zeta}\frac{dy}{(\zeta^2+y)\sqrt{\zeta^2-y^2}}\Big[v'(y)-\frac{3}{2}\frac{v(y)}{\zeta+y}\Big]\Bigg\}=0. \qquad (2.80)$$

Equation (2.80) provides less accuracy for calculation of the parameters of classic motion law than (2.79) does, but instead allows to present the results in an analytical form. The classical law of motion $x(t)$ can be defined on the basis of the approximate solution of the SE found above. Assuming the exact eigenfunctions $|\psi_m\rangle$ and corresponding eigenvalues E_m are known, the following wave packet can be constructed:

$$|\psi(t)\rangle=\sum_m C_m e^{-\frac{i}{\hbar}E_m(t-t_0)}|\psi_m\rangle;\qquad \sum_m|C_m|^2=1. \qquad (2.81)$$

The coefficients C_m determine the packet form at $t = t_0$. According to the general rules of quantum mechanics we find the average value of the coordinate $\hat{x}$ operator in a state (2.81) as:

$$x(t) = \sum_{mm'} C^*_{m'} C_m e^{\frac{i}{\hbar}(E_{m'}-E_m)(t-t_0)} \langle \psi_{m'} | \hat{x} | \psi_m \rangle, \tag{2.82}$$

or

$$x(t) = \sum_{mk} C^*_{m+k} C_m e^{\frac{i}{\hbar}(E_{m+k}-E_m)(t-t_0)} \langle \psi_{m+k} | \hat{x} | \psi_m \rangle. \tag{2.83}$$

The classical trajectory of a particle implies the wave packet in which the coefficients C_m as a function of m have a maxima close to the value $n \gg 1$, defined by (2.75) and (2.76):

$$E_n = E, \qquad \hbar n = \beta.$$

This definition demonstrates that the classical law of motion does not depend on the wave packet form and is defined by the following expression:

$$x(t) = \sum_{k=-\infty}^{\infty} e^{ik\Omega(t-t_0)} \lim_{\substack{\hbar \to 0 \\ n \to \infty}} \langle \psi_{n+k} | \hat{x} | \psi_n \rangle, \tag{2.84}$$

where

$$\Omega = \frac{\partial E}{\partial \beta} \tag{2.85}$$

is the frequency of the classical periodic motion. For the following calculations we use the approximate expressions, obtained by OM for wave vectors of stationary states $|\psi_n\rangle$ in (2.84). In the zeroth approximation these wave functions are defined by the formula (2.73):

$$|\psi_n^{(0)}\rangle = |n, \omega_n\rangle, \tag{2.86}$$

where from evidently follows that in the OM zeroth approximation every state with n excitations corresponds its own value ω_n due to the choice of the parameter ω described above. The states $|n, \omega_n\rangle$ and $|m, \omega_m\rangle$ are nonorthogonal at $m \neq n$ and that results in a certain change of the definition (2.84) for the classical law of motion :

$$x^{(0)}(t) = \frac{1}{N(t)} \sum_{k=-\infty}^{\infty} e^{ik\Omega(t-t_0)} \lim_{\substack{\hbar \to 0 \\ n \to \infty}} \langle n+k, \omega_{n+k} | \hat{x} | n, \omega_n \rangle, \tag{2.87}$$

where

$$N(t) = \sum_{k=-\infty}^{\infty} e^{ik\Omega(t-t_0)} \lim_{\substack{\hbar\to 0\\ n\to\infty}} \langle n+k, \omega_{n+k}|n, \omega_n\rangle. \tag{2.88}$$

The further calculations are based on the use of a unitary transformation (2.56), and applying the normal form (2.57) we find:

$$\langle n+k, \omega_{n+k}|n, \omega_n\rangle = \left(ch\, 2\zeta\right)^{-1/2} \langle n+k|e^{-\frac{A^{+}}{2} th\, 2\zeta} e^{-\hat{n}\ln ch\, 2\zeta} e^{\frac{A}{2} th\, 2\zeta}|n\rangle, \tag{2.89}$$

where

$$\zeta = \frac{1}{4}\ln\frac{\omega_{n+k}}{\omega_n} \simeq \frac{1}{2}\gamma\frac{k}{n}; \qquad \gamma = \frac{\beta}{2\omega}\frac{\partial\omega}{\partial\beta}, \tag{2.90}$$

and thus all the operators and wave vectors are reduced to the same value of the parameter ω_n. By performing the limit transition in formula (2.89):

$$\lim_{\substack{\hbar\to 0\\ n\to\infty}} \langle n+k, \omega_{n+k}|n, \omega_n\rangle = \begin{cases} J_{\frac{1}{2}k}(\gamma k), & k = 2l; \\ 0, & k = 2l+1, \end{cases} \tag{2.91}$$

where $J_\nu(x)$ is the ν-th order Bessel function. Then, applying the integral representation of the Bessel function [55]:

$$J_k(z) = \frac{1}{\pi}\int_0^{\pi} d\varphi\, \cos(k\varphi - z\sin\varphi), \tag{2.92}$$

the expression for normalizing factor $N(t)$ can be reduced to the following form ($t_0 = 0$):

$$N(t) = 1 + \frac{2}{\pi}\int_0^{\pi} \lim_{\epsilon\to 0}\sum_{k=1}^{\infty} e^{-\epsilon k}\cos(2\Omega tk)\cos(k\varphi - 2\gamma k\sin\varphi) \tag{2.93}$$

and takes its final form after summation and integration:

$$N(t) = \frac{1}{1-2\gamma\cos\varphi}, \tag{2.94}$$

with phase φ, which is determined by the equation:

$$\Omega t - \frac{1}{2}\varphi + \gamma\sin\varphi = 0. \tag{2.95}$$

By doing similar calculations in the numerator of the formula (2.87), the use of the OM zeroth approximation for eigenvalues and eigenfunctions of Hamiltonian in the range of high quantum numbers leads to the following expression for the classical law of motion:

$$x^{(0)}(t) = \sqrt{\frac{2\beta}{\omega}} \cos\frac{\varphi}{2}, \tag{2.96}$$

with the phase φ being defined by the Eqs. (2.95). The parameters β, γ, ω and Ω are determined by the classical motion energy E by formulas (2.78), (2.79) or (2.80), (2.85) and (2.90). The parameter ω equals to the classical motion frequency, and is calculated from the Eq. (2.80):

$$\omega(E) = \Omega(E). \tag{2.97}$$

Having finished with the bulky analytical computations in general case, we move to the quantitative comparison of the results obtained above with the exact solution for the classical law of motion for anharmonic oscillator, which is the main subject of this chapter. For anharmonic oscillator, the function $x(t)$ represents the solution of the differential equation in the form of (2.67):

$$\frac{d^2x}{dt^2} + x + 4\lambda x^3 = 0, \tag{2.98}$$

and is defined by the following formula:

$$x(t) = \left[\frac{\sqrt{1+16\lambda E}-1}{4\lambda}\right]^{1/2} cn(\tau, k), \tag{2.99}$$

where

$$\tau = t(1+16\lambda E)^{1/4}, \quad k^2 = \frac{1}{2}\left[1 - \frac{1}{\sqrt{1+16\lambda E}}\right];$$

E is the energy of the classical motion, $cn(\tau, k)$ is the elliptical Jacobi function. The classical motion frequency is defined by the following expression:

$$\Omega = \frac{\pi}{2K(k)}(1+16\lambda E)^{1/4}, \tag{2.100}$$

where $K(k)$ is the full elliptical integral of the first kind [55]. The averaging methods for the problem of QAO [54] deliver only the asymptotic expansions for functions (2.99) and (2.100), which are valid at $\lambda E \ll 1$. Within the limits of the OM zeroth approximation, the analytical formula for a classical oscillation frequency might be found by using the Eq. (2.80) and the relation (2.97):

$$\Omega_0(E) = \left(\frac{1 + 2\sqrt{1 + 18\lambda E}}{3}\right)^{1/2}.$$

The results of the calculation of the oscillation period $T^{(0)}(E) = 2\pi/\Omega_0(E)$ using above formula are shown in the Table 2.7 as a function of the dimensionless parameter λE. These results have been compared with the exact values of period $T(E)$ and function $T_2(E)$, obtained by the Eq. (2.79), which corresponds to the second order of the OM. In the same table the laws of the QAO motion obtained by the exact formula (2.99) and the approximate one (2.96) are compared in the case of $\lambda E \gg 1$, which is the most unfavorable case from the point of view of the result accuracy. Formulas (2.95) and (2.96) obtained above have a universal form, which does not depend on the form of the potential function. If anharmonicity parameter satisfies the condition $\lambda \ll 1$, they lead to the same results as the asymptotical method of averaging does [54]. However, the Table 2.7 shows that the formulas provide the motion laws similar to the exact ones in the case of $\lambda E \gg 1$. Thus, the results obtained indicate that OM provides evenly valid interpolation of the SE solution for QAO within the limits of asymptotically high quantum numbers.

Another useful result obtained from the formulas (2.100) and (2.97) is the analytical approximation of the full elliptical integral of the first kind. At $0 \le k \le 1/\sqrt{2}$ we obtain:

$$K(k) \approx \frac{\pi}{2}\left(\frac{3}{1 - 2k^2 + 2\sqrt{1 + \frac{1}{2}k^2(1 - k^2)}}\right)^{1/2}, \qquad (2.101)$$

and the precision of the approximations described above has the same order as for the estimation of the classical motion period.

Table 2.7 Oscillation period and motion law of a classical anharmonic oscillator

	λE				
T	0.1	0.5	1.0	10	$(\lambda E)^{1/4}T, \lambda E \gg 1$
$T(E)$	5.21198	4.00431	3.47306	2.04604	3.70815
$T^{(2)}(E)$	5.21195	4.00425	3.47302	2.04605	3.70820
$T^{(0)}(E)$	5.21992	4.02115	3.49105	2.06007	3.73600

	$\tau = 2(\lambda E)^{1/4}t, \lambda E \gg 1$				
x	0	0.4	1.0	1.4	1.6
$x(\tau)$	1	0.923	0.596	0.321	0.176
$x^{(0)}(\tau)$	1.017	0.905	0.551	0.293	0.164

2.6 Why Do the OM Successive Approximations Converge?

The set of the empirical results obtained above in using of the OM prescriptions for QAO is not a fully sufficient basis to expand this method to other systems. It is necessary to find the reason why the special choice of the zeroth approximation Hamiltonian with arbitrary parameter results in a radical reconstruction of the successive approximation series, which converge evenly at any quantum numbers within the entire range of the Hamiltonian parameters, and are not asymptotic with zeroth convergence radius as it is in the PT canonical scheme. For this purpose we compare once again the ways of the division of the QAO Hamiltonian into zeroth approximation Hamiltonian $\hat{H}_0{}'$ and perturbation operator $\hat{H}_1{}'$ in the case of canonical PT:

$$\hat{H}_0{}' = \frac{1}{2}(2\hat{n}+1); \quad \hat{n} = a^+a; \quad \hat{H}_1{}' = \frac{\lambda}{4}(a+a^+)^4, \tag{2.102}$$

and in OM ($\hat{H}_0$ and $\hat{H}_1$):

$$\begin{aligned} \hat{H}_0 &= \tfrac{1}{4\omega^2}\Big[\omega(\omega^2+1)(2\hat{n}+1) + 3\lambda(2\hat{n}^2+2\hat{n}+1)\Big]; \\ \hat{H}_1 &= \tfrac{1}{4\omega^2}a^{+2}\Big[2\lambda(2\hat{n}^2+3) - \omega(\omega^2-1)\Big] + \\ &\quad + \tfrac{1}{4\omega^2}\Big[2\lambda(2\hat{n}+3) - \omega(\omega^2-1)\Big]a^2 + \tfrac{\lambda}{4\omega^2}\Big(a^{+4} + a^4\Big). \end{aligned} \tag{2.103}$$

The formula (2.103) differs from (2.102) in two essential features: (i) it includes all addends commutating with the operator of excitation number $\hat{n}$ into $\hat{H}_0$; (ii) dependence of the zeroth approximation Hamiltonian $\hat{H}_0$ on free parameter (frequency ω). Parameter ω can be used either to choose the best zeroth approximation, or to increase the convergence rate of successive approximations. The Hamiltonian division similar to (2.103) can be carried out for any system on the basis of full set of the state vectors as it will be shown below for other applications of the OM [1–3]. Now we discuss the role of each of the mentioned factors in the PT series reconstruction in details. The factor (i) leads to the appearance of coupling constant λ and the quantum numbers of intermediate states in the denominators of the terms of the series for exact SE solution. They are located in the expansion of the operator $\hat{H}_1$ powers due to the propagator $(E_n{}^{(0)} - H_0)^{-1}$. The exact eigenvalues and eigenfunctions of Hamilton operator $\hat{H}$ can be represented in the form of the following operator series (see, e.g., [56]):

$$|\psi_n\rangle = \lim_{\alpha\to 0} \frac{\hat{U}_\alpha|n\rangle}{\langle n|\hat{U}_\alpha|n\rangle},$$

$$E_n = E_n^{(0)} + \lim_{\alpha\to 0} \frac{\langle n|\hat{H}_1\hat{U}_\alpha|n\rangle}{\langle n|\hat{U}_\alpha|n\rangle}, \tag{2.104}$$

where

$$\hat{U}_\alpha = 1 + \sum_{s=1}^{\infty} \frac{1}{E_n^{(0)} - \hat{H}_0 + is\alpha} \hat{H}_1 \frac{1}{E_n^{(0)} - \hat{H}_0 + i(s-1)\alpha} \hat{H}_1 \ldots$$

$$\ldots \frac{1}{E_n^{(0)} - \hat{H}_0 + i\alpha} \hat{H}_1, \tag{2.105}$$

and therefore the successive approximations are defined by the operator powers:

$$\hat{B}_n = \frac{1}{E_n^{(0)} - \hat{H}_0} \hat{H}_1, \tag{2.106}$$

which provides the following results for the arbitrary intermediate k-quantum state:

$$\hat{B}_n^{(PT)}|k\rangle = \tfrac{\lambda}{4} \frac{1}{E_n^{(0)} - (k+4+1/2)} \sqrt{\tfrac{(k+4)!}{k!}}|k+4\rangle +$$

$$+\tfrac{\lambda}{2} \frac{2k+3}{E_n^{(0)} - (k+2+1/2)} \sqrt{\tfrac{(k+2)!}{k!}}|k+2\rangle + \tfrac{3}{4}\lambda \frac{1+2k+2k^2}{E_n^{(0)} - k}|k\rangle + \tag{2.107}$$

$$+\tfrac{\lambda}{2} \frac{2k-1}{E_n^{(0)} - (k-2+1/2)} \sqrt{\tfrac{k!}{(k-2)!}}|k-2\rangle + \tfrac{\lambda}{4} \frac{1}{E_n^{(0)} - (k-4+1/2)} \sqrt{\tfrac{k!}{(k-4)!}}|k-4\rangle$$

in the case of PT for $H_0{}'$ and $H_1{}'$, defined by formulas (2.102), and

$$\hat{B}_n^{(OM)}|k\rangle =$$

$$\tfrac{\lambda}{4\omega^2} \frac{1}{E_n^{(0)} - \tfrac{1}{4\omega^2}[\omega(\omega^2+1)(1+2k+8)+3\lambda(1+2k+8+2k^2+16k+32)]} \sqrt{\tfrac{(k+4)!}{k!}}|k+4\rangle +$$

$$\tfrac{1}{4\omega^2} \frac{2\lambda(2k+3) - \omega(\omega^2-1)}{E_n^{(0)} - \tfrac{1}{4\omega^2}[\omega(\omega^2+1)(1+2k+4)+3\lambda(1+2k+4+2k^2+8k+8)]} \sqrt{\tfrac{(k+2)!}{k!}}|k+2\rangle +$$

$$\tfrac{1}{4\omega^2} \frac{2\lambda(2k-1) - \omega(\omega^2-1)}{E_n^{(0)} - \tfrac{1}{4\omega^2}[\omega(\omega^2+1)(1+2k-4)+3\lambda(1+2k-4+2k^2-8k+8)]} \sqrt{\tfrac{k!}{(k-2)!}}|k-2\rangle + \tag{2.108}$$

$$\tfrac{\lambda}{4\omega^2}\Big\{E_n^{(0)} - \tfrac{1}{4\omega^2}[\omega(\omega^2+1)(1+2k-8)+$$

$$+3\lambda(1+2k-8+2k^2-16k+8)]\Big\}^{-1} \sqrt{\tfrac{k!}{(k-4)!}}|k-4\rangle$$

for the operators (2.103) corresponding to the OM.

Apparently for any finite n in the limit of $k \gg 1$ the matrix element of the operator $\hat{B}_n^{(PT)}$ behaves as

$$\|\hat{B}_n^{(PT)}\| = c\lambda k \qquad (c = const),$$

while for the arbitrary finite values of n, λ and ω the matrix element of $\hat{B}_n^{(OM)}$ at $k \gg 1$ do not exceed:

$$\|\hat{B}_n^{(OM)}\| < \frac{2}{3},$$

which follows from the inclusion of the operator

$$\frac{3\lambda}{4\omega^2}(2\hat{n}^2 + 2\hat{n} + 1)$$

into $\hat{H}_0$ and results in a power decrease of all coefficients in the expansion (2.104) instead of their factorial increase when taking into account the whole anharmonicity by PT. To prove this fact, we consider the particular sequences (2.104) determining the nth steady state energy. The sequence of addends in the form:

$$n - \boxed{n+4} - \ldots - \boxed{n+4k-4} - \boxed{n+4k} - \boxed{n+4k-4} - \ldots - \boxed{n+4} - n \tag{2.109}$$

proves to be the one of the most quickly increasing within the limits of ordinary PT. Here we use a symbolic graphic presentation of intermediate states, following from the analytical formula (2.104) for the kth term of the series corresponding to diagram (2.109):

$$\begin{aligned} I_k = H_{n,n+4} & \frac{1}{E_n^{(0)} - H_{n+4,n+4}} H_{n+4,n+8} \frac{1}{E_n^{(0)} - H_{n+8,n+8}} \cdots \\ & \cdots \frac{1}{E_n^{(0)} - H_{n+4,n+4}} H_{n+4,n}. \end{aligned} \tag{2.110}$$

By substituting the matrix elements of the operators $\hat{B}_n^{(PT)}$ and $\hat{B}_n^{(OM)}$, defined by the formulas (2.107) and (2.108), into (2.110), we find the following expressions for I_k in the limits of PT and OM:

$$I_k^{(PT)} = -\left(\frac{\lambda}{4}\right)^{2k} \frac{1}{[(k-1)!]^2} \frac{(n+4k)!}{n!}; \tag{2.111}$$

$$I_k^{(OM)} = -\left(\frac{\lambda}{8}\right)^{2k} \frac{2k[\omega(\omega^2+1) + 3\lambda(4k+2n+1)]}{\omega^2 n!(k!)^2}. \tag{2.112}$$

For fixed m, λ and ω and at $k \gg 1$:

$$I_k^{(PT)} = -\left(\frac{4\lambda}{e}\right)^{2k} k^{2k+n+\frac{3}{2}} A_n^{(PT)}; \tag{2.113}$$

$$I_k^{(OM)} = -\frac{1}{6^{2k}} \frac{1}{k^{\frac{\omega(\omega^2+1)}{6\lambda}}} A_n^{(OM)}, \tag{2.114}$$

where $A_n^{(PT)}$ and $A_n^{(OM)}$ are the constants, which are independent on k. As follows from (2.113), the series formed of the addends (2.109), diverges at any $\lambda > 0$ in the case of PT. At the same time, the formula (2.114) shows that for OM a similar series converges even at $\lambda \gg 1$, with the rate which is not worse than the geometrical progression with denominator $q = \frac{1}{36}$. This radical change of the convergence is caused by including the intermediate state quantum numbers in propagator $[E_n{}^{(0)} - \hat{H}_0]^{-1}$. A similar treatment of the series formed of the addends corresponding to the diagrams:

$$n - \boxed{n+2} - \ldots - \boxed{n+2k-2} - \boxed{n+2k} - \boxed{n+2k-2} - \ldots - \boxed{n+2} - n \tag{2.115}$$

results (at $k \gg 1$) in the estimation:

$$J_k^{(OM)} = -\left(\frac{4}{9}\right)^k \frac{1}{k^{\frac{\omega(\omega^2+1)}{3\lambda} + \frac{\omega(\omega^2-1)}{4\lambda}}} A_n^{(1)}. \tag{2.116}$$

Thus the sequence of addends corresponding to the diagrams (2.115) converges not worse than the geometrical progression with the denominator $\frac{4}{9}$. The results obtained above allows us to conclude that the first specific feature of the OM (the including of all diagonal matrix elements in $\hat{H}_0$) results in an absolute convergence of different sequences appearing in the power series of the operator $\hat{H}_1$ in the whole range of the Hamiltonian parameters. This property of OM is universal and does not depend on the choice of the representation form [25–43].

The above choice of the Hamiltonian for zeroth approximation is very close to some methods widely used in the description of the systems with many particles. For example, in Hartree–Fock method the state vectors $|N_i\{\phi_i\}\rangle$ play the role of the basis set and correspond to N noninteracting particles, and one-particle wave function chosen from the variation analogue of the condition (2.14) [56, 57] act as arbitrary parameter. Another example is the method based on the separation of so-called coherent wave [58, 59] in describing the particle interaction with a system possessing an infinite number of degrees of freedom. In this case, the basis set corresponds to the plane waves with the refractive index different from unity and is an analogue of the parameter ω in the QAO problem.

However, the adequate choice of the Hamiltonian for zeroth approximation in the OM, resulting in the convergence of partial sequences of PT series over the operator $\hat{H}_1$, does not automatically mean that this series converges at arbitrary choice of the basic set of functions. The number of possible diagrams increases rapidly with the PT order increase. As an example we show in Fig. 2.7 all diagrams corresponding to the corrections of the 2–4th order for the QAO energy on nth level (analytical expression for diagram can be written using the same rules as in formula (2.110)). Thus, convergence of the series depends not only on convergence of partial sequences of diagrams, but on the rate of the number of such sequences S_k, and increases with the growth of the number k, which determines PT order. Therefore to limit the S_k growth and to accelerate a convergence of PT series, the additional degrees of freedom included into the given scheme of the OM can be used: (1) the choice of the full set of functions; (2) the use of the optimal values for the parameters like ω; (3) the calculations on the basis of iteration procedure providing the most rapid convergence. The way to use these opportunities depends on the specific peculiarities of the physical systems, however, several general recipes can be derived which are common for the majority of the applications.

Concluding this paragraph, we would like to briefly discuss how the additional degrees of freedom of the OM influence the value of S_k in the case of QAO. For example, all the uncoupled diagrams (diagrams 17–32 of the 4th-order correction in Fig. 2.7) disappear due to the wave function normalizing condition (2.31). A special choice of the parameter ω lessens the number of the diagrams, which are not equal to zero. In particular, by choosing ω from the condition for the matrix element of transition (in the nearest state for given level) to be zero [3], all the odd order diagrams disappear and the number of the diagrams in PT even order essentially decreases (for example, in Fig. 2.7 only the diagrams 1,4 in the second order and the diagrams 1,2 and 15,16 in the forth order are not equal to zero). By choosing ω_{opt} from the condition (2.42), the parameter takes a value, for which the sum of all diagrams beyond some given order turns to zero. Finally, the recurrent sequence defined by the iteration procedure (2.45) permits to calculate more precise values of energy, and this is a full analogue of Dyson equations arising from the summing diagrams which describe any quantum mechanical system.

The qualitative arguments listed in this paragraph are not a mathematical proof of the OM convergence, but to a certain extent they explain a high efficiency of the method in describing various physical systems which will be considered in the following chapters.

SECOND ORDER

$1)\ n-\boxed{n+4}-n$ $2)\ n-\boxed{n+2}-n$ $3)\ n-\boxed{n-2}-n$ $4)\ n-\boxed{n-4}-n$

THIRD ORDER

$1)\ n-\boxed{n+4}-\boxed{n+2}-n$ $2)\ n-\boxed{n+2}-\boxed{n+4}-n$

$3)\ n-\boxed{n+2}-\boxed{n-2}-n$ $4)\ n-\boxed{n-2}-\boxed{n+2}-n$

$5)\ n-\boxed{n-2}-\boxed{n-4}-n$ $6)\ n-\boxed{n-4}-\boxed{n-2}-n$

FOURTH ORDER

$1)\ n-\boxed{n-4}-\boxed{n-8}-\boxed{n-4}-n$ $2)\ n-\boxed{n-4}-\boxed{n-6}-\boxed{n-4}-n$

$3)\ n-\boxed{n-4}-\boxed{n-6}-\boxed{n-2}-n$ $4)\ n-\boxed{n-2}-\boxed{n-6}-\boxed{n-4}-n$

$5)\ n-\boxed{n-2}-\boxed{n-6}-\boxed{n-2}-n$ $6)\ n-\boxed{n-2}-\boxed{n-4}-\boxed{n-2}-n$

$7)\ n-\boxed{n-2}-\boxed{n+2}-\boxed{n-2}-n$ $8)\ n-\boxed{n-2}-\boxed{n+2}-\boxed{n+4}-n$

$9)\ n-\boxed{n+2}-\boxed{n-2}-\boxed{n-4}-n$ $10)\ n-\boxed{n+2}-\boxed{n-2}-\boxed{n+2}-n$

$11)\ n-\boxed{n+2}-\boxed{n+4}-\boxed{n+2}-n$ $12)\ n-\boxed{n+2}-\boxed{n+6}-\boxed{n+2}-n$

$13)\ n-\boxed{n+2}-\boxed{n+6}-\boxed{n+4}-n$ $14)\ n-\boxed{n+4}-\boxed{n+6}-\boxed{n+2}-n$

$15)\ n-\boxed{n+4}-\boxed{n+6}-\boxed{n+4}-n$ $16)\ n-\boxed{n+4}-\boxed{n+8}-\boxed{n+4}-n$

$$17)\ \begin{array}{c} n-\boxed{n-4}-n \\ \boxed{n-4} \\ n-\boxed{n-4}-n \end{array} \quad 18)\ \begin{array}{c} n-\boxed{n-2}-n \\ \boxed{n-2} \\ n-\boxed{n-4}-n \end{array} \quad 19)\ \begin{array}{c} n-\boxed{n+2}-n \\ \boxed{n+2} \\ n-\boxed{n-4}-n \end{array} \quad 20)\ \begin{array}{c} n-\boxed{n+4}-n \\ \boxed{n+4} \\ n-\boxed{n-4}-n \end{array}$$

$$21)\ \begin{array}{c} n-\boxed{n-4}-n \\ \boxed{n-4} \\ n-\boxed{n-2}-n \end{array} \quad 22)\ \begin{array}{c} n-\boxed{n-2}-n \\ \boxed{n-2} \\ n-\boxed{n-2}-n \end{array} \quad 23)\ \begin{array}{c} n-\boxed{n+2}-n \\ \boxed{n+2} \\ n-\boxed{n-2}-n \end{array} \quad 24)\ \begin{array}{c} n-\boxed{n+4}-n \\ \boxed{n+4} \\ n-\boxed{n-2}-n \end{array}$$

$$25)\ \begin{array}{c} n-\boxed{n-4}-n \\ \boxed{n-4} \\ n-\boxed{n+2}-n \end{array} \quad 26)\ \begin{array}{c} n-\boxed{n-2}-n \\ \boxed{n-2} \\ n-\boxed{n+2}-n \end{array} \quad 27)\ \begin{array}{c} n-\boxed{n+2}-n \\ \boxed{n+2} \\ n-\boxed{n+2}-n \end{array} \quad 28)\ \begin{array}{c} n-\boxed{n+4}-n \\ \boxed{n+4} \\ n-\boxed{n+2}-n \end{array}$$

$$29)\ \begin{array}{c} n-\boxed{n-4}-n \\ \boxed{n-4} \\ n-\boxed{n+4}-n \end{array} \quad 30)\ \begin{array}{c} n-\boxed{n-2}-n \\ \boxed{n-2} \\ n-\boxed{n+4}-n \end{array} \quad 31)\ \begin{array}{c} n-\boxed{n+2}-n \\ \boxed{n+2} \\ n-\boxed{n+4}-n \end{array} \quad 32)\ \begin{array}{c} n-\boxed{n+4}-n \\ \boxed{n+4} \\ n-\boxed{n+4}-n \end{array}$$

Fig. 2.7 The diagrams corresponding to the OM approximations of the 2-nd–4-th orders

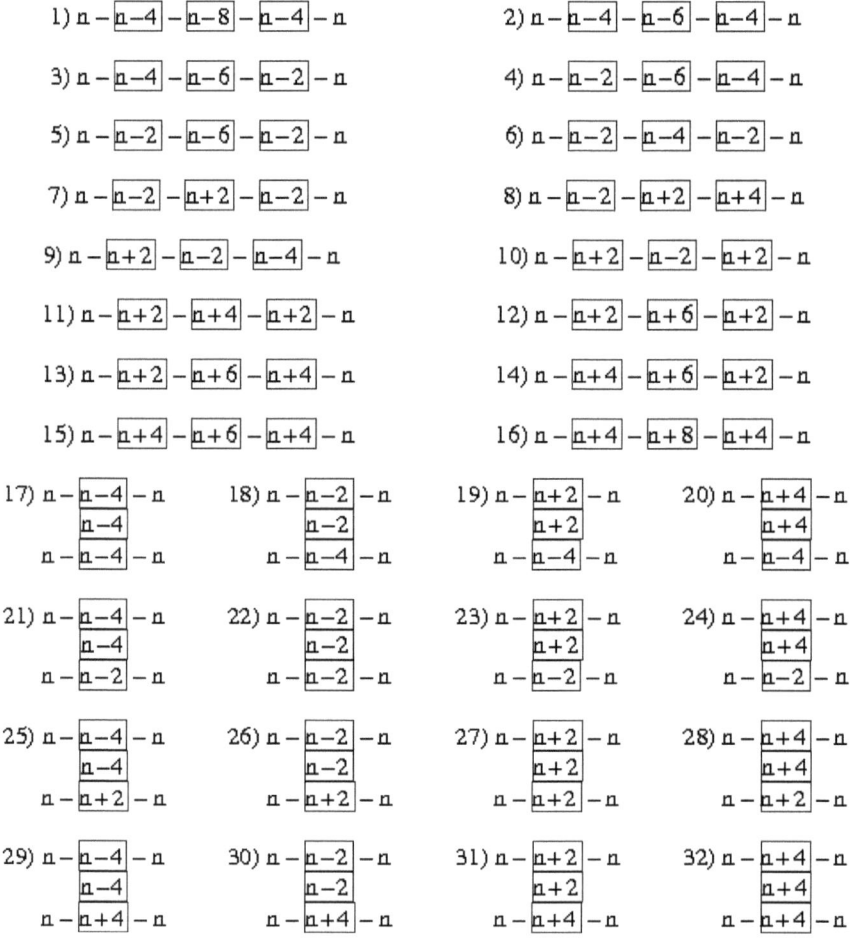

2.7 Calculation of Energy and Level Width of Quasi-Stationary States

The QAO model can be applied to various quantum systems, including the case with the states having a finite lifetime or quasi-stationary ones. These systems are described by an operator in the following form:

$$\hat{H} = \frac{1}{2}(\hat{p}^2 + \hat{x}^2) - \lambda \hat{x}^4 \equiv \frac{1}{2}\hat{p}^2 + V(x). \tag{2.117}$$

A quasi-stationary nature of these systems for arbitrary $\lambda > 0$ is shown in Fig. 2.8, which demonstrates the form of potential pit where the particle is moving. A particle is not able to stay an infinitely long time in the section I, which corresponds to the classical finite movement, due to the quantum mechanical tunnel-effect. This effect causes the particle moves into the section III after a certain period of time and transfers to infinity [56].

The probability P of the particle penetration through the barrier is known to decrease exponentially with the increase of the height and the width of the barrier. The magnitude P becomes $\sim exp(-\frac{a}{\lambda})$ at $\lambda \to 0$, where a is a certain constant [60]. Hence, the eigenvalues E_n of the Hamiltonian (2.117) are the complex values:

$$E_n = E_n' - i\frac{\Gamma_n}{2}.$$

The parameter $\Gamma_n \sim P$ and the level width $\Gamma \sim P$ is inversely proportional to the state lifetime, and the solution of SE is:

$$\hat{H}|\Psi_n\rangle = E_n|\Psi_n\rangle, \tag{2.118}$$

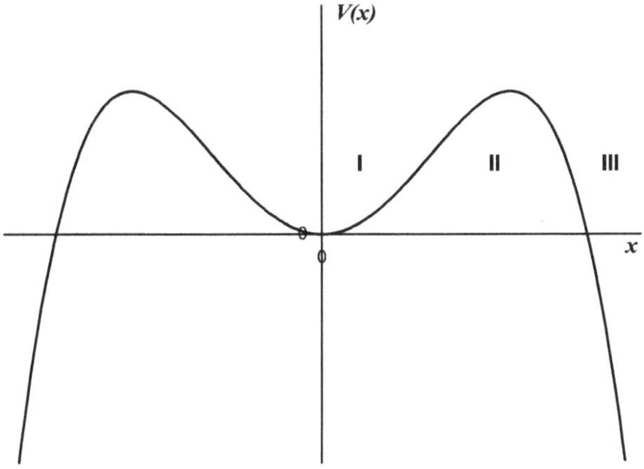

Fig. 2.8 Form of the potential pit for quasi-stationary states

which are considered as the functions of the parameter λ having essential singularity at $\lambda = 0$. Therefore, the use of the PT canonical form for this problem does not help to calculate the level width for small λ [61]. The method used to exceed the limits of PT and to estimate the value Γ is based on the quasi-classical approximation for calculation of the coefficient of barrier permeability [62]. However, this approach can be only applied for exponentially small values of Γ.

The method of coordinate rotation in the complex plane is used for numerical integration of the Eq. (2.118) for complex-valued E_n [63]. The substitution of the variables:

$$x = \rho e^{i\varphi} \tag{2.119}$$

allows to choose the method of integration of the Eq. (2.118) in a complex plane ρ at $\varphi > \varphi_0$, where the wave function $|\Psi_n\rangle$ becomes absolutely normalized. However, the Hamiltonian becomes non-hermitian and the finite-difference approximation of derivatives within the infinite integration interval becomes essentially complex.

According to the general theory of the differential equations [64], the solutions expressed in the complex-valued variables and parameters are the analytical continuation of the solutions defined in the real space. Therefore, the operator method is able to calculate not only the real energies for the states of the discrete spectrum but also the complex eigenvalues of the states with a finite lifetime. To investigate this new opportunity provided by OM, we consider the Eqs. (2.117) and (2.118). Within the limit of the zeroth approximation, the canonical transformation (2.2) can be applied to this problem. Because of the Hamiltonian (2.16) differs from the operator (2.117) only by the sign of the parameter λ, a simple transformation of the Eqs. (2.19) and (2.21) results in the following equation for zeroth approximation of energy ${E_n}^{(0)}(\omega, \lambda)$ for the state with the quantum number n and corresponding ω value:

$$E_n^{(0)} = \frac{1}{4}\left(\omega + \frac{1}{\omega}\right) - \frac{3\lambda}{4\omega^2}(1 + 2n + 2n^2); \qquad n = 0; 1; 2; \ldots \tag{2.120}$$

$$(\omega^3 - \omega)(2n + 1) + 6\lambda(1 + 2n + 2n^2) = 0. \tag{2.121}$$

In a coordinate representation, the wave function is:

$$|0, \omega\rangle \rightarrow \Psi_0(x) = \left(\frac{\omega}{\pi}\right)^{1/4} e^{-\frac{\omega}{2}x^2},$$

which is absolutely normalized only if the condition is fulfilled:

$$\omega' \equiv Re(\omega) > 0, \tag{2.122}$$

where $Re(\omega)$ is the real part of the parameter ω, corresponding to the state vector $|0, \omega\rangle$ which is the basic state of the OM basic set.

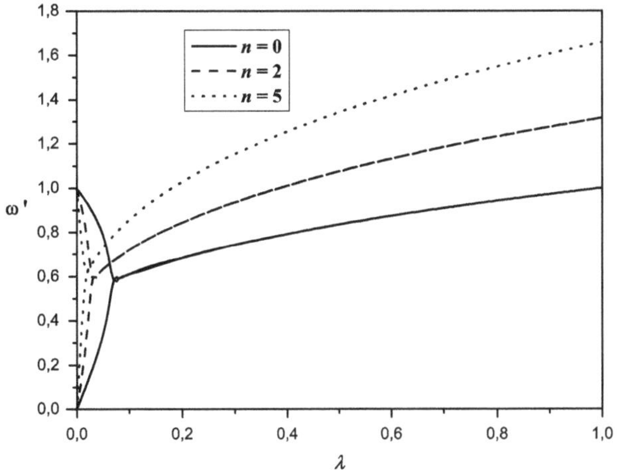

Fig. 2.9 Choice of the parameter ω for quasi-stationary states

Table 2.8 Comparison of the results of numerical E_n [63] and approximate $E_n^{(0)}$ calculations for the energy and the level width of QAO quasi-stationary states

$\lambda = 0.1$	E_n'	Γ_n	$ReE_n^{(0)}$	$\Gamma_n^{(0)}$
n = 0	0.397	0.045	0.384	0.042
n = 1	1.096	0.340	1.074	0.353
n = 2	1.753	0.969	1.769	0.971

In contrast to the Eq. (2.21) with a single root satisfying the condition (2.122) at any λ, the Eq. (2.121) depends on value λ (see Fig. 2.9). There are two real roots $\omega_{1,2}$ at $\lambda < {\lambda_b}^{(n)}$ with

$$\lambda_b^{(n)} = \frac{2n+1}{9\sqrt{3}(2n^2+2n+1)} \tag{2.123}$$

and there is a single complex root with $\omega' > 0$ at $\lambda > \lambda_b^{(n)}$. In the range $0 < \lambda < \lambda_b^{(n)}$, the root ω_2 has to be selected satisfying the adiabaticity condition and transforming to the solution $\omega = 1$ of the harmonic oscillator in the limit $\lambda \to 0$.

Thus, the OM zeroth approximation allows to find the complex-valued eigenvalues exclusively at $\lambda > \lambda_b^{(n)}$, where the energy level width becomes sufficiently high ${\Gamma_n}^{(0)} > 0.01\, Re\, E_n^{(0)}$. As follows from the results of Sect. 2.1, this approximation can provide accuracy of several percents for the energy values. For the values satisfying the condition (2.122), the simple analytical formulas (2.120) and (2.121) estimate the energy and the level width with a sufficiently high precision. Table 2.8 demonstrates this fact, presenting the values $\Gamma_n^{(0)}$ and $Re\, E_n^{(0)}$ calculated by Eqs. (2.120) and (2.121), compared with the results for Γ_n and E_n', obtained in [63] on the basis of a bulky numerical solution of the SE with Hamiltonian (2.117).

The value $\lambda = 0.1$ used in calculation satisfies the condition (2.123) even for basic state. With the increase of λ, the accuracy for calculated energy and level width is almost unchanged and varies around $\sim$ 2–3% which is typical for the OM zeroth approximation.

The algorithm of OM zeroth approximation requires the use of the complex-valued parameter ω, which lead to the strict definition of the basic set of the states $|n, \omega\rangle$ when the imaginary part ω'' ($Im\ \omega = \omega''$) is not equal to zero. It is easier to construct this definition using the operator $\hat{R}(\omega, \omega_1)$, which transforms the operators of creation and annihilation from the parameter ω_1 to the parameter ω, corresponding to the secondary quantization representation for an ordinary harmonic oscillator:

$$a_\omega = \hat{R}^{-1} a_1 \hat{R}; \qquad a_\omega^+ = \hat{R}^{-1} a_1^+ \hat{R}$$
$$|n; \omega\rangle = \hat{R}(\omega, 1)|n; 1\rangle; \qquad a_1|0; 1\rangle = 0. \tag{2.124}$$

According to the formula (2.55), the operator $R(\omega, 1)$ is defined by a simple analytical expression:

$$\hat{R}(\omega; 1) = exp\left[\frac{1}{4}(a^{+2} - a_1^2)\ln\omega\right]$$

and the full set of the eigenfunctions of harmonic oscillator is known, and therefore the formula (2.124) is used as analytical continuation of the basic sequence for the state vectors on the complex-valued ω. It is worth to note that the operators a_ω and $a_\omega{}^+$ and the state vectors $|n, \omega\rangle$ and $\langle\omega, n|$ are not Hermitian-conjugated at complex ω. The following expression can be derived using a closed form of the operator $\hat{R}(\omega)$:

$$(|n; \omega\rangle)^+ = \langle\omega^*; n|; \qquad (a_\omega)^+ = a_{\omega^*}^+.$$

The parameter ω, being a canonical parameter for transformation of (2.2):

$$\hat{x} = \frac{1}{\sqrt{2\omega}}(a_\omega^+ + a_\omega)$$

to the representation of the secondary quantization, assures the equivalence of the complex ω to the coordinate rotation in the complex plane which numerically integrates the SE for these states [63] in the case of application of OM to the quasi-stationary states.

Below the iteration procedure is applied for the OM successive approximations described in Sect. 2.2. The convergence for complex ω and the calculation of $E_n{}'$ and Γ_n with required accuracy is considered for the algorithm based on the following algebraic recurrent relations [31]:

$$E_n = \lim_{s\to\infty} E_n^{(s)}; \qquad C_{nk} = \lim_{s\to\infty} C_{nk}^{(s)}; \tag{2.125}$$

$$E_n^{(s)} = H_{nn} + \sum_{k\neq n} C_{nk}^{(s-1)} H_{nk};$$

$$C_{nk}^{(s)} = \left(E^{(s-1)_n} - H_{kk}\right)^{-1}\left(H_{nk} + \sum_{l\neq n,k} C_{nl}^{(s-1)} H_{lk}\right), \tag{2.126}$$

where C_{nk} are the coefficients of the state vector expansion:

$$|\Psi_n\rangle = |n,\omega\rangle + \sum_{k\neq n} C_{nk}|k,\omega\rangle$$

with the normalization condition:

$$\langle\omega, n|\Psi_n\rangle = 1.$$

The initial elements of the sequences (2.125) are defined by the OM zeroth approximation:

$$E_n^{(0)} = H_{nn}(\omega); \qquad C_{nk}^{(0)} = 0. \tag{2.127}$$

To complete the picture, the formulas for nonzero Hamiltonian matrix elements included in (2.126) are written below:

$$H_{nn} = \frac{1}{4}\left(\omega + \frac{1}{\omega}\right)(2n+1) - \frac{3\lambda}{2\omega^2}\left(n^2 + n + \frac{1}{2}\right);$$

$$H_{n,n+2} = H_{n+2,n} = -\frac{\sqrt{(n+1)(n+2)}}{4\omega^2}[\lambda(4n+6) + \omega(\omega^2 - 1)];$$

$$H_{n,n+4} = H_{n+4,n} = -\frac{\lambda}{4\omega^2}[(n+1)(n+2)(n+3)(n+4)]^{1/2}.$$

The parameter ω has to be fixed for the use in the formulas (2.126). As shown in the Sect. 2.4, the optimal value of ω_{opt} at which the sequence (2.126) quickly converges to the exact solution is different from the ones delivered by the Eq. (2.121) for ω_n, which corresponds to the best OM zeroth approximation. The parameter ω_n can not be used in (2.126) at $\lambda < {\lambda_b}^{(n)}$ because the condition $\Gamma_n \neq 0$ is not fulfilled. The same situation occurs in the method of the complex coordinate rotation, which implements the rotation of (2.119) at certain finite angle $\varphi > \varphi_0 \neq 0$ to normalize the state vector $|\Psi_n\rangle$ and to calculate Γ_n. The numerical computing using the formulas (2.125) shows that the OM successive approximations converge fast to

Table 2.9 The energy levels of QAO with $\hat{H} = \frac{1}{2}(\hat{p}^2 + \hat{x}^2) - \lambda \hat{x}^4$

	$\lambda = 0.1$			
$E_0^{(s)}(E', \frac{\Gamma}{2})$	$\omega = 0.1 + i\,0.8$		$\omega = 1.0 + i\,1.0$	
$E_0^{(0)}$	0.392400	0.122665	0.375000	0.162500
$E_0^{(6)}$	0.397473	0.044629	0.397682	0.044932
$E_0^{(10)}$	0.397441	0.044705	0.397682	0.044715
$E_0^{(s\geq 13)}$	0.397441	0.044706	0.397441	0.044706
$E_0^{\|70\|}$	0.397	0.045	0.	0.

	$\lambda = 0.025$		$\lambda = 0.5$	
$E_n^{(s)}(E', \frac{\Gamma}{2})$	$\omega = 1.0 + i\,0.2$		$\omega = 2.0 + i\,1.5$	
$E_0^{(s\geq 10)}$	0.479117	0.000007	0.373874	0.304990
$E_0^{\|70\|}$	0.479117	0.000007	0.37	0.31
$E_1^{(s\geq 20)}$	1.385667	0.000771	1.205286	1.257500
$E_1^{\|70\|}$	1.38564	0.00077	1.	1.3
$E_2^{(s\geq 20)}$	2.157234	0.021984	2.198786	2.68970
$E_2^{\|70\|}$	2.157	0.021	2.	2.5

the exact values of $E_n{}'$ and Γ_n in a wide range of the complex plane $\omega = \omega' + i\omega''$ where the following inequalities are satisfied:

$$\omega' > \omega_n'; \qquad \omega'' > \omega_n''. \tag{2.128}$$

Under the condition $\omega_n'' = 0$ at $\lambda < \lambda_b^{(n)}$, the value $\omega'' \geq 0.1\omega'$ is not firmly fixed because by satisfying the Eqs. (2.128) the convergence rate in (2.127) is sensitive to the values of ω' and ω''. Within the reasonable accuracy, the eigenvalues E_n do not change with the variation of ω, which can be set as iteration convergence criteria. At the same time, the coefficients C_{nk} depend on the choice of the parameter ω, which defines the basis set of the states $|n, \omega\rangle$ in the exact vector $|\Psi_n\rangle$. The results shown in the Table 2.9 demonstrate that for certain values of λ a high accuracy of results is obtained using fast numerical calculations by formulas (2.126) in comparison with the results obtained by other authors using complex numerical methods (see, e.g., [65]).

In the conclusion of this paragraph, we consider a calculation procedure for the eigenvalues of QAO Hamiltonian with a cubic nonlinearity:

$$\hat{H} = \frac{\hat{p}^2}{2} + V_3(x) \equiv \frac{1}{2}p^2 + \frac{1}{2}x^2 + \lambda x^3. \tag{2.129}$$

The states of this system are quasi-stationary at any signature of λ, and the operator $\hat{H}$ does not commutate with the parity operator in opposite to other systems considered in previous paragraphs. The potential energy is not symmetric with respect to the origin of the coordinate system, and therefore the functions have to be

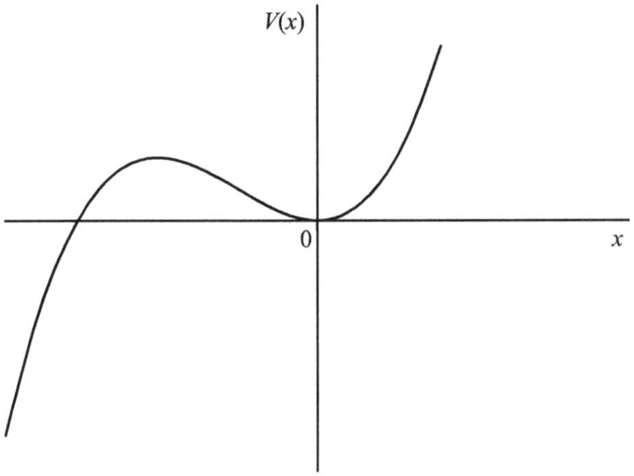

Fig. 2.10 Form of the potential $V_3(x)$

used, which are localized closely to certain value $x = u \neq 0$. For this purpose, the OM generalizes the canonical transformation (2.2) [3]:

$$\hat{x} = u + \frac{1}{\sqrt{2\omega}}(a^+ + a); \qquad \hat{p} = i\sqrt{\frac{\omega}{2}}(a^+ - a), \tag{2.130}$$

by introducing another arbitrary parameter u, additionally to ω, which defines the equilibrium position of the considered system (Fig. 2.10). The general scheme of calculations remains unchanged and is based on the use of the recurrent relations (2.126) with the matrix elements calculated by means of the substitution of operators (2.130) in Hamiltonian (2.129):

$$\begin{aligned} H_{nn} &= \frac{1}{4}\Big(\omega + \frac{1}{\omega}\Big)(2n+1) + \frac{1}{2}u^2 - \lambda\Big[u^3 + \frac{3u}{2\omega}(2n+1)\Big]; \\ H_{n,n+1} &= H_{n+1,n} = \frac{u}{\sqrt{2\omega}}\sqrt{n+1}\Big[1 - 3\lambda\Big(u^2 + \frac{n}{2\omega} + 1\Big)\Big]; \\ H_{n,n+3} &= H_{n+3,n} = -\frac{\lambda}{(2\omega)^{3/2}}[(n+1)(n+2)(n+3)]^{1/2}. \end{aligned}$$

The parameter u is found from the equation:

$$\frac{\partial E_n^{(0)}}{\partial u} = \frac{\partial H_{nn}}{\partial u} = u - 3\lambda\Big(u^2 + \frac{2n+1}{2\omega}\Big) = 0,$$

and ω is obtained using the same algorithm as in previous sections. Table 2.10 shows the effectiveness of operator method for the states of QAO.

Table 2.10 The energy levels of QAO with $\hat{H} = \frac{1}{2}(\hat{p}^2 + \hat{x}^2) - \lambda \hat{x}^3$

	$\lambda = 0.1$		$\lambda = 0.5$	
$E_n^{(s)}(E', \frac{\Gamma}{2})$	$\omega = 1.5 + i\,0.5$		$\omega = 2.7 + i\,1.9$	
$E_0^{(0)}$	0.494862	0.012833	0.385852	0.142919
$E_0^{(s\geq 20)}$	0.484316	0.000008	0.472129	0.252714
$E_0^{\|70\|}$	0.48432	0.00001	0.5	0.25
$E_1^{(s\geq 30)}$	1.378073	0.003141	1.631	1.051
$E_1^{\|70\|}$	1.3782	0.003	1.5	1.
$E_2^{(s\geq 30)}$	2.09437	0.09101	2.70	1.31
$E_2^{\|70\|}$	2.09	0.1	3.	2.

2.8 States with a Broken Symmetry (Integrals of Motion)

A modern quantum field theory deals frequently with the concept of the spontaneous symmetry breaking (see, e.g. [66, 67]). In general case, the solution of such kind of problems is reduced to the system states which do not possess a symmetry of the initial Hamiltonian. The general form of QAO Hamiltonian (2.1) proves to be sufficiently flexible for the modeling of mentioned systems:

$$H = \frac{1}{2}\hat{p}^2 + V_D(x) = \frac{1}{2}\hat{p}^2 - \frac{1}{2}x^2 + \lambda x^4. \tag{2.131}$$

In this equation, the function $V_D(x)$ consists of two symmetric potential pits, the depth and the distance between them increase with the decrease of λ (Fig. 2.11).

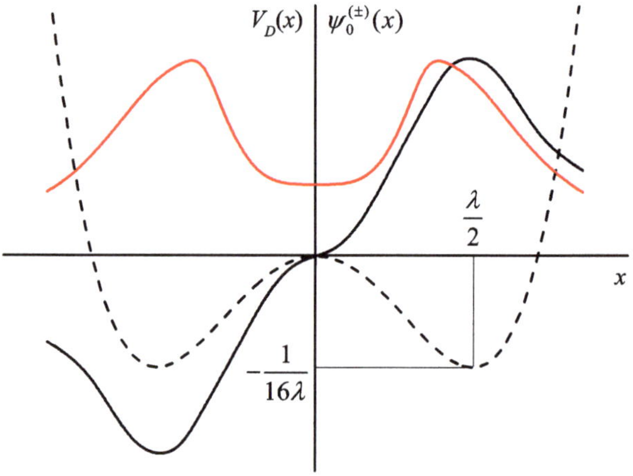

Fig. 2.11 The double-well potential and the function of the lowest symmetric (*red line*) and antisymmetric (*black line*) states

The analysis of the common properties of one-dimensional quantum system [68] shows that the eigenfunction of the operator (2.131) at energy $E < 0$ is concentrated near one of the potential energy minima. Again, as in previous paragraph, the canonical transformation (2.130) is applied:

$$\hat{x} = \frac{1}{\sqrt{2\omega}}(a^+ + a); \quad \hat{p} = i\sqrt{\frac{\omega}{2}}(a^+ - a); \quad a = u + b; \quad a^+ = u + b^+. \tag{2.132}$$

The state vector $|n\rangle$, being the vacuum state with respect to the operators b, is written as:

$$b|u\rangle = (a - u)|u\rangle = 0, \tag{2.133}$$

and acts as the zeroth approximation to the wave function of system basic state. In the coordinate representation, the state vector $|n\rangle$ corresponds to the wave function:

$$\langle x|u\rangle = \left(\frac{\omega}{\pi}\right)^{1/4} e^{-\frac{\omega(x-u)^2}{2}},$$

which describes the oscillator concentrated near the point $x = u$. On the other hand, $|n\rangle$ is the coherent state and the eigenvector of the operator a [69]:

$$\begin{aligned} a|u\rangle &= u|u\rangle; \\ |u\rangle &= e^{ua^+}|0\rangle e^{\frac{u^2}{2}}; \qquad a|0\rangle = 0. \end{aligned} \tag{2.134}$$

In contrast to the operator (2.129), the Hamiltonian (2.130) commutates with the parity operator $\hat{P}$, and therefore its eigenfunctions:

$$\hat{H}|\Psi_\mu\rangle = E_\mu|\Psi_\mu\rangle, \tag{2.135}$$

appear to be the eigenfunctions of $\hat{P}$ as well:

$$\hat{P}|\Psi_\mu\rangle = \mu|\Psi_\mu\rangle, \quad \mu = \pm 1; \quad [\hat{P}\hat{H}] = 0. \tag{2.136}$$

At the same time, the canonical transformation (2.142) does not commutate with the parity operator, which is presented in the terms of operators a and a^+ by the following expression [70]:

$$\hat{P} = e^{i\pi a^+ a}.$$

Therefore, in the function $|\Psi_\mu\rangle$ the exact integral of motion defined by the Eq. (2.136) has be taken into account before using any approximation for this function. For this purpose, we introduce here the operator of vector projection on the state with a definite value of the integral of motion, which has the following form with parity $\mu = \pm 1$:

$$T_\mu = \frac{1}{2}(1 + \mu e^{i\pi a^+ a}); \qquad T_\mu^2 = T_\mu. \tag{2.137}$$

Assuming $|\mu\rangle$ is the solution of the equation:

$$\hat{L}_\mu |\mu\rangle \equiv (\hat{H} - E_\mu)\hat{T}_\mu |\mu\rangle = 0, \tag{2.138}$$

the state vector $|\Psi_\mu\rangle = \hat{T}_\mu|\mu\rangle$ satisfies both Eqs. (2.135) and (2.136) at the same time. As the state vector $|\mu\rangle$ is not bound to any additional conditions, the regular OM procedure can be applied to solve (2.138) on the basis of canonical transformation (2.132) of the operator $\hat{L}_\mu$. In particular, to construct the OM zeroth approximation, the operator $\hat{L}_\mu$ has to be reduced to the normal form and then retain only the operators which commutate with the operator of excitation number b^+b. For example, the parity operator is reduced to the normal form as follows:

$$\hat{P} = e^{2u^2} e^{-2ub^+} e^{i\pi b^+ b} e^{-2ub}.$$

As a result, the expressions for the energies of two lowest states with different parities in the OM zeroth approximation are presented in the following way:

$$E_{0\mu}^{(0)}(\omega, u) = \frac{1}{4}\Big(\omega - \frac{1}{\omega} + \frac{3\lambda}{4\omega^2}\Big) + \frac{u^2}{\omega^2}(6\lambda - \omega + 4\lambda u^2) -$$

$$-\mu u^2 Z(u^2)\Big(\omega - \frac{1}{\omega} + \frac{6\lambda}{\omega^2} + \frac{4\lambda u}{\omega^2}\Big); \qquad Z(u^2) = \frac{e^{-2u^2}}{1 + \mu e^{-2u^2}}.$$

The parameters ω and u are calculated based on the independence principle of the eigenvalues of hermitian operator from the choice of representation, which results in the equations:

$$\frac{\partial E_{0\mu}^{(0)}}{\partial u} = \frac{\partial E_{0\mu}^{(0)}}{\partial \omega} = 0;$$

$$\omega^3 + \omega + 4\omega u^2 + 5e^{-2u^2}(\omega^3 + \omega - 4\omega^3 u^2) - \lambda(6 + 48u + 32u^2 + 6\mu e^{-2u^2}) = 0;$$

$$2\lambda(3 + 4u^2) - \omega + 2u^2 Z^2(u^2)[\omega(\omega^2 - 1) + 2\lambda(3 + 2u^2)] -$$

$$-\mu Z(u^2)[\omega(\omega^2 - 1) - 6\lambda - 2u^2\omega(\omega^2 - 1) - 4\lambda u^2(1 + 2u^2)] = 0.$$

The analytical solution for these equations can be found in both limits of large and small λ, and in the case $\lambda \to 0$ the solution is:

$$\begin{aligned} E_{0\mu}^{(0)} &\simeq -\frac{1}{16\lambda} + \frac{1}{\sqrt{2}} - \frac{3\mu}{16\lambda} e^{-\frac{1}{2\lambda\sqrt{2}}}; \\ \Delta E_0^{(0)} &= E_{0-}^{(0)} - E_{0+}^{(0)} \simeq \frac{3}{8\lambda} e^{-\frac{1}{2\lambda\sqrt{2}}}. \end{aligned} \tag{2.139}$$

Thus, the OM zeroth approximation identifies successfully the exponentially small splitting of levels [71].

The operator method is proved to have the outstanding capabilities for accurate calculation of the eigenvalues of the Hamiltonian (2.131), whereas this task is very challenging for regular numerical methods, especially for the range of small λ [72]. The iteration scheme of the OM for Eq. (2.135) can be applied in different ways, for example, the operator $\hat{H}$ can be transformed into the secondary quantization representation by a canonical transformation (2.2), which does not break the parity and contains only one arbitrary parameter ω. In this representation, the Hamiltonian $\hat{H}$ has a simple polynomial form with respect to a, a^+:

$$\begin{aligned} \hat{H} &= \frac{1}{4}\Big(\omega - \frac{1}{\omega}\Big)(2a^+a + 1) - \frac{1}{4}\Big(\omega + \frac{1}{\omega}\Big)(a^2 + a^{+2}) + \\ &+ \frac{\lambda}{4\omega^2}(6\hat{n}^2 + 6\hat{n} + 3 + a^{+4} + a^4 + 4a^{+2} + 4a^+a^2 + 6a^{+2} + 6a^2). \end{aligned} \tag{2.140}$$

Furthermore, the recurrent relations (2.126) are used with the matrix elements of the Hamiltonian (2.140), and the expansion of the eigenfunction $|\Psi_{ns}\rangle$ over the states with a certain excitation number:

$$\hat{n}|n\rangle = n|n\rangle;$$

$$|\Psi_{n\mu}\rangle = |n\rangle + \sum_{n \neq k} C_{nk}^{(\mu)} |k\rangle \tag{2.141}$$

has to include only the state vectors with even k at $\mu = +1$ and with odd k at $\mu = -1$. This approach has been used in [16], where in spite of the tangible difference between the zeroth approximation wave function $|n\rangle$ and the exact state vectors $|\Psi_s\rangle$, especially at small λ, the OM iteration sequence has been proved to converge to the exact SE solution and calculates the level splitting with any required accuracy. However, the convergence rate is found to be very slow in the range of exponentially small distance ΔE between the levels corresponding to symmetric and non-symmetric states, for example, it takes about 300 iterations to compute $\Delta E \leq 10^{-5}$. The use of the mentioned form of OM for the systems similar to (2.130) and with several degrees of freedom becomes ineffective. This fact is due to the qualitative change of the wave function of the system at $\lambda \to 0$, contrary to the wave functions of harmonic oscillator. Therefore, a certain modification of the recurrent relations (2.126) can be done, which allows to get more accurate wave functions for zeroth approximation and to provide the conservation of the state parity at the same time. First of all, we consider the iteration procedure of the SE solution:

$$\hat{H}|\Psi_0^{(\mu)}\rangle = E_0^{(\mu)}|\Psi_0^{(\mu)}\rangle \tag{2.142}$$

for the lowest symmetric $|\Psi_0^{(+)}\rangle$ and non-symmetric $|\Psi_0^{(-)}\rangle$ states. According to the procedures described above, the following normalized state vectors provide a sufficiently exact representation of the solution of the Eq. (2.141) with respect to the parity of the system:

$$\begin{aligned} |\Psi_0^{(\mu)}\rangle &\simeq |\varphi_0^{(s)}\rangle \equiv C(1+\mu\hat{P})|u\rangle = C(|u\rangle + \mu|-u\rangle); \\ C &= \frac{1}{\sqrt{2(1+|\langle u|-u\rangle|^2)}}. \end{aligned} \tag{2.143}$$

The state vector $|u\rangle$ is defined by the formula (2.133) with the parameters ω and u. The difference between the state vectors $|\Psi_0^{(\mu)}\rangle$ and $|\varphi_0^{(\mu)}\rangle$ is small at any λ and can be found in the form of the expansion into the series over a full set of states $|n\rangle$:

$$|\Psi_0^{(+)}\rangle = |\varphi_0^{(+)}\rangle + \sum_{k=0}^{\infty} C_{2k}^{(+)}|2k\rangle; \quad C_{2k+1}^{(+)} = 0;$$

$$|\Psi_0^{(-)}\rangle = |\varphi_0^{(-)}\rangle + \sum_{k=0}^{\infty} C_{2k+1}^{(-)}|2k+1\rangle; \quad C_{2k}^{(-)} = 0; \tag{2.144}$$

Substituting these expansions into the Eq. (2.142), the nonlinear algebraic equations for the eigenvalues E and coefficients C_n is obtained:

$$E_0^{(\mu)} = \left\{E_{00}^{\mu} + \sum_{k=0}^{\infty} C_k^{\mu}\langle k|\hat{H}|\varphi_0^{(\mu)}\rangle\right\}\left\{1 + \sum_{k=0}^{\infty} C_k^{\mu}\langle\varphi_0^{(\mu)}|k\rangle\right\}^{-1};$$

$$E_{00}^{\mu} = \langle\varphi_0^{(\mu)}|\hat{H}|\varphi_0^{(\mu)}\rangle; \tag{2.145}$$

$$C^{(\mu)} = -\left[H_{ll} - E_{00}^{(\mu)}\right]^{-1}\left\{E_0^{(\mu)}\langle l|\varphi_0^{(\mu)}\rangle - \langle l|H|\varphi_0^{(\mu)}\rangle - \sum_{p\neq l} C_p^{(\mu)} H_{pl}\right\}; \tag{2.146}$$

$$H_{kl} = \langle k|\hat{H}|l\rangle.$$

However, the solution of this equations by ordinary iterations scheme (2.126) faces the problem mentioned in Sect. 2.1: due to the nonorthogonality of the approximate state vector $|\varphi_0^{(\mu)}\rangle$ and the basis functions $\langle n|\varphi_0^{(\mu)}\rangle \neq 0$, the number of the accounted in the Eqs. (2.145) and (2.146) intermediate states becomes infinite in spite of a polynomial nature of Hamiltonian. This problem is not simplified even by choosing the basis functions in the form:

$$|n, u\rangle = \frac{(b^+)^n}{\sqrt{n!}}|0, u\rangle; \qquad b = a - u; \quad b|0, u\rangle = 0,$$

which corresponds to the excited states of the shifted harmonic oscillator without a definite parity. In order to simplify the solution of the Eqs. (2.145) and (2.146), we use the fact that the diagonal matrix element of the Hamiltonian $\langle\varphi_0^{(\mu)}|\hat{H}|\varphi_0^{(\mu)}\rangle$ defines the OM zeroth approximation evenly valid at any λ for proper choice of the parameters ω and u. Therefore, the procedure of the successive inclusion of the matrix elements $\langle k|\hat{H}|\varphi_0^{(\mu)}\rangle$ into the iteration scheme is utilized, which has been described in Sect. 2.2 with respect to the non-polynomial Hamiltonian (formula (2.49)). As a result, the OM recurrent relation system for the solution of SE can be presented in the following form:

$$E_0^{(s)} = \left[E_0^{(\mu)} + \sum_{n=n_0-s}^{n_0+s} C_n^{(s-1)}\langle n|H|\varphi_0^{(\mu)}\rangle\right]\left[1 + \sum_{k=n_0-s}^{n_0+s} C_k^{(s-1)}\langle k|\varphi_0^{(\mu)}\rangle\right]^{-1}; \tag{2.147}$$

$$C_l^{(s)} = -\left[E_0^{(\mu)} - E_0^{(s)}\right]^{-1}\left\{E_0^{(s)}\langle l|\varphi_0^{(\mu)} - \langle l|H|\varphi_0^{(\mu)}\rangle - \right. \tag{2.148}$$

$$\left. - \sum_{n=n_0-s}^{n_0+s} C_n^{(s)}\langle n|H|n_0\rangle\right\};$$

where

$$E_0^{(\mu)} = \lim_{s\to\infty} E_0^{(s)}; \quad C_l^{(\mu)} = \lim_{s\to\infty} C_l^{(s)}; \quad E_0^{(0)} = E_0^{(\mu)}; \quad C_l^{(0)} = 0. \tag{2.149}$$

Due to the nonorthogonality of the functions $|n\rangle$ and $|\varphi_0^{(\mu)}\rangle$, the main contribution to the corrections for the wave functions of zeroth approximation in the expansions (2.144) is determined by the state vector $|n_0\rangle$ with a quantum number n_0, which is not equal to zero even for the basic state. Therefore, in contrast to the recurrent formulas used before, the sequence (2.148) is calculated starting from the coefficient C_{n_0}. The parameter n_0 is an additional parameter and its optimal choice permits to increase the convergence rate of the iteration scheme (2.148). The value of this parameter is found from the fact that in the first iteration the coefficients C_n are defined by the overlap integral $\langle n|\varphi_0^{(\mu)}\rangle$, and therefore the maximum value of C_n reached at n_0 is found at the extremum condition:

$$\frac{\partial}{\partial n}\langle n|\varphi_0^{(\mu)}\rangle = 0. \tag{2.150}$$

Table 2.11 The energies of symmetric and antisymmetric states of the system with Hamiltonian (2.131)

λ	n_0	s_0	ω	u	$E_0^{(-)(0)}$	$E_0^{(-)(s\geq s_0)}$	$E_0^{(+)(s\geq s_0)}$
0.005	20	34	0.4	3.05	−11.1965301	−11.7979755	−11.7979755
0.012	15	19	0.6	2.35	−4.2749434	−4.5137280	−4.5137283
0.016	11	19	0.6	1.98	−2.9896903	−3.2160644	−3.2160702
0.02	9	14	0.6	1.73	−2.2255034	−2.4393456	−2.4394387
0.05	7	10	0.9	1.2	−0.5170587	−0.5765296	−0.6327464
0.5	3	10	2.0	0.2	1.5697561	1.4172688	0.3288265
10.	0	12	5.5	0.02	5.2279415	4.9895186	1.2836945
λ	n_0	s_0	ω	u	$E_1^{(-)(0)}$	$E_1^{(-)(s\geq s_0)}$	$E_1^{(+)(s\geq s_0)}$
0.01	15	48	1.5	4.25	−4.1287797	−4.2039848	−4.2039848

The described above scheme for the solution of SE can be generally used for any quantum system, where the exact symmetry of the wave functions in zeroth approximation can be constructed. For demonstration of the numerical results obtained for Hamiltonian (2.131), the expressions for matrix elements in Eqs. (2.147), (2.148) are explicitly derived:

$$\begin{aligned} \langle n|\varphi_0^{(\mu)}\rangle &= e^{-\frac{u^2}{2}}\tfrac{u^2}{n!}[1+\mu(-1)^n];\\ \langle\varphi_0^{(\mu)}|\hat{H}|\varphi_0^{(\mu)}\rangle &= \tfrac{1}{4}\left(\omega-\tfrac{1}{\omega}\right)+\tfrac{3\lambda}{4\omega^2}+\left(1+\mu e^{-2u^2}\right)^{-1}\times\\ &\times\left[\tfrac{\lambda u^2}{\omega^2}(4u^2+6)-\tfrac{u^2}{\omega}-\mu\omega u^2 e^{-2u^2}\right];\\ \langle l|H|k\rangle &= \sqrt{k(k+1)}\delta_{k+2,l}+\sqrt{k(k+1)(k+2)(k+3)}\delta_{l,k+4}. \end{aligned} \tag{2.151}$$

The calculation results show that the parameter n_0 influences the iteration convergence rate. The value $(n_0)_{opt}$, at which the convergence rate has a maximum, corresponds to the one derived from the formula (2.150). Table 2.11 displays the results for the basic state energy $E_0^{(\pm)}$ calculated at different values of λ, the values of the parameters ω and u at which the calculations are carried out, the value $(n_0)_{opt}$ for every λ and the number of the iterations s_0 required to obtain the mentioned results. These results, being compared to the ones of the paper [72], prove the modification of OM iteration scheme defined by (2.147) to considerably decrease a number of the iterations for calculation of the eigenvalues with required accuracy.

Summarizing this section, the choice of the wave function for excited states has been considered, and the best approximation is obtained with a symmetrized linear combination of the excited state wave functions of the shifted oscillator. The symmetric and non-symmetric functions of the nth excited state can be presented in the following form:

$$|\tilde{\varphi}_n^{(\mu)}\rangle = \frac{1}{\sqrt{n!}}\Big[(a^+-u)^n|0,u\rangle+(-1)^n\mu(a^++u)^n|0,-u\rangle\Big]. \tag{2.152}$$

This approximation can be further improved provided the orthogonality of this function to the wave functions of all lower laying states is preserved for calculation of the state vector $|\varphi_n^{(\mu)}\rangle$ by formulas:

$$|\varphi_n^{(\mu)}\rangle = |\tilde{\varphi}_n^{(\mu)}\rangle - \sum_{k=0}^{n-1} \langle \varphi_n^{(\mu)} || \tilde{\varphi}_n^{(\mu)} \rangle |\varphi_k^{(\mu)}\rangle. \tag{2.153}$$

The substitution of these functions in the iteration scheme defined by the formulas (2.146) and (2.147) delivers the solution of SE for the excited states. Table 2.11 shows the results of this calculation for two closest excited states.

References

1. I.D. Feranchuk, L.I. Komarov, Phys. Lett. A, **88**, 211 (1982)
2. I.D. Feranchuk, L.I. Komarov, J. Phys. C. Solid State Phys. **15**, 1965 (1982)
3. I.D. Feranchuk, L.I. Komarov, J. Phys. A: Math. Gen. **17**, 3111 (1984)
4. I.G. Holliday, P. Suranyi, Phys. Lett. B **85**, 421 (1979)
5. W.E. Caswell, Ann. Phys. (N.Y.) **123**, 153 (1979)
6. I.G. Holliday, P. Suranyi, Phys. Rev. D **21**, 1529 (1980)
7. I.K. Dmitrieva, G.I. Plindov, Phys. Lett. A **79**, 47 (1980)
8. K. Bhattacharyya, J. Phys. B: At. Mol. Phys. **14**, 783 (1981)
9. J. Austin, J. Killingbeck, J. Phys. A: Math. Gen. **15**, L433, (1982)
10. C.K. Au, G.W. Roders, J. Aharonov, Phys. Lett. A **95**, 287 (1983)
11. F.M. Fernandez, E.A. Castro, Phys. Lett. A **88**, 4 (1982)
12. F.M. Fernandez, E.A. Castro, Phys. Lett. A **91**, 339 (1982)
13. F.M. Fernandez, A.M. Meson, E.A. Castro, Phys. Lett. A **104**, 401 (1984)
14. F.M. Fernandez, A.M. Meson, E.A. Castro, Phys. Lett. A **111**, 104 (1985)
15. F.M. Fernandez, A.M. Meson, E.A. Castro, Phys. Lett. A **112**, 107 (1985)
16. F.M. Fernandez, A.M. Meson, E.A. Castro, Mol. Phys. **59**, 365 (1986)
17. C.C. Gerry, S. Silverman, Phys. Lett. A **97**, 481 (1983)
18. W. Witschel, Phys. Lett. A **97**, 315 (1983)
19. H. Mitter, K. Yamazaki, J. Phys. A: Math. Gen. **17**, 1215 (1984)
20. B. Bidyut, B. Sikha, R. Raykumar, Fizika **16**, 161 (1984)
21. K. Yamazaki, J. Phys. A: Math. Gen. **17**, 345 (1984)
22. C.-S. Hsue, J.L. Chern, Phys. Rev. D **29**, 643 (1984)
23. B. Rath, J. Math. Phys. **31**, 1175 (1991)
24. K.L. Chan, S.K. Lee, K.L. Liu, K. Young, C.K. Au, Phys. Lett. A **162**, 227 (1992)
25. I.D. Feranchuk, S.I. Fisher, L.I. Komarov, J. Phys. C: Solid State Phys. **17**, 4309 (1984)
26. C.Z. An, I.D. Feranchuk, Mol. Phys. **64**, 589 (1988)
27. I.D. Feranchuk, L.X. Hai, Phys. Lett. A **137**, 385 (1989)
28. L.I. Komarov, T.S. Romanova, J. Phys. B: At. Mol. Phys. **18**, 859 (1985)
29. I.D. Feranchuk, S.I. Fisher, L.I. Komarov, J. Phys. C: Solid State Phys. **18**, 5083 (1985)
30. C.Z. An, I.D. Feranchuk, L.I. Komarov, L.I. Nakhamchik, J. Phys. A: Math. Gen. **19**, 1583 (1986)
31. I.D. Feranchuk, L.I. Komarov, I.V. Nechipor, J. Phys. A: Math. Gen. **20**, 3849 (1987)
32. C.Z. An, I.D. Feranchuk, L.I. Komarov, Phys. Lett. A **125**, 123 (1987)
33. I.D. Feranchuk, L.I. Komarov, Phys. Lett. A **106**, 109 (1984)
34. I.D. Feranchuk, V.N. T'ok, Chem. Phys. Lett. **150**, 78 (1988)

35. I.D. Feranchuk, V.S. Kuz'min, A.P. Ulyanenkov, Chem. Phys. **157**, 61 (1991)
36. I.D. Feranchuk, A.P. Ulyanenkov, NIM B **88**, 369 (1994)
37. I.D. Feranchuk, L.I. Komarov, A.P. Ulyanenkov, J. Phys. A: Math. Gen. **29**, 4035 (1996)
38. I.D. Feranchuk, L.I. Komarov, A.P. Ulyanenkov, J. Phys. B: At. Mol. Opt. Phys. **35**, 3957 (2002)
39. I.D. Feranchuk, L.I. Gurskii, L.I. Komarov, O.M. Lugovskaya, F. Burgäzy, A.P. Ulyanenkov, Acta Crystallogr. A **58**, 370 (2002)
40. L.V. Hoang, L.I. Komarov, T.S. Romanova, J. Phys. A: Math. Gen. **22**, 1543 (1989)
41. L.I. Gurskii, L.I. Komarov, A.M. Solodukhin, Int. J. Quantum Chem. **72**, 499 (1999)
42. I.D. Feranchuk, A.V. Leonov, Phys. Lett. A **375**, 385 (2011)
43. I.D. Feranchuk, V.V. Triguk, Phys. Lett. A **375**, 2550 (2011)
44. I.D. Feranchuk, L.I. Komarov, I.V. Nechipor, A.P. Ulyanenkov, Ann. Phys. (N Y) **238**, 370 (1995)
45. A. Duncan, H.F. Jones, Nucl. Phys. B **320**, 189 (1989)
46. V.V. Kudrashov, V.I. Reshetnyak, E.A. Tolkachov, Dokl. Belarusian Acad. Sci. (in Russian) **31**, 135 (1987)
47. F.T. Hioe, D. MacMillen, E.W. Montroll, Phys. Rep. **335**, 307 (1978)
48. F.M. Fernandez, E.A. Castro, Phys. Lett. A **88**, 5083 (1982)
49. I.I. Gegusin, L.I. Leontieva, J. Phys. C: Condens. Matter **2**, 5869 (1990)
50. S. Srivastava, Vishamitter, Mol. Phys. **72**, 1285 (1991)
51. F.A. Kaempfer, *Concepts in Quantum Mechanics* (Academic, New York, 1965)
52. D.A. Kirzhnitz, *Field Methods of Many–Particle Theory (in Russian)* (Gosatomizdat, Moscow, 1963)
53. A.J. Barnes, W.J. Orville-Thomas, (ed.), *Vibrational Spectroscopy. Modern Trends* (Oxford, Amsterdam, Elsevier, New York, 1977)
54. C. Hayashi, *Nonlinear Oscillations in Physical Systems* (McGraw-Hill, New York, 1964)
55. I.S. Gradshtein, I.M. Ryzhik, *Tables of Integrals, Sums, Series and Products* (FIZMATGIZ, Moscow, 1963)
56. A. Messiah, *Quantum Mechanics* (North-Holland, Amsterdam, 1976)
57. N.H. March, W.H. Young, S. Sampanthar, *The Many-Body Problem in Quantum Mechanics* (University Press, Cambridge, 1967)
58. M.L. Goldberger, K.M. Watson, *Collision Theory* (Wiley, New York, 1964)
59. V.G. Baryshevsky, *High-Energy Nuclear Optics of Polarised Media* (World Scientific, Singapore, 2012)
60. P.E. Shanley, Phys. Lett. A **117**, 161 (1986)
61. R.J. Damburg, V.G. Kolosov, J. Phys. B: At. Mol. **9**, 3149 (1976)
62. H.K. Shepard, Phys. Rev. D **27**, 1288 (1983)
63. D. Farrelly, W.P. Reinhardt, J. Phys. B: At. Mol. **16**, 2103 (1983)
64. P.M. Morse, H. Feshbach, *Methods of Theoretical Physics* (McGraw-Hill, New York, 1953)
65. J.E. Drummond, J. Phys. A: Math. Gen. **14**, 1651 (1981)
66. N.N. Bogoliubov, Physica **26**, 1 (1960)
67. G. Ross, *Grand Unified Theories* (Westview Press, USA, 1984)
68. L.D. Landau, E.M. Lifshitz, *Quantum Mechanics* (Fizmatgiz, Moscow, 2004)
69. M.O. Scully, M.S. Zubairy, *Quantum Optics* (Cambridge University Press, Cambridge, 1997)
70. A. Bohm, *Quantum Mechanics: Foundations and Applications* (Springer, New York, 1986)
71. C.G. Callan, S. Coleman, Phys. Rev. D **16**, 1762, (1977)
72. F.M. Fernandez, E.A. Castro, A.M. Meson, J. Phys. A: Math. Gen. **7**, 1389 (1985)

Chapter 3
Applications of OM for One-Dimensional Systems

The extended analysis of various physical problems related to anharmonic oscillator performed in previous chapter makes it possible to sort out the capabilities of the operator method, which are not exclusively related to QAO and can be exploited for description of more complex quantum systems. These effective features of OM are

1. a special choice of the wave functions for zeroth approximation $\left|\Psi^{(0)}_{n\omega_n}\right\rangle$, in general case non-orthogonal, depending on the variational parameters ω_n, unique for every n;
2. an explicit receipt for calculation of the parameters ω_n in zeroth approximation of OM using the full Hamiltonian $\hat{H}$ without separation of the perturbation operator;
3. a construction of the iterative scheme for successive approximations for eigenvalues and eigenfunctions $|\Psi_n\rangle$ as the sequences uniformly converging to the exact solutions in the entire range of Hamiltonian's parameters.

For the physical calculations performed for the systems in previous sections, the wave functions in the occupation number representation $\left|\Psi^{(0)}_{n\omega_n}\right\rangle$ were chosen as the state vectors $|n, \omega_n\rangle$, corresponding to the excitation numbers. However, this choice of the functions $|n, \omega_n\rangle$ is not unique and predominantly conditioned by the simplicity of calculations for QAO. For more complex quantum systems, a certain complication of the calculations due to alternative choice of the basic set is worth in exchange to the improvement of the accuracy of zeroth approximation. This chapter considers several physical systems where the application of the extended operator method gives the effective results.

I. Feranchuk et al., *Non-perturbative Description of Quantum Systems*,
Lecture Notes in Physics 894, DOI 10.1007/978-3-319-13006-4_3

3.1 Anharmonic Oscillator with High Anharmonicity

We start here with the potential represented by a polynomial of high degree, that facilitates the investigation of the operator method for the potentials with qualitatively different behavior. The system is assumed to be described by the Hamiltonian [1]:

$$\hat{H} = \frac{1}{2}(\hat{p}^2 + \hat{x}^2) + \sum_{m=2}^{N} \lambda_m x^{2m}. \tag{3.1}$$

The systems with polynomial potentials, called also generalized QAO, are of wide interest for numerous applications. For example, the polynomial potential is used for the solution of inverse problem in molecular spectroscopy, because of the limitation by the fourth order of potential is insufficient for reconstruction of the interatomic interaction potential [2, 3]. The oscillators with high degree of anharmonicity [4, 5] have been investigated in numerous works, for example, in [6] by perturbation theory, or in [7–13] on the basis of various non-perturbative approaches. Below we use the technique of OM described in Chap. 2 for Hamiltonian (3.1), which implements the canonic transformation:

$$\hat{x} = \frac{1}{\sqrt{2\omega}}(\hat{a} + \hat{a}^+), \qquad \hat{p} = -i\sqrt{\frac{\omega}{2}}(\hat{a} - \hat{a}^+) \tag{3.2}$$

being applied to expression (3.1) modifies it to:

$$\hat{H} = \frac{1}{4}\left(\frac{1}{\omega}(\hat{a} + \hat{a}^+)^2 - \omega(\hat{a} - \hat{a}^+)^2\right) + \sum_{m=2}^{N} \lambda_m \frac{1}{(2\omega)^m}(\hat{a} + \hat{a}^+)^{2m}. \tag{3.3}$$

The anharmonic part of the potential (3.3) is then transformed to the normal form, using the transposition of the creation and annihilation operators:

$$(\hat{a} + \hat{a}^+)^{2m} = \sum_{k=0}^{m} \frac{(2m)!}{2^k k!} \sum_{j=0}^{2m-2k} \frac{(\hat{a}^+)^{2m-2k-j}\hat{a}^j}{j!(2m-2k-j)!}. \tag{3.4}$$

The Hamiltonian for zeroth approximation $\hat{H}_0$ is selected as a part of full operator $\hat{H}$, which commutates with the operator of excitation number $\hat{n} = \hat{a}^+\hat{a}$ and thus contains the terms with the products of the equal order of creation and annihilation operators:

$$\hat{H}_0 = \frac{2\hat{n}+1}{4}\left(\omega + \frac{1}{\omega}\right) + \sum_{m=2}^{N} \frac{\lambda_m}{(2\omega)^m} \sum_{k=0}^{m} \frac{(2m)!}{2^k k!((m-k)!)^2}(\hat{a}^+)^{m-k}\hat{a}^{m-k}. \tag{3.5}$$

As a next step, the formula for transition from the creation and annihilation operators to the particle number operator is used [14]:

$$(\hat{a}^{+})^{M}\hat{a}^{M} = \hat{n}(\hat{n}-1)\cdot\ldots\cdot(\hat{n}-M+1), \tag{3.6}$$

and the expression for the energy in zeroth approximation is found:

$$E_0(n,\omega_n) \equiv \langle n|\hat{H}|n\rangle = \frac{2n+1}{4}\left(\omega_n + \frac{1}{\omega_n}\right) + \sum_{m=2}^{N}\frac{\lambda_m(2m)!}{(2\omega_n)^m} \times\left[\sum_{k=0}^{m-1}\frac{1}{2^k k!((m-k)!)^2}\prod_{i=0}^{m-k-1}(n-i) + \frac{1}{2^m m!}\right]. \tag{3.7}$$

The parameter ω_n is derived from the expression (2.14) for the independence of eigenvalues from the parameter ω, where the substitution of (3.7) results in the algebraic equation of the order $(N+1)$:

$$\omega_n^{N+1} - \omega_n^{N-1} - \frac{1}{2n+1}\sum_{m=2}^{N}\frac{m(2m)!\lambda_m\omega_n^{N-m}}{2^{m-2}} \times\left[\sum_{k=0}^{m-1}\frac{1}{2^k k!((m-k)!)^2}\prod_{i=0}^{m-k-1}(n-i) + \frac{1}{2^m m!}\right] = 0. \tag{3.8}$$

The relationships (3.7) and (3.8) determine the zeroth approximation of OM for energy spectrum of the oscillator with polynomial potential. For particular case of the system with Hamiltonian (3.1) with $N = 3$:

$$\hat{H} = \frac{1}{2}(\hat{p}^2 + \hat{x}^2) + \lambda\hat{x}^4 + \eta\hat{x}^6, \tag{3.9}$$

the following expression is found from (3.7):

$$E_n^{(0)} = \frac{2n+1}{4}\left(\omega_n + \frac{1}{\omega_n}\right) + \frac{3\lambda}{4\omega_n^2}(2n^2+2n+1) + \frac{5\eta}{8\omega_n^3}(4n^3+6n^2+8n+3). \tag{3.10}$$

The equation for ω_n follows from (3.8):

$$\omega_n^4 - \omega_n^2 - \omega_n\frac{6\lambda(2n^2+2n+1)}{2n+1} - \frac{15\eta(4n^3+6n^2+8n+3)}{2(2n+1)} = 0. \tag{3.11}$$

Table 3.1 Energy spectrum of the quantum anharmonic oscillator of sixth order for $E_n^{(0)}(0, 1)$ and $E_n^{(0)}(0, 50)$ in comparison with the values $E_n(0, 1)$ and $E_n(0, 50)$ from [12]; n is a quantum number, $\xi_n^{(0)}$ is the relative accuracy of OM zeroth approximation

n	$E_n^{(0)}(0, 1)$	$E_n(0, 1)$	$\xi_n^{(0)}(0, 1)$ %	$E_n^{(0)}(0, 50)$	$E_n(0, 50)$	$\xi_n^{(0)}(0, 50)$ %
0	0,837797	0,804966	4,1	1,97349	1,85849	6,2
1	2,96566	2,87467	3,2	7,26237	6,97310	4,1
2	5,80729	5,77197	0,6	14,5779	14,4886	0,6
3	9,26189	9,32485	0,7	23,5802	23,7825	0,9
4	13,2404	13,4151	1,3	34,0057	34,5233	1,5
5	17,6816	17,9787	1,7	45,6801	46,5369	1,8
6	22,5411	22,9706	1,9	58,4801	59,6999	2,0
7	27,7853	28,3564	2,0	72,3131	73,9188	2,2
8	33,3877	34,1092	2,1	87,1062	89,1200	2,3
9	39,3265	40,2067	2,2	102,800	105,244	2,3
10	45,5835	46,6305	2,2	119,346	122,239	2,4
11	52,1433	53,3520	2,3	136,700	140,026	2,4
12	58,9922	60,3655	2,3	154,828	158,592	2,4
13	66,1187	67,6501	2,3	173,698	177,880	2,4
14	73,5122	75,1888	2,2	193,280	197,843	2,3
15	81,1633	82,9634	2,2	213,550	218,431	2,2
16	89,0637	90,9606	2,1	234,485	239,611	2,1
17	97,2057	99,1633	2,0	256,065	261,335	2,0

Table 3.1 shows the results for the energy $E_n^{(0)}(\lambda, \eta)$ of the QAO of sixth order [1]. For the comparison, the same energies are shown calculated by numerical method based on the generalized Bloch equation [12]. The iterative technique of OM provides the same results as in [12] after 5–6 iterations. This data confirms the uniform suitability of approximation obtained by OM. The parameter $\xi_n^{(0)}(\lambda, \eta)$ in Eq. (1.1), which determines the accuracy of the energy spectrum in OM zeroth approximation for arbitrary quantum number n and for entire range of Hamiltonian parameters, satisfies the condition $\xi_n^{(0)} \approx 0.06$. The conventional PT is not suitable for the description of this quantum system, because its applicability at small values of the anharmonic parameters only.

Table 3.2 presents the energy spectrum values for quantum anharmonic oscillator with general symmetric potential of sixth order at different parameters of the Hamiltonian (3.10) in zeroth approximation.

Table 3.2 Energy spectrum E_n of quantum anharmonic oscillator with Hamiltonian $\hat{H} = \frac{1}{2}(\hat{p}^2 + \hat{x}^2) + \lambda\hat{x}^4 + \eta\hat{x}^6$; n is a quantum number

n	$\lambda = 1,\ \eta = 1$	$\lambda = 1,\ \eta = 10$	$\lambda = 10,\ \eta = 1$	$\lambda = 10,\ \eta = 10$
0	0.963747	1.40218	1.55933	1.74713
1	3.39770	5.09920	5.50129	6.26672
2	6.60022	10.1483	10.5979	12.2809
3	10.4240	16.3176	16.4991	19.4450
4	14.7741	23.4339	23.0460	27.5691
5	19.5880	31.3823	30.1451	36.5339
6	24.8212	40.0814	37.7330	46.2556
7	30.4399	49.4698	45.7637	56.6717
8	36.4174	59.4991	54.2014	67.7327
9	42.7318	70.1299	63.0178	79.3989
10	49.3650	81.3293	72.1893	91.6367
11	56.3012	93.0694	81.6963	104.418
12	63.5271	105.326	91.5218	117.718
13	71.0308	118.078	101.651	131.515
14	78.8019	131.306	112.072	145.791
15	86.8309	144.994	122.772	160.528
16	95.1094	159.126	133.742	175.711
17	103.630	173.689	144.973	191.326
18	112.385	188.671	156.455	207.361
19	121.369	204.059	168.182	223.804
20	130.575	219.843	180.147	240.644
30	145.000	245.433	196.308	266.484
40	164.066	279.485	216.560	300.448
50	184.785	316.402	238.521	337.382
60	206.353	354.743	261.455	375.870
70	228.498	394.044	285.078	415.419
80	251.116	434.134	309.263	455.834
90	274.163	474.949	333.948	497.032
100	297.620	516.464	359.098	538.975
200	421.842	738.875	480.115	759.114

A similar data are also available for the QAO with the potential of eighth order described by Hamiltonian:

$$\hat{H} = \frac{1}{2}(\hat{p}^2 + \hat{x}^2) + \lambda\hat{x}^4 + \eta\hat{x}^6 + \theta\hat{x}^8. \tag{3.12}$$

Equations (3.7) and (3.8) in this case have the following form for energy of zeroth approximation:

$$E_n^{(0)} = \frac{2n+1}{4}\left(\omega_n + \frac{1}{\omega_n}\right) + \frac{3\lambda}{4\omega_n^2}(2n^2+2n+1) + \frac{5\eta}{8\omega_n^3}(4n^3+6n^2+8n+3) + \frac{7\theta}{16\omega_n^4}(10n^4+20n^3+50n^2+40n+15), \tag{3.13}$$

and for the parameter ω_n:

$$\omega_n^5 - \omega_n^3 - \omega_n^2\frac{6\lambda(2n^2+2n+1)}{2n+1} - \omega_n\frac{15\eta(4n^3+6n^2+8n+3)}{2(2n+1)} - \frac{7\theta(10n^4+20n^3+50n^2+40n+15)}{2n+1} = 0. \tag{3.14}$$

Equations (3.13) and (3.14) determine the energy spectrum for the oscillator with the potential of eighth order in OM zeroth approximation. Table 3.3 shows the results from [12] and by using OM iterative scheme (2.45). Similar to the case

Table 3.3 The energy spectrum for QAO of eighth order for $\theta = 1$ ($E_n^0(1)$) and $\theta = 50$ ($E_n^0(50)$) at $\lambda = 0$, $\eta = 0$ in comparison with the energy values $E_n(1)$ and $E_n(50)$; n is a quantum number, $\xi^{(0)}(1)$ and $\xi^{(0)}(50)$ are the relative accuracy of data

n	$E_n^0(1)$	$E_n(1)$	$\xi^{(0)}(1)$ %	$E_n^0(50)$	$E_n(50)$	$\xi^{(0)}(50)$ %
0	0,889691	0,820685	8,4	1,77826	1,59433	11,5
1	3,19731	2,99980	6,6	6,58611	6,09751	8,0
2	6,35835	6,21052	2,4	13,3490	13,0167	2,6
3	10,3210	10,3303	0,1	21,9069	21,9511	0,2
4	15,0132	15,2303	1,4	32,0826	32,6009	1,6
5	20,3726	20,8285	2,2	43,7308	44,7849	2,4
6	26,3497	27,0702	2,7	56,7386	58,3815	2,8
7	32,9051	33,9135	3,0	71,0173	73,2982	3,1
8	40,0065	41,3248	3,2	86,4952	89,4608	3,3
9	47,6273	49,2765	3,4	103,113	106,808	3,5
10	55,7449	57,7450	3,5	120,820	125,289	3,6
11	64,3396	66,7101	3,6	139,572	144,855	3,6
12	73,3943	76,1540	3,6	159,334	165,472	3,7
13	82,8939	86,0563	3,7	180,069	187,092	3,8
14	92,8249	96,4041	3,7	201,750	209,687	3,8

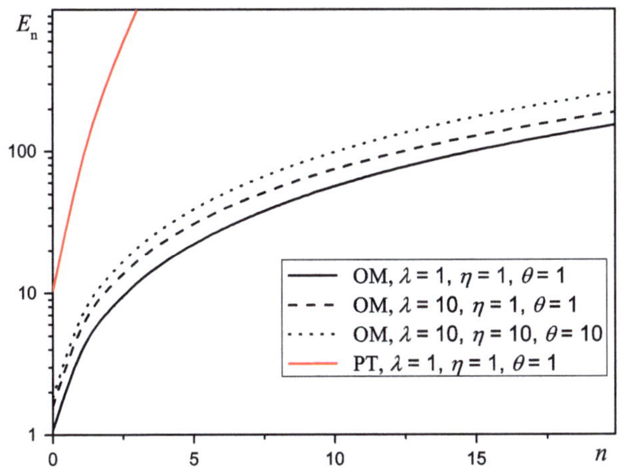

Fig. 3.1 The comparison of energy spectra of quantum anharmonic oscillator for various potentials

of QAO of sixth order, the uniformly suitable convergence exists over both the quantum number n and anharmonic parameter θ, and the parameter governing the uniform suitability is $\xi^{(0)} \approx 0.1$. The standard perturbation theory is not applicable for QAO of eighth order at small values of anharmonicity parameter θ even for low energy level (Fig. 3.1) .

Table 3.4 presents the energy of QAO for general potential of eighth order. The dependence of energy levels from the quantum number is non-monotonic with the increase of anharmonicity: the energy of levels at $\lambda = 1, \eta = 10, \theta = 1$ grows slower than the one with lower anharmonicity $\lambda = 10, \eta = 1, \theta = 1$, however, starting from $n = 4$ the energy of levels grows faster.

A comparison of energy levels for various coefficients and orders of anharmonicity is shown in Fig. 3.1. As follows from the Table 3.4 and Fig. 3.1, the role of anharmonicity is changing as the quantum number varies: the coefficient at the fourth order of perturbation λ is crucial at small quantum numbers n, the coefficient at eighth order θ dominates at $n \geq 30$, and for intermediate n the essential contribution is provided by the coefficient of sixth order η. This fact is very important for the determination of interatomic potentials from the oscillating spectra of molecules.

Table 3.4 The energy levels E_n of QAO with Hamiltonian $\hat{H} = \frac{1}{2}(\hat{p}^2+\hat{x}^2)+\lambda\hat{x}^4++\eta\hat{x}^6+\theta\hat{x}^8$; n is a quantum number

	$\lambda = 1$			$\lambda = 10$		
	$\eta = 1$		$\eta = 10$	$\eta = 1$		$\eta = 10$
n	$\theta = 1$	$\theta = 10$	$\theta = 10$	$\theta = 1$	$\theta = 10$	$\theta = 10$
0	1.06696	1.38823	1.57716	1.58103	1.71793	1.84817
1	3.82587	5.08402	5.78509	5.59915	6.18072	6.68534
2	7.55380	10.2140	11.6127	10.8381	12.1744	13.2239
3	12.1272	16.6434	18.8562	16.9751	19.4370	21.1632
4	17.4522	24.2421	27.3451	23.8665	27.8273	30.3242
5	23.4598	32.9058	36.9563	31.4278	37.2472	40.5867
6	30.0976	42.5541	47.5990	39.6018	47.6226	51.8629
7	37.3248	53.1231	59.2026	48.3461	58.8953	64.0852
8	45.1083	64.5614	71.7109	57.6275	71.0174	77.1995
9	53.4206	76.8261	85.0775	67.4192	83.9491	91.1608
10	62.2383	89.8809	99.2632	77.6986	97.6563	105.931
11	71.5414	103.695	114.234	88.4467	112.110	121.479
12	81.3123	118.240	129.962	99.6467	127.283	137.774
13	91.5353	133.492	146.420	111.284	143.153	154.793
14	102.197	149.430	163.585	123.345	159.700	172.512
15	113.284	166.035	181.437	135.819	176.904	190.912
16	124.786	183.288	199.957	148.694	194.750	209.974
17	136.691	201.173	219.128	161.962	213.220	229.682
18	148.992	219.676	238.934	175.612	232.302	250.020
19	161.678	238.784	259.361	189.638	251.981	270.974
20	174.742	258.483	280.396	204.030	272.246	292.531
30	199.882	298.311	320.770	228.215	311.304	332.386
40	234.758	353.679	376.696	261.397	365.600	387.510
50	272.704	413.612	437.497	298.067	424.763	447.723
60	312.227	475.827	500.786	336.708	486.451	510.611
70	352.894	539.712	565.877	376.781	549.978	575.432
80	394.566	605.094	632.549	418.065	615.118	641.926
90	437.196	671.928	700.732	460.457	681.791	709.997
100	480.772	740.212	770.407	503.905	749.974	779.608
200	740.757	1151.06	1185.04	758.210	1158.23	1191.94

3.2 Anharmonic Oscillator with Non-symmetric Potential

The interatomic potentials in real molecules and crystals are essentially asymmetric with respect to the reflection operation [15–18]. The accounting of this asymmetry is necessary for theoretical description of the thermal expansion of solids, phonon-phonon interaction, and other physical effects. To analyze the application of OM for

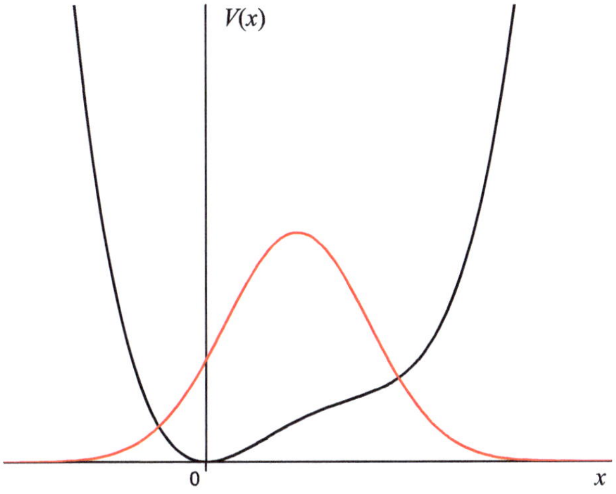

Fig. 3.2 The model potential (3.15)

these objects, the following QAO potential is considered here:

$$\hat{H} = \frac{1}{2}(\hat{p}^2 + \hat{x}^2) + \gamma\hat{x}^3 + \lambda\hat{x}^4, \tag{3.15}$$

which corresponds to the potential well presented in Fig. 3.2.

According to the procedure of OM, the supplementary parameters are introduced into Hamiltonian (3.15) by canonical transformation for coordinate and momentum transfer operators. Alternatively to transformation (3.2) for symmetric QAO, this operation has to be fulfilled taking into account the asymmetry of potential:

$$\hat{x} = u + \frac{1}{\sqrt{2\omega}}(\hat{a}^+ + \hat{a}), \qquad \hat{p} = i\sqrt{\frac{\omega}{2}}(\hat{a}^+ - \hat{a}), \tag{3.16}$$

where u is a parameter determining the non-zero mean value of coordinate operator $\bar{x}$. This parameter will be further considered as variational one, and the value of it is found from the condition for best approximation in zeroth order of OM. For zeroth approximation of the energy, the diagonal matrix elements of full Hamiltonian (3.15) are selected, that results in:

$$E_n^{(0)} = E_n^{(0)}(u_n, \omega_n) = \frac{2n+1}{4}\left(\omega_n + \frac{1}{\omega_n}\right) + \frac{u_n^2}{2} + \gamma\left[u_n^3 + \frac{3u_n}{2\omega_n}(2n+1)\right]$$
$$+\lambda\left[u_n^4 + \frac{3u_n^2}{\omega_n}(2n+1) + \frac{3}{4\omega_n^2}(2n^2+2n+1)\right]. \tag{3.17}$$

and the best approximation condition is written as:

$$\frac{\partial E_n^{(0)}}{\partial u_n}=0, \quad u_n^3+\frac{3\gamma}{4\lambda}u_n^2+\frac{\omega_n+6\lambda(2n+1)}{4\lambda\omega_n}u_n+\frac{3\gamma(2n+1)}{8\lambda\omega_n}=0,$$
$$\frac{\partial E_n^{(0)}}{\partial \omega_n}=0, \quad \omega_n^3-[1+6u_n(\gamma+\lambda u_n)]\,\omega_n-\frac{6\lambda(2n^2+2n+1)}{2n+1}=0. \tag{3.18}$$

The expressions (3.17) and (3.18) define the uniformly suitable approximation for the energy of the system with Hamiltonian (3.15). Depending on the relationship between the parameters, two qualitatively different forms of potential are possible: in case the equation for the extrema points of potential

$$x + 3\gamma x^2 + 4\lambda x^3 = 0 \tag{3.19}$$

possesses a single root $x = 0$, the potential well has a single minimum (Fig. 3.2). This condition is fulfilled if the denominator of the quadratic equation followed from (3.19) is less or equal to zero:

$$D = 9\gamma^2 - 16\lambda \leq 0, \tag{3.20}$$

that implements the limitation for Hamiltonian parameters, which is satisfied for majority of the molecular systems:

$$|\gamma| \leq \frac{4}{3}\sqrt{\lambda}. \tag{3.21}$$

However, there are exist physical systems with extremely large potential asymmetry, for example, the atomic oscillations in the crystals with complex elementary unit cell. In such cases, the problem is treated as a one with two asymmetric potential wells, which causes the peculiarities in thermodynamical features of the system (see [19] citations therein). This case is described by inequality:

$$|\gamma| > \frac{4}{3}\sqrt{\lambda}. \tag{3.22}$$

The simple algebraic relationships (3.17) and (3.18) derived from zeroth approximation of the operator method are found to describe all qualitative properties of the system. Assuming the validity of the inequality (3.21), the mean value of the coordinate operator $\bar{x}$ over the wave functions of OM zeroth approximation:

$$\bar{x} = \langle n|\,\hat{x}\,|n\rangle = u, \tag{3.23}$$

is not equal to zero for any non-zero values of the parameter γ, which corresponds to the shifted position of the localization of wave function and makes possible the description of the thermal expansion of solids. Table 3.5 presents the numerical data

Table 3.5 Energy levels of the oscillator with Hamiltonian $\hat{H} = \frac{1}{2}(\hat{p}^2 + \hat{x}^2) + \gamma\hat{x}^3 + \lambda\hat{x}^4$ by zeroth approximation of OM $E_n^{(0)}$ and by group-theoretic approach [20] $E_n^{(TG)}$; n is a quantum number

	$\lambda = 1$					
	$E_n^{(0)}$			$E_n^{(TG)}$		
n	$\gamma = 0.1$	$\gamma = 0.7$	$\gamma = 1.3$	$\gamma = 0.1$	$\gamma = 0.7$	$\gamma = 1.3$
0	0.8118	0.7760	0.6657	0.7875	0.7535	0.6484
1	2.7578	2.6508	2.3541	2.7164	2.6164	2.3281
2	5.1691	5.0083	4.5749	5.1588	5.0033	4.5681
3	7.9037	7.6966	7.1464	7.9590	7.7504	
4	10.895	10.646	9.9906	11.080	10.866	
5	14.103	13.815	13.063	14.371		
6	17.501	17.177	16.333	17.851		
7	21.069	20.710	19.780	21.490		
8	24.795	24.400	23.386	25.287		
9	28.657	28.233	27.140	29.250		
10	32.654	32.200	31.030	33.331		
20	78.5099	77.7885	75.9507			

	$\lambda = 25$					
	$E_n^{(0)}$			$E_n^{(TG)}$		
n	$\gamma = 1$	$\gamma = 3$	$\gamma = 6.5$	$\gamma = 1$	$\gamma = 3$	$\gamma = 6.5$
0	2.0379	2.0271	1.9808	1.9563	1.9483	1.9055
1	7.2021	7.1741	7.0562	7.0548	7.0306	6.9220
2	13.865	13.824	13.654	13.825	13.789	13.624
3	21.528	21.477	21.263	21.735	21.669	
4	29.974	29.914	29.660	30.550		
5	39.077	39.008	38.717	40.094		
6	48.755	48.674	48.349	50.148		
7	58.938	58.852	58.494			
8	69.586	69.493	69.103			
9	80.661	80.560	80.140			
10	92.130	92.023	91.574			
20	224.306	224.137	223.435			

for the energy of QAO with Hamiltonian (3.15) for several values of the parameters γ and λ, satisfying the condition (3.21) and calculated in zeroth approximation of the operator method. For comparison, the results by group-theoretic approach from [20] are displayed, too.

Provided the condition (3.22) is fulfilled, there are several solutions of (3.18) available for parameter u_n at fixed quantum number n. This leads to sophisticated sequence of the energy levels due to the overlap of the wave functions localized in different potential wells. Nevertheless, the results from the Table 3.5 demonstrate the ability of OM to deliver the uniformly suitable approximation for both energy levels and their relative positions.

3.3 Morse Potential

A real interatomic potential possesses the finite number of bound states, and therefore the use of perturbation theory, which exploits the harmonic oscillator with infinite number of bound states for initial approximation, yields the incorrect results for the energy spectra. To analyze the effectiveness of the operator method for such systems, the model with the finite number of the discrete energy levels is applied, which corresponds to Morse potential (Fig. 3.3).

The Schrödinger equation with this potential has an analytical solution, which is widely used to model the two-atomic molecules [16, 21–23]. The exact energy spectrum for the system with Hamiltonian

$$\hat{H} = \frac{\hat{p}^2}{2} + D_c(e^{-2\alpha x} - 2e^{-\alpha x}) \tag{3.24}$$

is determined by the following expression [14]:

$$E_n = -D_c\left[1 - \frac{\alpha}{\sqrt{2D_c}}\left(n + \frac{1}{2}\right)\right]^2, \tag{3.25}$$

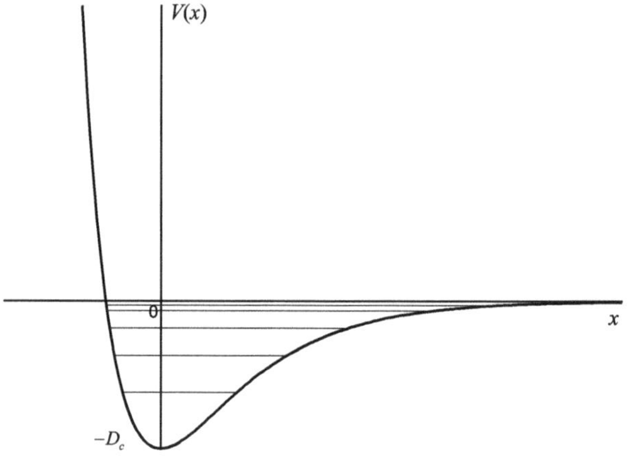

Fig. 3.3 Morse potential (3.24)

where the quantum number n varies in the limits:

$$0 \leq n < \frac{\sqrt{2D_c}}{\alpha} - \frac{1}{2}. \tag{3.26}$$

The Morse oscillator is frequently used for approbation of both PT and non-perturbative approaches [8, 24–27]. According to general OM prescription for calculation of the energy spectrum, we select the diagonal elements of the Hamiltonian (3.24) with respect to the full set of functions, which can be the wave functions of harmonic oscillator with undefined frequency ω. The following mean values have to be calculated:

$$E_n^{(0)} = \langle n | \frac{\hat{p}^2}{2} + D_c(e^{-2\alpha x} - 2e^{-\alpha x}) | n \rangle , \tag{3.27}$$

where $|n\rangle$ are the eigenfunctions of the operator of particle number $\hat{n} = \hat{a}^+\hat{a}$. The averaging of the operator of kinetic energy results in:

$$\langle n | \frac{\hat{p}^2}{2} | n \rangle = \frac{\omega}{2}\left(n + \frac{1}{2}\right). \tag{3.28}$$

To calculate the contribution of the potential energy, the expression for the averaged value of the exponential operator over the wave functions of the harmonic oscillator with arbitrary frequency is used:

$$\langle n | e^{-\alpha x} | n \rangle = \frac{\sqrt{\omega}}{\sqrt{\pi} 2^n n!} \int_{-\infty}^{+\infty} e^{-\alpha x} e^{-\omega x^2} \mathrm{H}_n^2(\sqrt{\omega} x) dx = \mathrm{L}_n\left(-\frac{\alpha^2}{2\omega}\right) e^{\frac{\alpha^2}{4\omega}}, \tag{3.29}$$

where $\mathrm{L}_n(x)$ are the orthogonal Laguerre polynomials [28], and the result is:

$$\langle n | e^{-2\alpha x} - 2e^{-\alpha x} | n \rangle = \mathrm{L}_n\left(-\frac{2\alpha^2}{\omega}\right) e^{\frac{\alpha^2}{\omega}} - 2\mathrm{L}_n\left(-\frac{\alpha^2}{2\omega}\right) e^{\frac{\alpha^2}{4\omega}}, \tag{3.30}$$

that with the use of (3.28) leads to the expression for the energy in OM zeroth approximation:

$$E_n^{(0)} = \frac{\omega_n}{2}\left(n + \frac{1}{2}\right) + D_c\left\{\mathrm{L}_n\left(-\frac{2\alpha^2}{\omega_n}\right) e^{\frac{\alpha^2}{\omega_n}} - 2\mathrm{L}_n\left(-\frac{\alpha^2}{2\omega_n}\right) e^{\frac{\alpha^2}{4\omega_n}}\right\}, \tag{3.31}$$

and the optimal choice of the parameter ω_n follows from the solution of the non-linear algebraic equation:

$$(1+2n)\omega_n^2 - 4e^{\frac{\alpha^2}{\omega_n}}\alpha^2 D_c \left[\mathrm{L}_n\left(-\frac{2\alpha^2}{\omega_n}\right) + 2\mathrm{L}_{n-1}^1\left(-\frac{2\alpha^2}{\omega_n}\right)\right]$$
$$+2e^{\frac{\alpha^2}{4\omega_n}}\alpha^2 D_c \left[\mathrm{L}_n\left(-\frac{\alpha^2}{2\omega_n}\right) + 2\mathrm{L}_{n-1}^1\left(-\frac{\alpha^2}{2\omega_n}\right)\right] = 0, \qquad (3.32)$$

where $\mathrm{L}_n^a(x)$ are the generalized orthogonal Laguerre polynomials and the expression for the derivative of Laguerre polynomials [28] has been used:

$$\frac{d}{dx}\mathrm{L}_n(x) = -\mathrm{L}_{n-1}^1(x). \qquad (3.33)$$

In opposite to perturbation approaches and direct summation of asymptotic series [8], the zeroth approximation of OM takes into account the existence of the upper border of the discrete energy spectrum. The spectrum of the system with Hamiltonian (3.24) is continuous at $E \geq 0$, whereas the discrete non-degenerate levels exist at $E < 0$ [14] only. From the expression (3.31), the following inequality follows:

$$n < \frac{2D_c}{\omega_n}\left\{2\mathrm{L}_n\left(-\frac{\alpha^2}{2\omega_n}\right)e^{\frac{\alpha^2}{4\omega_n}} - \mathrm{L}_n\left(-\frac{2\alpha^2}{\omega_n}\right)e^{\frac{\alpha^2}{\omega_n}}\right\} - \frac{1}{2}, \qquad (3.34)$$

which permits to find the maximal number n_m of molecule bound states for different parameters of the potential. By solving the Eq. (3.32) for each n and substituting the results in (3.31), we obtain the energy spectrum of Morse oscillator in OM zeroth approximation. Table 3.6 demonstrates the uniform suitability of the approximation obtained for both parameters of the physical system.

The relatively poor accuracy ($\sim 10\,\%$) corresponds to the highly excited states at small values of $\sqrt{D_c}/\alpha$, when the total number of coupled states is small. For the range $\sqrt{D_c}/\alpha > 30$, which corresponds to the vibrational terms for the majority of real molecules, the accuracy stays within the 3 % range.

Table 3.6 The energy levels E_n of the Morse oscillator; n_m and $n_m^{(ex)}$ are the maximal number of coupled states for OM and exact number of coupled states, respectively; ξ is the accuracy of the energy levels

	$D_c = 10$					
	$E_n^{(0)}$		Exact values E_n		Accuracy ξ, %	
	$\alpha = 0,01$	$\alpha = 0,1$	$\alpha = 0,01$	$\alpha = 0,1$	$\alpha = 0,01$	$\alpha = 0,1$
n	$n_m = 497$	$n_m^{(ex)} = 446$	$n_m = 51$	$n_m^{(ex)} = 44$		
0	−9.97762	−9.77423	−9.97765	−9.77764	0.0003	0.03
1	−9.93281	−9.31846	−9.93303	−9.34043	0.002	0.24
2	−9.88791	−8.85445	−9.88851	−8.91322	0.006	0.66
3	−9.84293	−8.38248	−9.84409	−8.49600	0.01	1.3
4	−9.79786	−7.90281	−9.79977	−8.08879	0.02	2.3
5	−9.75271	−7.41569	−9.75555	−7.69158	0.03	3.7
6	−9.70747	−6.92132	−9.71142	−7.30436	0.04	5.2
7	−9.66214	−6.41990	−9.66740	−6.92715	0.05	7.3
8	−9.61673	−5.91161	−9.62348	−6.55993	0.07	9.9
9	−9.57124	−5.39661	−9.57966	−6.20272	0.09	13.0
10	−9.52566	−4.87508	−9.53594	−5.85551	0.11	16.7

	$D_c = 100$					
	$E_n^{(0)}$		Exact values E_n		Accuracy ξ, %	
	$\alpha = 0,05$	$\alpha = 0,2$	$\alpha = 0,05$	$\alpha = 0,2$	$\alpha = 0,05$	$\alpha = 0,2$
n	$n_m = 314$	$n_m^{(ex)} = 282$	$n_m = 79$	$n_m^{(ex)} = 70$		
0	−99.6459	−98.5771	−99.6468	−98.5908	0.0009	0.014
1	−98.9366	−95.7142	−98.9422	−95.8024	0.006	0.09
2	−98.2252	−92.8176	−98.2400	−93.0539	0.015	0.25
3	−97.5116	−89.8881	−97.5404	−90.3455	0.03	0.51
4	−96.7958	−86.9265	−96.8433	−87.6771	0.05	0.86
5	−96.0779	−83.9334	−96.1487	−85.0487	0.07	1.3
6	−95.3580	−80.9095	−95.4566	−82.4602	0.10	1.9
7	−94.6359	−77.8554	−94.7670	−79.9118	0.14	2.6
8	−93.9117	−74.7718	−94.0799	−77.4034	0.18	3.4
9	−93.1855	−71.6590	−93.3953	−74.9349	0.22	4.4
10	−92.4572	−68.5176	−92.7132	−72.5065	0.28	5.5

3.4 Solution of the Mathieu Equation

The Mathieu equation is a canonic equation of the mathematical physics, which describes numerous physical system (see, for example, [29]). The properties of the solutions for Mathieu equation, so called Mathieu functions, are well studied, however, these functions are expressed either through other special functions or as infinite fractions [28], which limits their use in real applications. The use of operator method for the solution of Mathieu equation may deliver the convenient analytical form for the approximation of Mathieu functions, useable for physical applications. We use here the dimensionless form of the Mathieu equation, which corresponds to the Schroödinger equation for the particle of mass $m = \frac{1}{2}$ moving in the field of one-dimensional potential:

$$\left(\hat{H} - E\right)\Psi = \left(-\frac{d^2}{dx^2} + h\cos 2x - E\right)\Psi = 0, \tag{3.35}$$

where $h \geq 0$ is a dimensionless amplitude of the potential; E is an eigenvalue of Hamiltonian $\hat{H}$. We look for the periodic solutions of the Eq. (3.35):

$$\Psi(x + 2\pi) = \Psi(x), \qquad \Psi(x + \pi) = \pm\Psi(x), \tag{3.36}$$

without any preliminary assumptions about parameter h. To construct the wave functions of OM zeroth approximation for Mathieu equation, the Hamiltonian $\hat{H}$ is presented in the form, which simplifies the approximate factorization:

$$\hat{H} = \hat{p}^2 + \frac{h}{2}\cos^2 x + \frac{h}{2} \tag{3.37}$$

by using two adjoint operators:

$$\hat{a} = \hat{p} - i\sqrt{\frac{h}{2}}\cos x, \qquad \hat{a}^+ = \hat{p} + i\sqrt{\frac{h}{2}}\cos x, \tag{3.38}$$

which define the basic functions of two types:

$$\left|\Psi^{\pm}\right\rangle = \exp\left(\pm\sqrt{\frac{h}{2}}\sin x\right), \quad \hat{a}\left|\Psi^{-}\right\rangle = 0, \quad \hat{a}^+\left|\Psi^{+}\right\rangle = 0. \tag{3.39}$$

The commutator of the operators $\hat{a}$ and $\hat{a}^+$ is more complex then for harmonic oscillator, and therefore these operators do not factorize exactly the Hamiltonian $\hat{H}$ and the functions (3.39) are not the exact solutions of the Eq. (3.35). Nevertheless, they can be used to build a basic set of functions Ψ_n for operator method. In previous chapter, the wave functions of the harmonic oscillator have been used to approximate the wave functions of the anharmonic oscillator (2.16) by using the single-parameter unitary transformation (2.2) for the operators of creation and annihilation with the parameter ω in the basic set of functions. The similar operation can be performed for the operators (3.38), which corresponds to the introduction of the parameter ν in considered functions [30]:

$$\left|\Psi^{\pm}\right\rangle \to \left|\Psi_{\nu}^{\pm}\right\rangle = \exp\left(\pm\frac{\nu}{2}\sin x\right). \tag{3.40}$$

To build the whole spectrum of the periodic solutions of Eq. (3.35), the symmetry of wave functions (3.36) has to be taken into account and the algorithm for the construction of the excited states has to be created. The translations of the state vectors (3.36) for translations can be provided by linear combinations of the functions (3.40), and then the approximate normalized equations (3.35), corresponding to two minimal values E_0 and E_1, are written as:

$$|\Psi_{0,\nu}\rangle = \frac{1}{\sqrt{1+I_0(\nu)}}\,\mathrm{ch}\left(\frac{\nu}{2}\sin x\right), \tag{3.41}$$

$$|\Psi_{1,\nu}\rangle = \frac{1}{\sqrt{-1+I_0(\nu)}}\,\mathrm{sh}\left(\frac{\nu}{2}\sin x\right). \tag{3.42}$$

Here $I_0(\nu)$ is the Bessel function of zeroth order with imaginary argument, and the orthogonality and normalization conditions follow from the scalar product:

$$\langle\Psi_{\alpha,\nu}|\,\Psi_{\beta,\nu}\rangle = \int_0^{2\pi}\Psi^*_{\alpha,\nu}(x)\Psi_{\beta,\nu}(x)dx = \pi\delta_{\alpha\beta}. \tag{3.43}$$

These simple analytical functions $|\Psi_{0,\nu}\rangle$ and $|\Psi_{1,\nu}\rangle$ give a good approximation for Mathieu functions ce_0 and se_0 [29] uniformly suitable for ν and x provided the parameter ν is chosen properly. The cumbersome form of the commutator $\hat{a}$ and $\hat{a}^+$ complicates the algebraic calculation of both the excited states of the system and the matrix elements of the perturbation operator, as in the case of harmonic oscillator. Nevertheless, using the linear combinations of the functions (3.41) and (3.42) multiplied by polynomials over $\cos x$, and orthogonality procedures [28], the wave functions of OM zeroth approximation can be constructed for required number of the excited states $|\Psi_{n,\nu}\rangle$. The optimal value ν_n varies for different n, and thus the functions $|\Psi_{n,\nu_n}\rangle$, being orthogonal at certain ν, become non-orthogonal when the parameter ν is being optimized. Nevertheless, the preliminary orthogonalization of state vectors $|\Psi_{n,\nu}\rangle$ is necessary to provide the correct symmetry and proper number of nodes in wave functions for the excited states. We derive here the normalized functions $|\Psi_{2,\nu}\rangle$ and $|\Psi_{3,\nu}\rangle$ for two excited states, which give an analytical approximation for the Mathieu functions ce_1 and se_1:

$$\begin{aligned}|\Psi_{2,\nu}\rangle &= \left[\frac{1}{\nu}I_1(\nu)+\frac{1}{2}\right]^{-1/2}\cos x\,\mathrm{ch}\left(\frac{\nu}{2}\sin x\right),\\ |\Psi_{3,\nu}\rangle &= \left[\frac{1}{\nu}I_1(\nu)-\frac{1}{2}\right]^{-1/2}\cos x\,\mathrm{sh}\left(\frac{\nu}{2}\sin x\right)\end{aligned} \tag{3.44}$$

and correspond to the eigenvalues $E_{2,3}$. The general algorithm to obtain the approximate solutions $|\Psi_{n,\nu}\rangle$ for arbitrary n is built in the following way: the set of the functions $|\Psi_{n,\nu}\rangle$ is introduced according to the following rule:

$$|\Psi_{n,\nu}\rangle = \begin{cases} \cos kx \cdot \mathrm{ch}\left(\frac{\nu}{2}\sin x\right); & n = 2k; \quad k = 0, 1, \dots \\ \cos kx \cdot \mathrm{sh}\left(\frac{\nu}{2}\sin x\right); & n = 2k+1; \quad k = 0, 1, \dots \end{cases} \tag{3.45}$$

The state vectors $|\Psi_{n,\nu}\rangle$ possessing the required number of nodes and orthogonal to the wave functions of all states with smaller n are found from the recurrent equations:

$$|\Psi_{n,\nu}\rangle = C_{n,\nu}\left\{|\Psi_{n,\nu}\rangle - \sum_{l=0}^{n-1}|\Psi_{l,\nu}\rangle\langle\Psi_{l,\nu}|\,\Psi_{n,\nu}\rangle\right\}, \tag{3.46}$$

where the constants $C_{l,\nu}$ are determined from the normalization conditions. The integrals in formulas (3.45) and (3.46) are expressed through the Bessel functions, and the choice of the parameters ν_n in OM zeroth approximation is defined by the independence condition of eigenvalues E_n on the parameters, which manage the representation of the wave functions, as stated in Sect. 2.1:

$$\frac{\partial}{\partial\nu_n}E_n^{(0)}(\nu) = 0. \tag{3.47}$$

At $n = 0$, only one of these equations coincides with the form of the variational principle on the functions class $|\Psi_{n,\nu}\rangle$. To apply the variational principle for calculation of the energy of excited states ($n \neq 0$), the trial functions depending on n parameters are used, which are required for orthogonality of the state vector $|\Psi_{n,\nu}\rangle$ to the wave functions of the states, corresponding to smaller quantum numbers [28]. The energy of the system in OM zeroth approximation at certain ν is calculated as:

$$E_n^{(0)}(\nu) = \int_0^{2\pi}\Psi_{n,\nu}^*(x)\hat{H}\Psi_{n,\nu}(x)dx, \tag{3.48}$$

where the state vectors $|\Psi_{n,\nu}\rangle$ are determined from (3.45) and (3.46). For example, for the states $n = 0, 1, 2, 3$ the equations are:

$$E_{0,1}^{(0)} = \frac{1}{I_0 \pm 1}\left[\mp\frac{1}{8}\nu^2 + \frac{1}{4}\nu I_1 + h\left(\frac{2}{\nu}I_1 - I_0\right)\right], \tag{3.49}$$

$$h = \frac{\nu}{4}\frac{I_0^2 - 1 \pm I_1\left(\frac{1}{2}\nu \mp I_1\right)}{I_1\left(\frac{2I_1}{\nu} - I_0\right) + (I_0 \pm 1)\left[I_1\left(1 + \frac{4}{\nu^2}\right) - \frac{2I_0}{\nu}\right]},$$

$$E_{2,3}^{(0)} = \frac{1}{\frac{2I_1}{\nu} \pm 1} \left\{ \pm 1 \mp \frac{3}{16}\nu^2 + \frac{3}{2}I_0 - \frac{I_1}{\nu} - \right.$$
$$\left. - h\left[\mp\frac{1}{2} - \frac{12I_0}{\nu^2} + \frac{2I_1}{\nu}\left(1 + \frac{12}{\nu^2}\right)\right]\right\}, \quad (3.50)$$
$$h = \frac{I_1^2 - I_0^2 + \frac{2}{\nu}I_0 I_1 - \frac{1}{8}\nu^2 \pm \frac{2}{\nu}I_1 \pm I_0\left(\frac{1}{8}\nu^2 - 1\right)}{\frac{8}{\nu^2}I_0^2 \pm I_0\left(1 + \frac{16}{3\nu^2}\right) - \frac{8}{\nu^2}I_1^2\left(1 + \frac{4}{\nu^2}\right) \mp \frac{2}{\nu}I_1\left(3 + \frac{16}{\nu^2}\right)}.$$

These equations express the dependence of the eigenvalues of the Mathieu equation on the potential amplitude. The use of the perturbation theory or strong coupling approximation for solution of Mathieu equations [31] results in asymptotic series over h^2 or $h^{-1/2}$, respectively. The OM zeroth approximation $E_n^{(0)}(h)$ by formulas (3.47) and (3.48), however, calculates the system energy with the accuracy, which is independent on h. For example, the solutions of the Eqs. (3.49) at $n = 0$ in the limits $h \ll 1$ or $h \gg 1$ are found in analytical form:

$$h \simeq \frac{1}{4}\nu^2; \quad E_0^{(0)}(h) \simeq \frac{1}{16}\left(\frac{1}{8}\nu^4 - h\nu^2\right) \simeq -\frac{1}{8}h^2; \quad h \ll 1;$$
$$h \simeq \frac{1}{8}\nu^2; \quad E_0^{(0)}(h) \simeq \frac{\nu}{4}\frac{I_1}{I_0} + h\left(\frac{2}{\nu}\frac{I_1}{I_0} - 1\right) \simeq -h + \sqrt{2h}; \quad h \gg 1, \quad (3.51)$$

and this result is equivalent to the asymptotic expansion for exact eigenvalues $E_0(h)$ [32]. Further terms in the expansion (3.51) contain the same orders of h as asymptotic series do, but with different coefficients, and similar results are obtained for the excited states, too. Table 3.7 shows the accuracy of OM zeroth approximation for eigenvalues and eigenfunctions of the Mathieu equation for intermediate values h.

For real applications of the approximation for Mathieu functions, the use of the parameter ν is more convenient than the parameter h. This replacement makes a use of the Eq. (3.47) for calculation of functions $h(\nu)$, instead of the solution of transcendental equations. This approach is especially beneficial for the applications with a wide range of the parameter h.

The zeroth approximation of the operator method gives an accurate approximation for the periodic Mathieu functions, which are only particular solutions of the Eq. (3.35). According to the Floquet theorem (or Bloch theorem), which is valid for all periodic potentials [28], the general solution of the Eq. (3.35) is:

$$\left|\Psi_n^k(x)\right\rangle = e^{ikx}\varphi_n(x); \qquad \varphi_n(x+\pi) = \varphi_n(x), \quad (3.52)$$

which depends on the quasimomentum k varying in the range:

$$-1 \leq k \leq 1,$$

Table 3.7 Comparison of the exact periodic solutions and eigenvalues of the Mathieu equation [32] with OM zeroth approximation

	h	E_0	$ce_0\left(0, \frac{h}{2}\right)$	$ce\left(\frac{\pi}{2}, \frac{h}{2}\right)$
Exact	10	−5.800	4.48×10^{-2}	1.335
solution	30	−22.513	1.93×10^{-3}	1.550
=8.331	10	−5.800	4.14×10^{-2}	1.333
=14.944	30	−22.513	1.76×10^{-3}	1.550
	h	E_1	$Sl_1'\left(0, \frac{h}{2}\right)$	$Sl_1\left(\frac{\pi}{2}, \frac{h}{2}\right)$
Exact	10	−5.790	1.75×10^{-1}	1.337
solution	30	−22.513	1.39×10^{-2}	1.550
=8.371	10	−5.790	1.70×10^{-1}	1.337
=14.944	30	−22.513	1.32×10^{-2}	1.550
	h	E_2	$Cl_1\left(0, \frac{h}{2}\right)$	$Cl_1'\left(\frac{\pi}{2}, \frac{h}{2}\right)$
Exact	10	1.858	2.57×10^{-1}	−3.469
solution	30	−8.101	1.50×10^{-2}	−5.764
=6.784	10	1.867	2.28×10^{-1}	−3.388
=13.772	30	−8.100	1.17×10^{-2}	−5.740
	h	E_3	$Sl_2'\left(0, \frac{h}{2}\right)$	$Sl_2\left(\frac{\pi}{2}, \frac{h}{2}\right)$
Exact	10	2.099	7.33×10^{-1}	−3.641
solution	30	−8.099	9.18×10^{-2}	−5.766
=7.212	10	2.100	7.09×10^{-1}	−3.615
=13.779	30	−8.099	8.06×10^{-2}	−5.742

and specifying the zone spectrum $E_n(k)$ of the particle in the periodic potential. The approximate calculation of $E_n(k)$ in OM zeroth approximation is executed in a standard way as in previous paragraphs: the expression (3.52) is substituted into (3.35) and the equation for the periodic function $\varphi_n(x)$ is obtained:

$$\left(\hat{H}_k - E\right)\varphi_n(x) \equiv \left[-\frac{d^2}{dx^2} - 2ik\frac{d}{dx} + k^2 + h\cos 2x - E\right]\varphi_n(x) = 0. \tag{3.53}$$

The Hamiltonian $\hat{H}_k$ for this equation does not possess any certain parity, therefore the combination of even and odd vectors has to be used for construction of the basic set of the wave functions, in opposite to (3.45):

$$\varphi_n^{(0)}(x) = A\,|\Psi_{n,\nu}\rangle + B\,|\Psi_{n+1,\nu}\rangle\, e^{-ix}. \tag{3.54}$$

The functions $|\Psi_{n,\nu}\rangle$ follow from the Eq. (3.46), and the coefficients A and B are related each to other by normalization condition:

$$|A|^2 + |B|^2 = 1.$$

Table 3.8 The energy spectrum of the lower levels for Mathieu equation in OM zeroth approximation at $h = 30$

k	0.0	0.5	1.0
$E_0^{(0)}(k)$	−22.513026	−22.512980	−22.512993
$E_1^{(0)}(k)$	−8.098766	−8.099101	−8.100342

The energy $E_n^{(0)}(k)$, corresponding to the level with quasimomentum k in nth energy zone is found from the formula:

$$E_n^{(0)}(k) = \int_0^{2\pi} \varphi_n^{(0)}(x)^* \hat{H}_k \varphi_n^{(0)}(x) dx, \tag{3.55}$$

and the coefficients A and B as well as parameter ν follow from:

$$\frac{\partial E_n^{(0)}}{\partial A} = \frac{\partial E_n^{(0)}}{\partial B} = \frac{\partial E_n^{(0)}}{\partial \nu} = 0. \tag{3.56}$$

The results of numerical simulations for zone spectrum using formula (3.55) are presented in Table 3.8.

As a next step of the application of operator method to Mathieu equation, we analyze the convergence of successive approximations for numerical solution of Mathieu equation by OM. Along with the algorithm (3.45) for calculation of the wave functions of basic set for zeroth approximation, the iterative scheme (2.49) can be applied for further approximations of non-polynomial Hamiltonians. However, this calculation faces the following problem: due to complicated algebra of the operators (3.38), the matrix elements $\hat{H}_k$ of the basic set (3.54) required for high iterations are extremely cumbersome. Assuming the wave function of zeroth approximation $\left|\varphi_n^{(0)}\right\rangle$ gives a good results in a wide range of x, this obstacle can be overcame. The Fourier series converges quickly for all arguments and for small difference $|\varphi_n\rangle - \left|\varphi_n^{(0)}\right\rangle$, which is a periodic function. Therefore, for the basic set of the exact solution, not the functions of zeroth approximation (3.54) are used like in iterative scheme (2.49), but the coefficients of the expansion C_{nl} of the Fourier series:

$$|\varphi_n(x)\rangle = \left|\varphi_n^{(0)}(x)\right\rangle + \sum_{l=-\infty}^{+\infty} C_{nl} |l\rangle ; \qquad |l\rangle = \frac{1}{\sqrt{2\pi}} e^{ilx}. \tag{3.57}$$

In this case and opposite to (2.49), the function for zeroth approximation is not orthogonal to the basic vectors:

$$\langle \varphi_n^{(0)} | l \rangle \neq 0,$$

that results in the modification of the iterative scheme (2.49) by accounting of the overlapping integral, and the iterative scheme becomes:

$$
\begin{aligned}
E_n^{(s)} &= \left[1 + \sum_{l=-(s-1)}^{s-1} C_{nl}^{(s-l-1)} \langle \varphi_n^{(0)} | l \rangle \right]^{-1} \\
&\times \left\{ \langle \varphi_n^{(0)} | \hat{H}_k | \varphi_n^{(0)} \rangle + \sum_{l=-(s-1)}^{s-1} C_{nl}^{(s-l-1)} \left\langle \varphi_n^{(0)} \middle| \hat{H}_k \middle| l \right\rangle \right\}, \\
C_{nl}^{(s)} &= \left[E_n^{(s-1)} - \langle l | \hat{H}_k | l \rangle \right]^{-1} \\
&\times \left\{ \left\langle l \middle| \hat{H}_k \middle| \varphi_n^{(0)} \right\rangle - E_n^{(s-1)} \langle l | \varphi_n^{(0)} \rangle + \sum_{m=-(s-1)}^{s-1} C_{nm}^{s-m-1} \langle m | \hat{H}_k | l \rangle \right\}.
\end{aligned}
\tag{3.58}
$$

Here $\left\langle l \middle| \hat{H}_k \middle| \varphi_n^{(0)} \right\rangle$ and $\langle l | \hat{H}_k | m \rangle$ are the matrix elements of the Hamiltonian, and the exact solution of the Schrödinger equation is defined by the limit of the sequence:

$$
\begin{aligned}
E_n(k) &= \lim_{s \to \infty} E_n^{(s)}(k); & C_{nl} &= \lim_{s \to \infty} C_{nl}^{(s)}; \\
E_n^{(0)} &= \left\langle \varphi_n^{(0)} \middle| \hat{H}_k \middle| \varphi_n^{(0)} \right\rangle; & C_{nl}^{(0)} &= 0.
\end{aligned}
\tag{3.59}
$$

Table 3.9 demonstrates the convergence of the numerical simulations of the successive approximations for $E_0^{(s)}(k)$ based on (3.58) for lower energy zone:

$$
\begin{aligned}
\langle l | \hat{H}_k | m \rangle &= (l + k)^2 \delta_{l,m} + \frac{h}{2} \left[\delta_{l,m-2} + \delta_{l,m+2} \right], \\
\left\langle \varphi_0^{(0)} \middle| \hat{H}_k \middle| l \right\rangle &= (l + k)^2 \langle \varphi_0^{(0)} | l \rangle + h \left\langle \varphi_0^{(0)} \middle| \cos 2x \middle| l \right\rangle \\
&= (-1)^l \left\{ \frac{(l + k)^2}{\sqrt{2}} \left[A I_l \left(\frac{\nu}{2} \right) + B I_{l+1} \left(\frac{\nu}{2} \right) \right] \right.
\end{aligned}
$$

Table 3.9 Convergence of OM successive approximations for lower energy zone of the Mathieu equation at $h = 30$

k	0.0	0,5	1.0
$E_0^{(0)}$	−22.513026	−22.512980	−22.512993
$E_0^{(1)}$	−22.513031	−22.512991	−22.512993
$E_0^{(10)}$	−22.513031	−22.512999	−22.512998
$E_0^{(s \geq 20)}$	−22.513037	−22.513021	−22.513004

$$-\frac{h}{2}\left[A\left(I_{l+2}\left(\frac{\nu}{2}\right)+I_{l-2}\left(\frac{\nu}{2}\right)\right)\right.$$
$$\left.\left.+\ B\left(I_{l+3}\left(\frac{\nu}{2}\right)+I_{l-1}\left(\frac{\nu}{2}\right)\right)\right]\right\}, \tag{3.60}$$

where the coefficients A and B in linear combination (3.54) follow from (3.56).

In the conclusion of this section, we discuss shortly a further possible generalization of OM for complex systems. As follows from the Chap. 1, the eigenfunctions of the harmonic oscillator can be efficiently used in OM zeroth approximation for the solution of Schrödinger equation with polynomial Hamiltonian. A similar conclusion can be made from this chapter for the particle moving in the field of one-dimensional periodic potential represented as Fourier series:

$$V(x) = \sum_{n=-\infty}^{+\infty} V_n e^{inx}. \tag{3.61}$$

Despite of the basic function set (3.52) and (3.54) obeying the Bloch theorem has been introduced for a simple form of the potential (3.61), it is found to be very effective for both zeroth approximation and iterative scheme for arbitrary potential. This fact is successfully used for the analysis of electromagnetic radiation in the channeling of charged particles inside the crystals [33, 34]. The zeroth approximation of OM based on (3.52) and (3.54), being applied to this problem, has certain advantages over other methods and provides the accurate values for energy spectrum (relative accuracy is below 1 %) and analytical expressions for matrix elements of transitions, population coefficients and other important physical characteristics of electromagnetic radiation for both one-dimensional (plane channeling) and two-dimensional (axial channeling) cases [35].

3.5 Quasienergies and Wave Functions of the Two-Level System in a Classical Monochromatic Field

The method of quasienergies is an effective technique for the description of quantum systems undergoing the influence of the external temporal periodic fields. This method has been introduced for the first time in [36] and is widely used for the analysis of various problems (see, for example, [37] and citations therein). In this section, the capability of the operator method will be demonstrated for the description of the evolution of the two-levels atom in a monochromatic linear polarized field. This physical system describes very well the behavior of many real phenomena [38]. There are numerous numerical and approximate analytical calculations done for the quasienergies in various limiting cases. However, the analytical expressions for the wave functions and quasienergies of two-level system are of a special interest for the use as a basis for the description of complex atomic states.

As usual, we start with the construction of OM zeroth approximation for Schrödinger equation for two-levels atom in a classic monochromatic field [38]:

$$\hat{H}\,|\Psi(\tau)\rangle \equiv \left[-i\omega\frac{\partial}{\partial\tau} + \frac{1}{2}E\sigma_3 - F\sigma_1\cos\tau\right]|\Psi(\tau)\rangle = \xi\,|\Psi(\tau)\rangle\,. \tag{3.62}$$

Here $\tau = \omega t$ is a dimensionless time, ω is a frequency of the field, E is a distance between the energy levels of the atom, ξ is a quasienergy, σ_i are Pauli matrices, parameter F is proportional to the amplitude of the field and to the dipole matrix element of the transition between the atomic states. The calculation of the exact values of the quasienergy ξ [37] requires the computing of the infinite Hill determinant.

Equation (3.62) is the system of two differential equations of the first order, with a mathematical structure similar to the equations obtained from the factorization of the Mathieu equation (Sect. 3.4). Therefore, the basic set of functions (3.40) is used for zeroth approximation, taking into account the spinor structure of the Eq. (3.62):

$$\left|\Psi^{\pm}\right\rangle = \exp\left[\pm i\lambda\sin\tau\right]\chi_{\pm};\quad \chi_{+} = \begin{pmatrix}1\\0\end{pmatrix};\quad \chi_{-} = \begin{pmatrix}0\\1\end{pmatrix}, \tag{3.63}$$

where λ is arbitrary parameter. The combination of spinors (3.63) for two lower quasienergy zones $\xi_n^{(0)}$ $(n = 1,2)$ is built to reconstruct the exact solution of the Eq. (3.62) in quasi-classic limit ($F \gg 1$) and in the limit of weak coupling ($F \ll 1$). The state vectors obeying these conditions are:

$$\left|\Psi_{n,\lambda}^{(0)}(\tau)\right\rangle = C_1\chi_1 + C_2\chi_2 e^{i\tau}; \tag{3.64}$$

$$\chi_1 = \begin{pmatrix}\cos\varphi\\ -i\sin\varphi\end{pmatrix};\quad \chi_2 = \begin{pmatrix}\sin\varphi\\ i\cos\varphi\end{pmatrix};\quad \varphi = \lambda\sin\tau,$$

where coefficients $C_{1,2}$ are connected through the normalization [39] for periodic functions:

$$\langle\langle\Psi_\alpha(\tau)|\ \Psi_\beta(\tau)\rangle\rangle = \int_0^{2\pi}\langle\Psi_\alpha(\tau)|\ \Psi_\beta(\tau)\rangle\,d\tau = \delta_{\alpha\beta}.$$

Other branches of the quasienergy spectrum can be calculated by using the functions analogous to (3.45) represented through the polynomials over $e^{i\tau}$ as the multipliers at spinors (3.64). Because of the limited space for this chapter, we consider further only the states, corresponding to quasienergies $\xi_n^{(0)}$. The zeroth approximation for these functions is a diagonal part of the Hamiltonian $\hat{H}$ with respect to the functions (3.64):

$$\xi(C_1, C_2, \lambda) = \langle\langle \Psi_{n,\lambda} \,|\, \hat{H} \,|\, \Psi_{n,\lambda} \rangle\rangle \,. \tag{3.65}$$

By variation of the expression (3.65) over the coefficients $C_{1,2}$, the following equation system is obtained:

$$\begin{aligned} \xi C_1 &= \frac{1}{2} E \left[J_0(2\lambda) C_1 + i J_1(2\lambda) C_2 \right] + \frac{1}{2} i (F - \omega\lambda) C_2, \\ \xi C_2 &= \omega C_2 - \frac{1}{2} F \left[J_0(2\lambda) C_2 + i J_1(2\lambda) C_1 \right] - \frac{1}{2} i (F - \omega\lambda) C_1, \end{aligned} \tag{3.66}$$

where J_0 and J_1 are the Bessel functions of real argument. The zeroth approximation for quasienergy follows from the existence of the non-trivial solution of the Eq. (3.66):

$$A_{1,2} = \frac{\xi_{1,2}^{(0)}(\lambda)}{E} = \frac{1}{2} \left\{ \frac{\omega}{E} \pm \sqrt{\left[\frac{\omega}{E} - J_0(2\lambda) \right]^2 + \left[\frac{\omega\lambda}{E} - \frac{F}{E} - J_1(2\lambda) \right]^2} \right\}, \tag{3.67}$$

provided the optimal value of λ is a root of the equation:

$$\frac{\partial}{\partial\lambda} \xi_{1,2}^{(0)}(\lambda) = 0,$$

$$2J_1(2\lambda) \left[\frac{\omega}{E} - J_0(2\lambda) \right] + \left[\frac{\omega\lambda}{E} - \frac{F}{E} - J_1(2\lambda) \right] \left[\frac{\omega}{E} + \frac{J_1(2\lambda)}{\lambda} - 2J_0(2\lambda) \right] = 0, \tag{3.68}$$

where the following formulas are used:

$$\frac{\partial J_0}{\partial\lambda} = J_0'(2\lambda) = -J_1(2\lambda); \qquad J_0''(2\lambda) = \frac{J_1(2\lambda)}{2\lambda} - J_0(2\lambda).$$

The coefficients $C_{1,2}$ can be found for above defined $\xi^{(0)}$ and λ, using the normalization condition for wave function (3.64):

$$C_1 = i \frac{\frac{F}{E} + J_1(2\lambda) - \frac{\omega\lambda}{E}}{\Delta}; \quad C_2 = \frac{2A - J_0(2\lambda)}{\Delta}; \tag{3.69}$$

$$\Delta = \sqrt{\left[2A - J_0(2\lambda) \right]^2 + \left[\frac{F}{E} + J_0(2\lambda) - \frac{\omega\lambda}{E} \right]^2}.$$

Equations (3.67)–(3.69) reconstruct properly the behavior of asymptotic expansions for quasienergy in different limiting cases [38]. For example, the Eq. (3.68) in the limit $F \gg \omega$ results in:

$$\lambda \simeq \frac{F}{\omega} + \frac{E}{\omega} J_1\left(\frac{2F}{\omega}\right) + O\left(\sqrt{\frac{\omega}{F}}\right),$$

and from (3.67) follows:

$$\xi_1^{(0)} \simeq \frac{F}{2}\sqrt{\frac{2}{\pi x_0}} \cos\left(x_0 - \frac{\pi}{4}\right),$$

$$\xi_2^{(0)} \simeq \omega - \frac{F}{2}\sqrt{\frac{2}{\pi x_0}} \cos\left(x_0 - \frac{\pi}{4}\right),$$

$$x_0 = \frac{2F}{\omega}.$$

In the resonant limit ($F \ll E, \omega \simeq E$), the following expressions are obtained:

$$\lambda \simeq \frac{F}{\omega + E},$$

$$\xi_{1,2}^{(0)} \simeq \frac{1}{2}\left\{\omega \pm \sqrt{(\omega - F)^2 + \frac{1}{4}F^2}\right\},$$

that agrees well with know results by alternative methods.

The important advantage of the operator method, the uniformly suitable interpolation of the exact solutions in a wide range of arguments, is illustrated in the Fig. 3.4. The plots obtained from the formulas (3.67) and (3.68), are compared with the results of the numerical calculations at intermediate values of the parameters

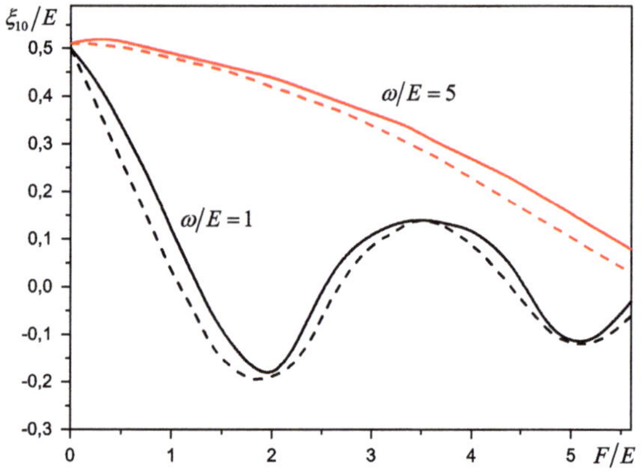

Fig. 3.4 Comparison of the numerical calculation of quasienergies ξ_{10} [40] (*dash line*) and OM zeroth approximation $\xi_{10}^{(0)}$ (*solid line*) for the same values as a function of the external field magnitude

[40]. The properly chosen zeroth approximation along with the iterative scheme for sequential approximations makes possible the expansion of the temporal periodic spinor into quickly converging Fourier series (similar to Mathieu equation):

$$|U(\tau)\rangle = |\Psi_n(\tau)\rangle - \left|\Psi_{n,\lambda}^{(0)}(\tau)\right\rangle$$

and to find an exact solution of the Eq. (3.62) in the following form:

$$|\Psi_n(\tau)\rangle = \left|\Psi_{n,\lambda}^{(0)}(\tau)\right\rangle + \sum_{l=-\infty}^{\infty} \begin{pmatrix} C_{l_1} \\ C_{l_2} \end{pmatrix} e^{il\tau}. \tag{3.70}$$

Taking into account $2n+1$ terms of this series (from $-n$ to n), the approximation of nth order is defined. The iterative scheme for the solution of the Eq. (3.62) is built similarly to (3.58), assuming the sth approximation is determined by the limited number $l \leq s$ of the transitions $\left\langle l \left| \hat{H} \right| \Psi_{n,\lambda}^{(0)} \right\rangle$. The equations for the coefficients C_{l_1} and C_{l_2} are found from the projections:

$$\left\langle\left\langle m_i \left| \hat{H} - \xi \right| \Psi_n \right\rangle\right\rangle; \qquad i = 1, 2; \qquad |m_{1,2}\rangle = e^{im\tau}\chi_{1,2},$$

which lead to the following recurrent relationships:

$$\begin{aligned} C_{k_1}^{(s)} = {} & \left(\xi_{1,2}^{(s-1)} - \langle k|\, \hat{H}_0^+ \,|k\rangle\right)^{-1} \Bigg\{ \left\langle k \left| \hat{H}_0^+ \right| \Psi_1^{(0)} \right\rangle \\ & + \sum_{l \neq k} C_{l_1}^{(s-1)} \langle k|\, \hat{H}_0^+ \,|l\rangle - \xi_{1,2}^{(s-1)} \left\langle k \mid \Psi_1^{(0)} \right\rangle \\ & - F \left[\left\langle k \left| \cos\tau \right| \Psi_2^{(0)} \right\rangle + \sum_{l} C_{l_2}^{(s-1)} \langle k \left| \cos\tau \right| l\rangle \right] \Bigg\}, \\ C_{k_2}^{(s)} = {} & \left(\xi_{1,2}^{(s-1)} - \langle k|\, \hat{H}_0^- \,|k\rangle\right)^{-1} \Bigg\{ \left\langle k \left| \hat{H}_0^- \right| \Psi_2^{(0)} \right\rangle \\ & + \sum_{l \neq k} C_{l_2}^{(s-1)} \langle k|\, \hat{H}_0^- \,|l\rangle - \xi_{1,2}^{(s-1)} \left\langle k \mid \Psi_2^{(0)} \right\rangle \\ & - F \left[\left\langle k \left| \cos\tau \right| \Psi_1^{(0)} \right\rangle + \sum_{l} C_{l_1}^{(s-1)} \langle k \left| \cos\tau \right| l\rangle \right] \Bigg\}, \end{aligned} \tag{3.71}$$

where

$$H_0^{\pm} = -i\omega\frac{\partial}{\partial\tau} \pm \frac{E}{2},$$

and the matrix elements in (3.71) are:

$$\left\langle \Psi_1^{(0)} \mid l \right\rangle = \begin{cases} B, & l = 2k, \\ 0, & l = 2k+1, \end{cases}$$

$$\left\langle \Psi_2^{(0)} \mid l \right\rangle = \begin{cases} 0, & l = 2k, \\ -B, & l = 2k+1, \end{cases}$$

$$\left\langle \Psi_{1,2}^{(0)} \left| \hat{H}_0^{\pm} \right| l \right\rangle = \left\langle \Psi_{1,2}^{(0)} \mid l \right\rangle \left(\omega l \pm \frac{E}{2} \right),$$

$$\langle k | \, \hat{H}_0^{\pm} \, | l \rangle = \left(\omega l \pm \frac{E}{2} \right) \delta_{kl}, \tag{3.72}$$

$$\left\langle \Psi_1^{(0)} \left| \cos\tau \right| l \right\rangle = \begin{cases} 0, & l = 2k, \\ \frac{\omega l}{\lambda} B - i \frac{C_2^*}{\lambda} J_{l-1}(\lambda), & l = 2k+1, \end{cases}$$

$$\left\langle \Psi_2^{(0)} \left| \cos\tau \right| l \right\rangle = \begin{cases} -\frac{\omega l}{\lambda} B + i \frac{C_2^*}{\lambda} J_{l-1}(\lambda), & l = 2k, \\ 0, & l = 2k+1, \end{cases}$$

$$\langle k | \cos\tau \, | l \rangle = \frac{1}{2} \left(\delta_{k,l+1} + \delta_{k,l-1} \right),$$

$$B = C_1^* J_l(\lambda) + i C_2^* J_{l-1}(\lambda).$$

The series of the successive approximations for quasienergy is built from the equation:

$$\left\langle \Psi_{1,2}^{(0)} \left| \hat{H} - \xi \right| \Psi_{1,2} \right\rangle = 0,$$

which in explicit form is written as:

$$\xi_{1,2}^{(s)} = \left\{ 1 + \sum_{l=-s}^{s} \left[C_{l_1}^{(s)} \left\langle \Psi_1^{(0)} \mid l \right\rangle + C_{l_2}^{(s)} \left\langle \Psi_2^{(0)} \mid l \right\rangle \right] \right\}^{-1}$$
$$\times \left\{ \xi_{1,2}^{(0)} + \sum_{l=-s}^{s} \left[C_{l_1}^{(s)} \left\langle \Psi_1^{(0)} \left| \hat{H}_0^{+} \right| l \right\rangle + C_{l_2}^{(s)} \left\langle \Psi_2^{(0)} \left| \hat{H}_0^{-} \right| l \right\rangle \right.\right.$$
$$\left.\left. - F \left(C_{l_2} \left\langle \Psi_1^{(0)} \left| \cos\tau \right| l \right\rangle + C_{l_1} \left\langle \Psi_2^{(0)} \left| \cos\tau \right| l \right\rangle \right) \right] \right\}, \tag{3.73}$$

where the exact values of the coefficients $C_{l_{1,2}}$ and quasienergies $\xi_{1,2}$ are the limits of the sequences:

$$C_{l_{1,2}} = \lim_{s \to \infty} C_{l_{1,2}}^{(s)}; \qquad \xi_{1,2} = \lim_{s \to \infty} \xi_{1,2}^{(s)}.$$

Table 3.10 The quasienergies of two-level system at different parameters

	$F = 0.4$		$F = 0.4$	
	$E = 0.1$	$E = 0.3$	$E = 0.1$	$E = 0.3$
$\xi_{0;1}^{(2)}$	4.2146×10^{-2}	1.0390×10^{-1}	4.7658×10^{-2}	1.3564×10^{-1}
$\xi_{0;1}^{(10)}$	4.1420×10^{-2}	1.0075×10^{-1}	4.7631×10^{-2}	1.3516×10^{-1}
$\xi_{0;1}^{(\geq 20)}$	4.1140×10^{-2}	1.0060×10^{-1}	4.7589×10^{-2}	1.3514×10^{-1}

The results of the calculations for the quasienergy of two-level system using the OM described above are presented in the Table 3.10, and illustrate the convergence of the successive approximations of OM.

3.6 More Applications of Operator Method

To illustrate the power of the operator method in analytical investigation of complex physical systems, we consider in this section the Schrödinger equation with diverse one-dimensional potentials, which are widely used in various applications, and in these examples we limit ourselves to zeroth approximation for eigenvalues of Hamiltonian. Let us start with the non-polynomial potential, which has been studied in numerous works [41]:

$$V(x) = x^2 + \frac{\lambda x^2}{1 + g x^2}. \tag{3.74}$$

The method of the summation of asymptotic series is not applicable in this situation, however, the Schrödinger equation with this potential has an exact solution [42], if the parameters λ and g obey certain conditions. The OM is able to calculate the eigenvalues $E(\lambda, g)$ with high accuracy for any values of λ and g. The Schrödinger equation is written as:

$$\left[\hat{p}^2 + \hat{x}^2 + \frac{\lambda \hat{x}^2}{1 + g\hat{x}^2}\right] |\Psi\rangle = E\, |\Psi\rangle\,. \tag{3.75}$$

To implement the OM procedure, the particle number representation for operators of coordinate $\hat{x}$ and momentum $\hat{p}$ are introduced:

$$\hat{x} = \frac{1}{\sqrt{2\omega}}\left(\hat{a}^+ + \hat{a}\right); \qquad \hat{p} = i\sqrt{\frac{\omega}{2}}\left(\hat{a}^+ - \hat{a}\right); \qquad \left[\hat{a}; \hat{a}^+\right] = 1,$$

which contains the arbitrary parameter ω. Using the expression

$$(1 + gx^2)^{-1} = \int_0^\infty \exp\left[-u(1 + gx^2)\right] du,$$

the Eq. (3.75) is transformed to the normal form as in (2.55):

$$\left\{-\frac{g\omega}{2}\left((\hat{a}^+)^2+\hat{a}^2-1-2\hat{a}^+\hat{a}\right)+\frac{1}{2g\omega}\left((\hat{a}^+)^2+\hat{a}^2+1+2\hat{a}^+\hat{a}\right)\right.$$
$$\left.+\frac{\lambda}{g}\left[1-\omega\int\limits_0^\infty e^{\varphi_1(\hat{a}^+)^2}e^{\varphi_2\hat{a}^+\hat{a}}e^{\varphi_1\hat{a}^2}e^{\frac{1}{2}\varphi_2-\omega u}du\right]-E\right\}|\Psi\rangle=0, \tag{3.76}$$

where

$$\varphi_1=\frac{u}{2(1+u)};\qquad \varphi_2=-\ln(1+u).$$

The operator $\hat{H}_0$ corresponds to the zeroth approximation of OM and commutates with the operator of the excitation number $\hat{n}=\hat{a}^+\hat{a}$, and its eigenvector $|n\rangle$ obeys the equation:

$$\hat{a}^+\hat{a}\,|n\rangle=\hat{n}\,|n\rangle=n\,|n\rangle\,.$$

The eigenvalues of the operator $\hat{H}_0$ determine the zeroth approximation for the energy spectrum of the system with potential (3.74), which has a cumbersome form due to non-polynomial Hamiltonian:

$$E_n^{(0)}=\left(n+\frac{1}{2}\right)\left(g\omega+\frac{1}{g\omega}\right) \tag{3.77}$$
$$+\frac{\lambda}{g}\left[1-\omega\sum_{k=0}^{\frac{n}{2}}\frac{n!}{(k!)^2(n-2k)!}\int\limits_0^\infty e^{\frac{1}{2}\varphi_2-\omega u}\varphi_1^{2k}e^{\varphi_2(n-2k)}du\right].$$

All the integrals in the Eq. (3.77) are expressed through the error function [43] and its derivatives. For each quantum number n, the parameter ω_n is found from the equation obtained from (3.77) by differentiation over ω:

$$\left(n+\frac{1}{2}\right)\left(g-\frac{1}{g\omega^2}\right)-\frac{\lambda}{g}\sum_{k=0}^{\frac{n}{2}}\frac{n!}{(k!)^2(n-2k)!}$$
$$\times\int\limits_0^\infty\varphi_1^{2k}e^{\varphi_2\left(n+\frac{1}{2}-2k\right)}(1-\omega u)e^{-\omega u}du. \tag{3.78}$$

Table 3.11 Comparison of the approximate and exact values for energy levels of particle in the potential $V(x) = x^2 + \frac{\lambda x^2}{1+gx^2}$

g	λ	$E_0^{(0)}$	E_0	$E_1^{(0)}$	E_1
0.5	0.1	1.036	1.031	3.080	3.074
	0.5	1.177	1.157	3.379	3.371
	1.0	1.345	1.315	3.759	3.742
1.0	0.1	1.027	1.024		
	0.5	1.125	1.121		
	1.0	1.259	1.242		
2.0	0.1	1.018	1.017		
	0.5	1.091	1.086		
	1.0	1.178	1.172		
100	0.1	1.00085	1.00084	3.00114	3.00098
	10	1.085	1.084	3.102	3.098
	100	1.850	1.841	3.985	3.983
500	0.1	1.00019	1.00018		
	100	1.187	1.185		
	500	1.924	1.922		

As a result, the equation system for energy of the ground state is written as:

$$
\begin{aligned}
&E_0^{(0)} = \frac{1}{2}\left(g\omega + \frac{1}{g\omega}\right) + \frac{\lambda}{g}\left(1 - \int_0^\infty \frac{e^{-u}}{\sqrt{\omega + u}} du\right); \\
&\left|\Psi_0^{(0)}\right\rangle = |0\rangle; \\
&-g\omega + \frac{1}{g\omega} + \frac{\lambda}{g}\sqrt{\omega}\int_0^\infty \frac{e^{-u}}{\sqrt{\omega + u}}\left(1 - \frac{\omega}{\omega + u}\right) du = 0,
\end{aligned}
\tag{3.79}
$$

and is equivalent to the equations obtained in [41] using variational principle. However, in opposite to variational approach the OM finds the energy spectrum in a consistent way, and further corrections to zeroth approximation can be calculated. To illustrate the accuracy of the approximation for $E_n(\lambda.g)$ by OM, the Table 3.11 shows the values $E_0^{(0)}(\lambda, g)$ and $E_1^{(0)}(\lambda, g)$ in comparison with $E_0(\lambda, g)$ and $E_1(\lambda, g)$, found in [41] from the numerical solution of the Schrödinger equation with the potential (3.74).

The zeroth approximation of the OM is proved in the table to be able to calculate the eigenvalues with the accuracy better than 3 % in the entire range of the parameters λ and g. The uniformity of this approximation is confirmed by the principle terms of the expansion $E_0^{(0)}(\lambda, g)$ for small and large g, which equal to the known exact solutions:

$$E_0^{(0)} \simeq \sqrt{1+\lambda}\left(1 - \frac{3\lambda g}{4\sqrt{1+\lambda}}\right), \qquad g \ll 1,$$

$$E_0^{(0)} \simeq 1 + \frac{\lambda}{g} - \frac{\lambda\sqrt{\pi}}{g^{3/2}}, \qquad g \gg 1.$$

The accuracy of the zeroth approximation of OM can be also estimated from the first correction to $E_0^{(0)}$ over the operator $\hat{H}_1 = \hat{H} - \hat{H}_0$. This correction is expressed as:

$$E_0^{(2)} = -\langle 0|\,\hat{H}_1\left(\hat{H}_0 - E_0^{(0)}\right)^{-1}\hat{H}_1\,|0\rangle\,,$$

and due to non-polynomial potential $V(x)$ the action of the operator $\hat{H}_1$ on the state vector of zeroth approximation results in the infinite series over the operator:

$$\hat{H}_1\,|0\rangle = \sum_{k=4}^{\infty} C_k \frac{(\hat{a}^+)^k}{\sqrt{k!}}\,|0\rangle\,.$$

As a result, the correction $E_0^{(2)}$ is determined by the sum of the infinite series:

$$E_0^{(2)} = -\sum_{k=4}^{\infty} C_k^2 \frac{1}{E_0(k) - E_0^{(0)}}, \tag{3.80}$$

where $E_0(k)$ is a result of the action after operator $\hat{H}_0$ on the state $|k\rangle$ with certain number of excitations and for parameter ω, corresponding to ground state.

By applying the described in the Sect. 2.3 iterative scheme for OM based on the successive inclusion of matrix elements of non-polynomial Hamiltonian, the correction $E_0^{(2)}$ takes into account only the first term in (3.80), which corresponds to the transition of the system in four-quantum intermediate state. The numerical calculations demonstrate that by choosing the parameter ω from the Eq. (3.79), the series (3.80) converge quite fast and the contribution from four-quantum intermediate state is much higher than the one from the next to it eight-quantum state, which is of the same order as correction of third order to operator $\hat{H}_1$. As an example, the values of the energy of ground state E_0 at $\lambda = 1$ and $g = 1$, found numerically in [41] and calculated by formulas for zeroth and second approximations of OM are equal:

$$E_0 = 1.24013; \qquad E_0^{(0)} = 1.24032;$$

$$E_0^{(2)}(4) = -0.00074; \qquad E_0^{(2)}(8) = -0.00002,$$

where $E_0^{(2)}(k)$ is a contribution to the second order correction due to excitation of the state with k quanta.

The initial Schrödinger equation (3.74) can be formally transformed to polynomial one, and thus the application of OM is simplified after the the multiplication by the operator $1 + g\hat{x}^2$. However, the effective Hamiltonian becomes non-hermitian and the OM zeroth approximation gives less accurate results than for the initial Hamiltonian. As shown for Mathieu equation in Sect. 3.5, the accurate estimation of the eigenvalues in zeroth approximation requires the choice of the basic wave functions, which are more complex than ones for eigenvectors of the harmonic oscillator. To formalize the approach for the selection of these functions, we consider again the one-dimensional system with Hamiltonian:

$$\hat{H} = \frac{1}{2}\left(\hat{p}^2 + \hat{x}^2\right) - \lambda\hat{x}^4, \tag{3.81}$$

which has been already considered in Sect. 2.5. The choice of the waves functions of harmonic oscillator for the basic set of OM zeroth approximation makes it possible to calculate the width of the levels Γ of quasi-stationary states for restricted values of the coupling constant (2.123):

$$\lambda > \lambda_{min} = \frac{2n+1}{9\sqrt{3}\left(2n^2+2n+1\right)},$$

and for the smaller λ the successive approximations of OM have to be used, which require the numerical simulations. However, the analytical approximation of the dependence of the level width $\Gamma(\lambda)$ on the coupling constant is of essential interest for numerous applications, especially because of the function $\Gamma(\lambda)$ has a singularity at $\lambda \to 0$. Here we introduce two operators as was done for the solution of the Mathieu equation above:

$$\hat{A}_{\pm} = \frac{1}{\sqrt{2}}\hat{p} \pm i\sqrt{\frac{1}{2}\hat{x}^2 - \lambda\hat{x}^4}. \tag{3.82}$$

By neglecting the commutator of these operators, the operators factorize approximately the initial Hamiltonian (3.81). All further derivations are performed for the ground state. Two state vectors $\left|\Psi_{0\nu}^{\pm}\right\rangle$ are further introduced, which contain the arbitrary parameter ν and are the solutions of the following differential equations of the first order:

$$\left[\frac{1}{\sqrt{2}}\hat{p} \pm i\nu\sqrt{\frac{1}{2}\hat{x}^2 - \lambda\hat{x}^4}\right]\left|\Psi_{0\nu}^{\pm}\right\rangle = 0. \tag{3.83}$$

The arbitrary parameter ν has been introduced as a linear combination of the operators $\hat{A}_{\pm}$ in a similar way as for anharmonic oscillator in (2.2). To evaluate the energy of the ground state, the wave function is chosen as a linear superposition of the vectors:

$$|\Psi_{00}\rangle = \begin{cases} \sqrt{2}\,\mathrm{ch}\left[i\frac{\pi}{4} + \frac{\nu}{\lambda}\left(1 - 2\lambda x^2\right)^{3/2}\right]; & |x| \leq \frac{1}{2\lambda} \\ \exp\left[i\frac{\nu}{\lambda}\left(2\lambda x^2 - 1\right)^{3/2}\right]; & |x| > \frac{1}{2\lambda}. \end{cases} \tag{3.84}$$

The coefficients in the function (3.84) are selected in such a manner that it contains only diverging at $|x| \to \infty$ wave and is continuous along with its first and second derivatives in singular point $x = \frac{1}{\sqrt{2\lambda}}$. The vectors $|\Psi_{n\nu}\rangle$ for the excited states are constructed as products of the wave function $|\Psi_{0\nu}\rangle$ and Hermit polynomials $H_n\left(\sqrt{\frac{\nu}{3}}x\right)$. Then the energy in zeroth approximation $E_0^{(0)}$ is determined by the equation:

$$\langle\Psi_{0\nu}|\,\hat{H} - E_0^{(0)}(\nu)\,|\Psi_{0\nu}\rangle = 0, \tag{3.85}$$

where the optimal value ν is found from the equation:

$$\frac{\partial}{\partial\nu}E_0^{(0)}(\nu) = 0. \tag{3.86}$$

In a general case of quasi-stationary states, the solution of the Eq. (3.86) is a complex value, at which the state vector $\langle\Psi_{0\nu}|$ is defined as [44]:

$$\langle\Psi_{0\nu}| = \left\{|\Psi_{0\nu}\rangle^{+}\right\}^{*}.$$

Equation (3.85) is then transformed to:

$$\begin{aligned}
&\frac{1}{\sqrt{2\lambda}}\int_0^1 \frac{t dt}{\sqrt{1-t^2}}\left\{\frac{i}{2}\left[\exp\left(\frac{2\nu t^3}{\lambda}\right)\times\left(\frac{1-36\nu^2}{4\lambda}t^2(1-t^2) + \frac{3\nu}{t}(2t^2-1) - E_0\right)\right.\right.\\
&\left.-\exp\left(-\frac{2\nu t^3}{\lambda}\right)\left(\frac{1-36\nu^2}{4\lambda}t^2(1-t^2) - \frac{3\nu}{t}(2t^2-1) - E_0\right)\right]\\
&\left.+\frac{1-36\nu^2}{4\lambda}t^2(1-t^2) - E_0\right\}\\
&+\frac{1}{\sqrt{2\lambda}}\int_0^\infty \frac{z dz}{\sqrt{1+z^2}}\exp\left(i\frac{2\nu z^3}{\lambda}\right)\\
&\times\left(\frac{1-36\nu^2}{4\lambda}z^2(1+z^2) + \frac{3i\nu}{z}(2z^2+1) - E_0\right) = 0.
\end{aligned} \tag{3.87}$$

The technique described above has an exclusive application area at small $\lambda \leq \lambda_{min} \leq 0.1$, where the simple form of OM is not applicable (see Sect. 2.5), and therefore we use a limit ($\lambda \ll 1$) for the Eq. (3.87), which simplifies essentially all simulations and delivers $E_0^{(0)}(\nu)$ in analytical form:

$$E_0^{(0)}(\nu) \simeq F(\nu) - 4i\sqrt{\frac{3\nu}{\pi\lambda}}\exp(-\frac{2\nu}{\lambda})\left(\frac{1-36\nu^2}{30\lambda} - F(\nu)\right)$$
$$-4\sqrt{\frac{6\nu}{\pi}}\exp\left(-\frac{2\nu}{\lambda}\right)\Gamma\left(\frac{1}{3}\right)\frac{\nu^{2/3}}{2^{5/4}\lambda^{1/4}}\exp\left(i\frac{\pi}{6}\right)\left[1+O(\lambda^{1/6})\right],$$
$$F(\nu) = \frac{1}{24\nu} + \frac{3\nu}{2} - \frac{\lambda}{48\nu^2}. \tag{3.88}$$

The solution of the Eq. (3.86) in the same limit results in:

$$\nu_0 \equiv \nu_0(\lambda) = \frac{1}{6}\left(1 - \frac{3}{2}\lambda\right) + O(\lambda^2),$$

and finally the following formula for the energy of ground state is obtained:

$$E_0^{(0)}(\lambda) \simeq \frac{1}{2} - \frac{3}{4}\lambda - \frac{9}{4}\lambda^2 + O(\lambda^3) \tag{3.89}$$
$$+i\sqrt{\frac{e}{2\pi}}\exp\left(-\frac{1}{3\lambda}\right)\left(\frac{8}{5\sqrt{\lambda}} - \frac{\Gamma\left(\frac{1}{3}\right)}{3^{2/3}\lambda^{1/6}} + O(\lambda^{1/6})\right).$$

The parameter ν_0 contains the exponentially small correction $\Delta\nu \sim \exp\left(-\frac{1}{3\lambda}\right)$, which can be neglected in the Eq. (3.88) because of the corresponding change of the energy $E_0^{(0)}$ is proportional to $(\Delta\nu)^2$ due to the extremum of the function $E_0^{(0)}(\lambda)$ in the point $\nu = \nu_0$. Table 3.12 presents the results for the real and imaginary parts of the ground state energy for the system with Hamiltonian (3.81), computed by formula (3.89), and the results of the numerical calculations from [44] for $E_0^{'}$ and $E_0^{''}$.

The approximate factorization of the linear Schrödinger equation by operators (3.82) used above is also effective for the approximate solution of non-linear equations of the second order in the case when the equations of the first order obtained after factorization have an analytical representation. To illustrate this opportunity, the approximate solution of Thomas-Fermi equation [14] is discussed

Table 3.12 The energy of the quasi-stationary states for the system with Hamiltonian (3.81) at $\lambda \ll 1$

λ	$E_0^{'}$	$\Re E_0^{(0)}(\lambda)$	$E_0^{''}$	$\Im E_0^{(0)}(\lambda)$
0.025	0.479	0.480	7.8×10^{-6}	8.2×10^{-6}
0.030	0.474	0.475	6.2×10^{-5}	6.8×10^{-5}
0.050	0.450	0.456	3.3×10^{-3}	4.2×10^{-3}
0.100	0.397	0.402	4.5×10^{-2}	7.4×10^{-2}

here, which determines the self-consistent potential $\varphi(x)$ in multiple-electron atom:

$$\frac{d^2\varphi}{dx^2} - \frac{\sqrt{\varphi^3}}{\sqrt{x}} = 0, \qquad 0 \le x < \infty \tag{3.90}$$

with boundary conditions:

$$\varphi(0) = 1, \qquad \varphi(\infty) = 0. \tag{3.91}$$

The asymptotic behavior of the solution for Eq. (3.90) at $x \to \infty$ is determined by the formula [14]:

$$\varphi(x) \simeq \frac{144}{x^3}, \tag{3.92}$$

and we write a non-linear equation (3.90) in the form of Schrödinger equation:

$$\hat{H}\varphi(x) = 0$$

with zeroth eigenvalue and Hamiltonian depending on function:

$$\hat{H}(\varphi) \equiv \hat{p}^2 + \sqrt{\frac{\varphi}{x}}. \tag{3.93}$$

The analogue of the operators (3.82) for approximate factorization of the Hamiltonian (3.93) has the following form:

$$\hat{B}_\pm = \hat{p} \pm i\left(\frac{\varphi}{x}\right)^{1/4}. \tag{3.94}$$

The linear combination of these operators with arbitrary parameter makes the equation of the first order:

$$\left[\hat{p} - i\nu\left(\frac{\varphi_0}{x}\right)^{1/4}\right]\varphi_0 = 0, \tag{3.95}$$

and its solution approximates the function $\varphi(x)$ at the proper choice of the parameters. This solution is easy to find:

$$\varphi_0(x) = \frac{1}{\left(\frac{\nu}{3}x^{3/4} + C\right)^4},$$

where the constants ν and C follow from the boundary condition:

$$C = 1, \qquad \left(\frac{3}{\nu}\right)^4 = 144.$$

Table 3.13 Comparison of the approximate and the exact solution for Thomas-Fermi equation

x	$\varphi_0(x)$	$\varphi(x)$
0	1	1
1	0.362599	0.388633
2	0.205361	0.227292
3	0.132319	0.137382
4	0.0918463	0.0948242
5	0.0670401	0.0703753
6	0.0507696	0.0537674
7	0.0395559	0.0408496
8	0.0315267	0.0321377
9	0.0256000	0.0258943
10	0.0211146	0.0212346
100	9.50131×10^{-5}	9.60678×10^{-5}

Finally, the analytical solution of the Thomas-Fermi equation is:

$$\varphi_0(x) = \frac{144}{(x^{3/4} + \sqrt{12})^4}, \tag{3.96}$$

which gives a good approximation to the exact solution in the entire range of the parameter x (Table 3.13).

3.7 Operator Method for Uniformly Suitable Approximation of Integrals and Sums

A fast development of the computers does not underestimate the analytical methods for description of the physical processes. The analytical solutions expose the qualitative features of the physical system and choose the proper initial approximation for further numerical simulations. The advantage of the analytical methods is an ability to provide the correct functional dependencies of the physical properties in a broad range of the parameters, and the absolute accuracy plays the secondary role. As shown above by multiple examples, the zeroth approximation of OM successfully solves the former task, and we demonstrate below that this ability is not limited to the differential equations for quantum physics. In many cases, the solutions of the physical problems are expressed through special functions or reduced to the integrals and sums, which are not analytically calculated. We apply here operator method for analytical approximations of integrals and sums, which can not be calculated by using asymptotic methods [45]. The first example is an integral:

$$J(\lambda) = \int_{-\infty}^{+\infty} e^{-x^2-\lambda x^4} dx, \tag{3.97}$$

the expansions of which for large and small λ are:

$$J(\lambda) \simeq \sqrt{\pi}\left[1 - \frac{3}{2}\lambda + \ldots\right], \qquad \lambda \ll 1, \tag{3.98}$$

$$J(\lambda) \simeq \frac{1}{2}\lambda^{-1/4}\left[\Gamma\left(\frac{3}{4}\right) - \frac{1}{\sqrt{\lambda}}\Gamma\left(\frac{5}{4}\right) + \ldots\right], \qquad \lambda \gg 1.$$

Both expansions are asymptotic and cannot describe the function $J(\lambda)$ in the whole range $0 \le \lambda < \infty$. To apply OM for calculation of the integral (3.97), the following normalization vector is introduced:

$$|\omega\rangle = \left(\frac{\omega}{\pi}\right)^{1/4} e^{-\frac{1}{2}\omega x^2} \tag{3.99}$$

with the parameter ω, which will be determined later. The initial integral is transformed to the averaged expression with the state vector for some exponential operator:

$$J(\lambda) = \sqrt{\frac{\pi}{\omega}}\,\langle\omega|\, e^{(\omega-1)\hat{x}^2 - \lambda\hat{x}^4}\,|\omega\rangle \equiv \sqrt{\frac{\pi}{\omega}}\,\langle\omega|\, e^{\hat{H}(\omega,\lambda)}\,|\omega\rangle\,, \tag{3.100}$$

where the average is defined by standard quantum mechanics rules. The idea of OM procedure for calculation of this average value is based on the choice of the parameter ω to make the state vector $|\omega\rangle$ being the approximate eigenfunction of the operator $\hat{H}$:

$$\hat{H}(\omega,\lambda)\,|\omega\rangle \simeq \mathcal{E}(\omega,\lambda)\,|\omega\rangle\,, \qquad \omega = \omega(\lambda). \tag{3.101}$$

Then the estimate for integral is written as:

$$J(\lambda) \simeq \sqrt{\frac{\pi}{\omega}} e^{\mathcal{E}(\omega,\lambda)}. \tag{3.102}$$

Thus the calculation of the integral is reduced to the approximate solution of the Eq. (3.101), which is constructed by the procedure of OM. Using the second quantization representation:

$$\hat{x} = \frac{1}{\sqrt{2\omega}}\left(\hat{a} + \hat{a}^{+}\right), \qquad \left[\hat{a};\hat{a}^{+}\right] = 1,$$

and

$$\hat{a}\,|\omega\rangle = 0,$$

and split in $\hat{H}$ the operator of zeroth approximation $\hat{H}_0$ and the operator of excitation $\hat{H}_1$:

$$\begin{aligned}
\hat{H} &= \hat{H}_0 + \hat{H}_1, \\
\hat{H}_0 &= \frac{\omega - 1}{2\omega}\left(2\hat{a}^+\hat{a} + 1\right) - \frac{3\lambda}{4\omega^2}\left(2\left(\hat{a}^+\right)^2\hat{a}^2 + 4\hat{a}^+\hat{a} + 1\right), \\
\hat{H}_1 &= \frac{1}{2\omega^2}\left[\omega(\omega - 1) - 3\lambda\right]\left[\left(\hat{a}^+\right)^2 + \hat{a}^2\right] \\
&\quad - \frac{\lambda}{\omega^2}\left[\left(\hat{a}^+\right)^3\hat{a} + \hat{a}^+\hat{a}^3\right] - \frac{\lambda}{4\omega^2}\left[\left(\hat{a}^+\right)^4 + \hat{a}^4\right].
\end{aligned} \tag{3.103}$$

In zeroth approximation of OM, the state vector $|\omega\rangle$ is an approximate eigenfunction of the operator $\hat{H}$ provided the matrix element of the transition to the nearest excited state equals zero, i.e. the term with $\left(\hat{a}^+\right)^2|\omega\rangle$ in expression $\hat{H}_1|\omega\rangle$, which lead to the following equation for the parameter ω:

$$\omega(\omega - 1) - 3\lambda = 0. \tag{3.104}$$

The equivalent equation is obtained if the Bogolyubov-Feynman [46] inequality is applied to the expression (3.100):

$$J(\lambda) = \sqrt{\frac{\pi}{\omega}}\langle\omega|\, e^{\hat{H}}\,|\omega\rangle \geq J_0(\lambda, \omega) = \sqrt{\frac{\pi}{\omega}} e^{\langle\omega|\hat{H}|\omega\rangle} \tag{3.105}$$

and the parameter ω is found from the extremum condition for the right part of the inequality (3.105):

$$\frac{\partial}{\partial\omega} J_0(\lambda, \omega) = 0. \tag{3.106}$$

This integral is expressed through the Bessel function of the order 1/3 [43], however, substituting the solution of the Eq. (3.104) into formula (3.102), the analytical approximation is obtained:

$$J(\lambda) \simeq J_0(\lambda, \omega)\Big|_{\omega=\omega(\lambda)} = \sqrt{\frac{2\pi}{1+\sqrt{1+12\lambda}}}\exp\left[\frac{3\lambda}{\left(1+\sqrt{1+12\lambda}\right)^2}\right]. \tag{3.107}$$

The first terms of the expansion of this function at $\lambda \ll 1$ coincide with first equation in (3.98), whereas the central term in (3.107) at $\lambda \gg 1$ is:

$$J_0(\lambda) \simeq \sqrt{\frac{2\pi\sqrt{e}}{\sqrt{12\lambda}}} \simeq \frac{1.73}{\lambda^{1/4}}, \qquad \lambda \gg 1, \tag{3.108}$$

and is comparable with asymptotic expansion in (3.98):

$$J(\lambda) \simeq \frac{1.81}{\lambda^{1/4}}, \qquad \lambda \gg 1.$$

Thus, the simple formula (3.107) demonstrates the proper functional dependency $J(\lambda)$, and numerical calculations show the deviations from the exact solution less than 5 % in the entire range $0 \leq \lambda < \infty$. Whereas the zeroth approximation of OM is satisfactory for required accuracy of the estimates for the integrals, the corrective iterative procedure for improvement of the accuracy of zeroth approximation can be realized. Assuming the exact eigenfunctions of the operator $\hat{H}$ (3.103) are known:

$$\hat{H}\left|\Psi_n\right\rangle = E_n\left|\Psi_n\right\rangle, \tag{3.109}$$

the solution of the Eq. (3.109) can be expanded into the series over the eigenfunctions of the harmonic oscillator:

$$\left|\Psi_n\right\rangle = \sum_{k=0}^{\infty} C_{nk}\left|k\omega\right\rangle, \qquad \left|k\omega\right\rangle = \frac{\left(\hat{a}^+\right)^k}{\sqrt{k!}}\left|\omega\right\rangle$$

and re-write the formula (3.100) as follows:

$$J(\lambda) = \sqrt{\frac{\pi}{\omega}}\sum_{n=0}^{\infty} C_{n0}^2 e^{E_n}. \tag{3.110}$$

The complete set of eigenvalues E_n and eigenvectors is found by OM from the Eq. (3.109) using the expansion over the perturbation operator $\hat{H}_1$ from (3.103), and zeroth approximation results in:

$$\begin{aligned}
&\left|\Psi_n^{(0)}\right\rangle \simeq \left|n\omega\right\rangle, \\
&C_{n0} = \delta_{n0}, \\
&E_n^{(0)} \simeq \mathcal{E}_n = \frac{\omega-1}{2\omega}(2n+1) + \frac{3\lambda}{4\omega^2}\left(2n^2+4n+1\right),
\end{aligned} \tag{3.111}$$

and the first order correction for wave function is:

$$\begin{aligned}
\left|\Psi_n^{(1)}\right\rangle &= -\left(\hat{H}_0 - E_n^{(0)}\right)^{-1}\hat{H}_1\left|n\omega\right\rangle \\
&= \frac{\lambda}{8\omega^2}\left[\frac{\omega^2}{2\omega(\omega-1)-3\lambda(2n+5)}\sqrt{\frac{(n+4)!}{n!}}\left|n+4,\omega\right\rangle\right. \\
&\quad \left. - \frac{1}{2}\frac{\omega(\omega-1)-\lambda(2n+3)}{2\omega(\omega-1)-3\lambda(2n+3)}\sqrt{\frac{(n+2)!}{n!}}\left|n+2,\omega\right\rangle\right].
\end{aligned} \tag{3.112}$$

The equation for parameter ω is not equivalent to the extremum condition (3.106), because of the coefficient at operator $\left(\hat{a}^{+}\right)^2$ in expression $\hat{H}_1\left(\left|\Psi_0^{(0)}\right\rangle + \left|\Psi_0^{(1)}\right\rangle\right)$ must be zero. The latter expression arises from the action of the perturbation operator onto the ground state vector with accounting a first order correction, which results in equation:

$$\omega^2 - \omega - 3\lambda + \frac{3\lambda}{2}\frac{\omega(\omega-1)-7\lambda}{2\omega(\omega-1)-15\lambda} = 0, \tag{3.113}$$

with analytical solution:

$$\omega = \frac{1}{2}\left(\sqrt{1+4\xi}+1\right), \qquad \xi = \frac{\lambda}{8}\left(39 - \sqrt{39^2 - 16\cdot 69}\right).$$

Using the function $\left|\Psi_0^{(0)}\right\rangle + \left|\Psi_0^{(1)}\right\rangle$ in formula (2.104), the following expression is found:

$$J(\lambda) \simeq J_1(\lambda) = \sqrt{\frac{\pi}{\omega}}\left\{\left[1 - \frac{3\lambda^2}{8(2\xi-15\lambda)} - \frac{1}{2}\left(\frac{\xi-3\lambda}{2\xi-9\lambda}\right)^2\right]e^{E_0^{(0)}+E_0^{(2)}}\right.$$
$$\left. + \frac{1}{2}\left(\frac{\xi-3\lambda}{2\xi-9\lambda}\right)^2 e^{E_2^{(0)}+E_2^{(2)}} + \frac{3}{8}\frac{\lambda^2}{(2\xi-15\lambda)^2}e^{E_4^{(0)}+E_4^{(2)}}\right\}, \tag{3.114}$$

where

$$\begin{aligned} E_n^{(2)} = &-\frac{\lambda^2}{32\omega^2}\frac{(n+1)(n+2)(n+3)(n+4)}{[2\xi-3\lambda(2n+5)]} \\ &-\frac{[\xi-\lambda(2n+3)]^2(n+1)(n+2)}{4\omega^2[2\xi-3\lambda(2n+3)]} \\ &+\frac{[\xi-\lambda(2n-1)]^2 n(n-1)}{4\omega^2[2\xi-3\lambda(2n-1)]} \\ &+\frac{\lambda^2}{32\omega^2}\frac{(n-1)(n-2)(n-3)(n-4)}{[2\xi-3\lambda(2n-3)]}, \end{aligned}$$

and the parameters ω and ξ are determined from the Eq. (3.113). The formula (3.114) gives more accurate approximation for the function $J(\lambda)$ in comparison with (3.107). In the limit of $\lambda \gg 1$:

$$J_1(\lambda) \simeq \frac{1.80}{\lambda^{1/4}},$$

which has a relative accuracy better than 0.5 %. For some integrals, a more general than (3.103) canonic transformation of the operators has to be used. For example, the integral defining the gamma function:

$$\Gamma(x) = \int_0^\infty t^{x-1} e^{-t} dt = \frac{1}{x} \int_0^\infty t^x e^{-t} dt. \tag{3.115}$$

The operator form for $\Gamma(x)$, similar to (3.100), can be obtained by the substitution of the variable extending the integration interval to the entire numerical axis:

$$t = e^\rho, \qquad dt = e^\rho d\rho, \qquad -\infty < \rho < +\infty, \tag{3.116}$$

and by transition to the second quantization and further separation of classic component from the operators, because of the maximum of the integrand is not located in the origin:

$$\rho = u + \frac{1}{\sqrt{2\omega}} \left(\hat{a} + \hat{a}^+\right), \qquad \left[\hat{a}; \hat{a}^+\right] = 1. \tag{3.117}$$

As a result, we obtain:

$$\Gamma(x) = \frac{e^{u(x+1)}}{x} \sqrt{\frac{\pi}{\omega}} \left\langle \omega \right| e^{\hat{H}} \left| \omega \right\rangle, \tag{3.118}$$

where

$$\hat{H} = \frac{(x+1)}{\sqrt{2\omega}} \left(\hat{a} + \hat{a}^+\right) - e^{u+\frac{1}{4\omega}} e^{\frac{\hat{a}^+ + \hat{a}}{\sqrt{2\omega}}} + \frac{1}{2} \left(\left(\hat{a}^+\right)^2 + \hat{a}^2 + 2\hat{a}^+\hat{a} + 1\right),$$
$$\hat{a} \left|\omega\right\rangle = 0.$$

According to OM procedure, the parameters ω and u for integral are chosen to nullify the coefficients at operators $\left(\hat{a}^+\right)^2$ and $\hat{a}^+$ in the expansion $\hat{H}$, which define the transition in the nearest excited states:

$$e^{u+\frac{1}{4\omega}} = x + 1, \qquad e^{u+\frac{1}{4\omega}} = 2\omega. \tag{3.119}$$

Substituting the solutions of (3.119) into $\hat{H}$ and using the Bogolyubov-Feynman inequality for (3.118), the following analytical formula is found for gamma function:

$$\Gamma(x) \simeq \Gamma_1(x) = \frac{e^{u(x+1)}}{x} \sqrt{\frac{\pi}{\omega}} e^{\langle \omega | \hat{H} | \omega \rangle} = \sqrt{2\pi} \frac{(x+1)^{x+\frac{1}{2}}}{x} e^{-(x+1)}. \tag{3.120}$$

Despite the function $\Gamma_1(x)$ is expressed through the elementary functions, it deviates from $\Gamma(x)$ less than 8 % in the entire range $0 < x < +\infty$:

$$\Gamma_1(x) \simeq \frac{0.922}{x}, \quad x \ll 1, \quad \Gamma_1(1) \simeq 0.960, \quad \Gamma_1(2) \simeq 0.972,$$
$$\Gamma(x) \simeq \frac{1}{x}, \quad x \ll 1, \quad \Gamma(1) = 1, \quad \Gamma(2) = 1,$$
$$\Gamma_1(x) \simeq \sqrt{2\pi x} x^{x-\frac{1}{2}} e^{-x}, \quad x \gg 1,$$

which coincides in the limit with the asymptotic Stirling's formula. A certain complication for evaluation of integrals may occur in the case if the equation for variational parameters does not have an analytical solution. We consider below two special functions: the error function [32]:

$$\operatorname{erfc} z = \frac{2}{\sqrt{\pi}} \int_z^{\infty} e^{-t^2} dt, \tag{3.121}$$

and the function $\mathrm{H}(b, a)$, which describes the broadening of spectrum lines due to the radiation damping and the Doppler effect [47]:

$$\mathrm{H}(b, a) = \frac{a}{\pi} \int_{-\infty}^{+\infty} \frac{e^{-t^2}}{(t-b)^2 + a^2} dt. \tag{3.122}$$

The error function approximation is obtained in a similar way as (3.117):

$$\begin{aligned} \operatorname{erfc} z &= \frac{2}{\sqrt{\omega}} \left\langle \omega \left| \exp\left\{ \omega \hat{y}^2 + u + \hat{y} - \left(e^{u+\hat{y}} + z \right)^2 \right\} \right| \omega \right\rangle, \\ \hat{y} &= \frac{1}{\sqrt{2\omega}} \left(\hat{a} + \hat{a}^+ \right). \end{aligned} \tag{3.123}$$

Restricting ourselves to the OM zeroth approximation, the inequality of Bogolyubov-Feynman is applied to (3.123):

$$\operatorname{erfc} z \simeq J(\omega, u, z) = \frac{1}{\sqrt{\omega}} \exp\left[u + \omega - z^2 - 1 \right] \tag{3.124}$$

and the extrema of the function with respect to the parameters ω and u are:

$$z^2 = \frac{(1-\omega)^2}{\omega - \frac{1}{2}} e^{\frac{1}{2\omega}}, \qquad u = \frac{1}{2}\left[\ln\left(\omega - \frac{1}{2} \right) - \frac{1}{\omega} \right]. \tag{3.125}$$

Table 3.14 Comparison of the exact values of the error function erfc$z$ with the results of zeroth approximation by operator method

ω	$\sqrt{\frac{e}{2}}J(z)$	erfc$z$
0.04	0.7726	0.7679
0.25	0.4878	0.4804
0.92	0.1802	0.1751
1.00	0.1620	0.1573
3.68	0.0070	0.0067

In this case, the equation for ω doesn't have analytical solution, however, the dependence

$$z = \varphi(\omega), \qquad J = f(\omega)$$

can be used instead as an approximating function $J(z)$, which gives the uniformly suitable approximation for erfc$z$. In the asymptotic limits:

$$z \ll 1, \quad \text{erfc}z \simeq 1, \quad J(z) \simeq \sqrt{\frac{2}{e}}, \tag{3.126}$$

$$z \gg 1, \quad \text{erfc}z \simeq \frac{1}{\sqrt{\pi}}\frac{e^{-z^2}}{z}, \qquad J(z) \simeq \frac{\sqrt{2}}{e}\frac{e^{-z^2}}{z}.$$

In both limits, the function $J(z)$ produces the one-side deviation from the exact solution, and therefore the function $J(z)$ can be multiplied by factor $\sqrt{\frac{e}{2}} \simeq 1.167$, which is selected from the condition of the coincidence of asymptotics for $J(z)$ and erfc$z$ at $z \to 0$. Table 3.14 shows the values of the function $J(z)$ approximating the error function with a high accuracy for all z.

The OM zeroth approximation of the function $\text{H}(b, a)$ is obtained as:

$$\text{H}(b, a) \simeq W(b, a, u, \omega) = \frac{a}{\sqrt{\pi\omega}} \frac{\exp\left[\frac{1}{2} - b^2 - \frac{1}{2\omega}\right]}{(u - b)^2 + a^2 + \frac{1}{2\omega}}, \tag{3.127}$$

where the parameters ω and u are found from the equation:

$$\omega = \frac{b}{b - u}, \qquad \frac{1}{2}\left(1 - \frac{u}{b}\right) = \frac{b}{u} - 1 - a^2 - (u - b)^2.$$

In the limiting cases, the formula (3.127) demonstrates a good agreement with the exact formulas:

$$b \gg 1, \quad a \gg 1, \qquad W \simeq \frac{1}{2a\sqrt{\pi}}, \qquad \text{H} \simeq \frac{1}{2a\sqrt{\pi}},$$

$$b \gg 1, \quad a \ll 1, \qquad W \simeq \frac{a}{\sqrt{\pi}b^2}, \qquad \text{H} \simeq \frac{a}{\sqrt{\pi}b^2},$$

Table 3.15 Comparison of the exact values of the function $\mathrm{H}(a, b)$ with the results by zeroth approximation of operator method

a	b	$W(a,b)$	$\mathrm{H}(a,b)$
0.5	0.2	0.5331	0.6015
0.5	10	0.0028	0.0030
1.0	0.4	0.3835	0.4038
1.0	1.0	0.2745	0.3047
10	2.0	0.0539	0.0541
10	40	0.0033	0.0033

$$b \ll 1, \quad a \gg 1, \qquad W \simeq \frac{1}{a\sqrt{\pi}}, \qquad \mathrm{H} \simeq \frac{1}{a\sqrt{\pi}},$$

$$b \ll 1, \quad a \ll 1, \quad W \simeq \sqrt{\frac{e}{2\pi}}e^{-b^2}, \quad \mathrm{H} \simeq \left(1 - \frac{2a}{\sqrt{\pi}}\right)e^{-b^2+a^2},$$

and the results for intermediate arguments are shown in the Table 3.15.

Feynman and Kleinert [48] proposed the approximate method for calculation of the integrals based on the trajectory integrals for non-quadratic Lagrangian, and the results of this approach agree well with the ones from the OM. The operator method can also be extended to the approximation of the functions represented by sums, which is important for several problems of quantum statistics. As an example, the uniformly suitable interpolation for Weierstrass function (θ-function) is derived below, which is used in the solution of the diffusion equation or for the calculation of statistical sums of quantum systems. The canonic form of this function is written as [43]:

$$\theta_3(\lambda, v) = \sum_{n=-\infty}^{+\infty} \exp\left[-\pi\lambda n^2 + i2\pi v n\right]. \tag{3.128}$$

In a similar way as for (3.100), the θ-function is represented as a quantum-mechanical average using the following state vector different than for (3.100):

$$|u\rangle = \sqrt{1-u^2}\sum_{n=0}^{\infty} u^n \, |n\rangle\,, \tag{3.129}$$

where u is an arbitrary parameter, and $|n\rangle$ is an eigenvector of the operator of excitation number. Then the function $\theta_3(\lambda, v)$ is transformed to the following form:

$$\theta_3(\lambda, v) = \frac{2}{1-u^2}\left\langle u \left| \exp\left[i2\pi v\hat{n} - \pi\lambda\hat{n}^2 - 2\hat{n}\ln u^2\right] \right| u \right\rangle - 1, \tag{3.130}$$

where $\hat{n}$ is an operator of the excitation number. Using the Bogolyubov-Feynman inequality in (3.130), the result for the mean value is:

$$\langle u\,|\hat{n}|\,u\rangle = \frac{u^2}{2(1-u^2)}, \qquad \left\langle u\,\left|\hat{n}^2\right|\,u\right\rangle = \frac{u^2(1+u^2)}{(1-u^2)^2},$$

and

$$\theta_3(\lambda, v) \geq S(\alpha) \equiv 2\exp\left[\frac{2\pi v\alpha}{1-\alpha}i - \frac{\pi\lambda\alpha(1+\alpha)}{(1-\alpha)^2} - \frac{\alpha\ln\alpha}{1-\alpha} - \ln(1-\alpha)\right] - 1,$$
$$\alpha = u^2. \tag{3.131}$$

The estimation (3.131) delivers the best results if the parameter α is determined from the extremum condition $S(\alpha)$:

$$\ln\alpha = 2\pi vi + \pi\lambda\frac{1+3\alpha}{\alpha-1}. \tag{3.132}$$

At α satisfying the Eq. (3.132) the approximation has a simple analytical form:

$$S = \frac{2}{1-\alpha}\exp\left[2\pi\lambda\frac{\alpha^2}{(1-\alpha)^2}\right]. \tag{3.133}$$

This formula gives the results equivalent to the asymptotic for θ-function in the limiting cases:

$$\lambda \gg 1, \qquad \alpha \simeq e^{2\pi vi - \pi\lambda}, \qquad S \simeq 1 + e^{2\pi vi - \pi\lambda},$$
$$\lambda \ll 1, v = 0, \qquad \alpha \simeq 1 - 2\sqrt{\pi\lambda}, \qquad S \simeq \sqrt{\frac{e\lambda}{\pi}} - 1,$$
$$\lambda \ll 1, \qquad \alpha \simeq e^{2\pi vi}, \qquad S \simeq \left[1 - e^{2\pi vi}\right]^{-1}.$$

Table 3.16 shows that the formula above can calculate the values of θ-function with absolute accuracy $\sim 95\,\%$ for any values of the arguments.

Table 3.16 Comparison of OM zeroth approximation for θ-function with exact results

λ	v	OM	[43]
0	1.6	1.007	1.013
0	0.74	1.098	1.195
1.0	1.6	0.993	0.987
1.0	1.0	0.957	0.914

References

1. A.A. Ivanov, I.D. Feranchuk, Proc. Natl. Acad. Sci. Belarus, Ser. Phys. Math. Sci (in Russian) **4**, 64 (2004)
2. I.D. Feranchuk, V.N. T'ok, Chem. Phys. Lett. **150**, 78 (1988)
3. J. Killingback, Phys. Lett. A **84**, 95 (1981)
4. A.I. Ivanov, J. Phys. A Math. Gen. **31**, 5697 (1998)
5. F. Pan, J.R. Klauder, J.P. Draayer, Phys. Lett. A **262**, 131 (1999)
6. W. Janke, H. Kleinert, Phys. Lett. A **199**, 287 (1995)
7. V. Popescu, Phys. Lett. A **299**, 197 (2002)
8. M. Jafarpour, D. Afshar, J. Phys. A Math. Gen. **35**, 87 (2002)
9. A. Pathak, J. Phys. A Math. Gen. **33**, 5607 (2000)
10. S. Mandal, Phys. Lett. A **305**, 37 (2002)
11. L.K. Sharma, J.O. Fiase, Eur. J. Phys **21**, 167 (2000)
12. H. Meissner, E.O. Steinborn, Phys. Rev. A **56**, 1189 (1997)
13. Y. Meurice, J. Phys. A Math. Gen. **35**, 8831 (2002)
14. L.D. Landau, E.M. Lifshitz, *Quantum Mechanics* (Fizmatgiz, Moscow, 2004)
15. A. Barrat, E. Trizac, Phys. Rev. E **66**, 051303 (2002)
16. E. Delabaere, D.T. Trinh, J. Phys. A Math. Gen. **33**, 8771 (2000)
17. U. Back, M.G. Dondi, U. Valbusa, Phys. Rev. A **8**, 2409 (1973)
18. A. Voros, J. Phys. A Math. Gen. **32**, 1301 (1999)
19. I.D. Feranchuk, V.S. Kuz'min, A.P. Ulyanenkov, Chem. Phys. **157**, 61 (1991)
20. J.-L. Chen, L.C. Kwek, C.H. Oh, Phys. Rev. A **67**, 012101 (2003)
21. M. Carvajal, J.M. Arias, J. Gomez-Camacho, Phys. Rev. A **59**, 1852 (1999)
22. T. Shreecharan, P.K. Panigrahi, J. Banerji, Phys. Rev. A **69**, 012102 (2004)
23. G. Faure, P. Andre, A. Lefort, J. Phys. D Appl. Phys. **32**, 2376 (1999)
24. B. Molner, M.G. Benedict, J. Bertrand, J. Phys. A Math. Gen. **34**, 3139 (2001)
25. H.J. Korsch, B. Schnellhaas, Eur. J. Phys. **21**, 73 (2000)
26. X.G. Hu, Q.-S. Li, J. Phys. A Math. Gen. **32**, 139 (1999)
27. F.M. Fernandez, J.F. Olilvie, J. Phys. B At. Mol. Opt. Phys. **25**, 1375 (1992)
28. P.M. Morse, H. Feshbach, *Methods of Theoretical Physics* (McGraw-Hill, New York, 1953)
29. N.W. McLachlan, *Theory and Application of Mathieu Functions* (Clarendon Press, Oxford, 1951)
30. I.D. Feranchuk, L.I. Komarov, J. Phys. A. Math. Gen. **17**, 3111 (1984)
31. A.H. Nayfeh, *Perturbation Methods* (Wiley, New York, 1973)
32. M. Abramowitz, I.A. Stegun, *Handbook of Mathematical Functions* (Dover, New York, 1964)
33. V.G. Baryshevsky, *Channeling, Radiation and Decays in Crystals at High Energies (in Russian)* (Belarusian University, Minsk, 1982)
34. B.A. Chevganov, I.D. Feranchuk, J. de Phys. Paris **43**, 1687 (1982)
35. I.D. Feranchuk, A.P. Ulyanenkov, Nuclear Inst Methods Phys. Res. B **88**, 369 (1994)
36. Y.B. Zeldovich, Sov. Phys. JETP **24**, 1066 (1967)
37. S.-I. Chua, D.A. Telnov, Phys. Rep. **390**, 1 (2004)
38. L. Allen, *Optical Resonance and Two-Level Atoms* (Dover, New York, 1987)
39. Y.B. Zeldovich, Sov. Phys. Usp. **18**, 79 (1975)
40. A.G. Fainshtein, N.L. Manakov, L.P. Rapoport, J. Phys. B At. Mol. Opt. Phys. **11**, 2561 (1978)
41. R.N. Chandhuri, B. Mikherjee, J. Phys. A Math. Gen. **16**, 4031 (1983)
42. A.V. Turbiner, Uspekhi Fiz. Nauk (in Russian) **144**, 35 (1984)
43. I.S. Gradshtein, I.M. Ryzhik, *Tables of Integrals, Sums, Series and Products* (Fizmatgiz, Moscow, 1963)
44. D. Farrelly, W.P. Reinhardt, J. Phys. B Atom. Mol. **16**, 2103 (1983)
45. I.D. Feranchuk, L.I. Komarov, Phys. Lett. A **106**, 109 (1984)

46. R.P. Feynman, A.R. Hibbs, *Quantum Mechanics and Path Integrals* (McGraw-Hill, New York, 1965)
47. I.I. Sobelman, *Introduction to the Theory of Atomic Spectra (in Russian)* (Nauka, Moscow, 1963)
48. H. Kleinert, *Path Integrals in Quantum Mechanics, Statistics, Polymer Physics, and Financial Markets*, 5th edn. (World Scientific, Singapore, 2009)

Chapter 4
Operator Method for Quantum Statistics

The construction of the uniformly available approximation (UAA) for thermodynamical functions has two specific features, which differ it from the solution of the Schrödinger equation for stationary states. First of all, these functions depend both on Hamiltonian parameters and on the temperature T. In second, these functions are expressed through the partition function (PF) Z of the system, which requires the UAA for the whole energy spectrum and the summation to be carried out:

$$Z(\beta) = \sum_{n} g_n e^{-\beta E_n}, \tag{4.1}$$

where $\beta = 1/kT$ is an inverse temperature, k is Boltzmann constant, E_n is the energy level corresponding to the quantum number n (for the systems with multiple degrees of freedom it is a set of quantum numbers $\{n\}$) and to the degeneration multiplicity g_n. Thus, the algorithms for both finding the energy spectrum and for summation over the energy states have to be established to find an approximation, which is uniform over the Hamiltonian parameters and temperature. This chapter illustrates the application of the operator method for such kind of tasks.

4.1 General Algorithm for Calculation of PF

To find an algorithm for the summation over the states, the partition function is represented as a quantum-mechanical mean of the exponential operator [1–3]. As a first step, the basic set of the state vectors is considered, which corresponds to the eigenfunctions of the operator, which is formally equivalent to the operator of the excitation numbers with eigenvalue corresponding to the quantum number n:

$$\hat{n} \left| n \right\rangle = n \left| n \right\rangle . \tag{4.2}$$

I. Feranchuk et al., *Non-perturbative Description of Quantum Systems*,
Lecture Notes in Physics 894, DOI 10.1007/978-3-319-13006-4_4

By using these states, the definition (4.1) is re-written as:

$$Z(\beta) = \langle\beta^* \left|\exp\left[-\beta E(\hat{n}) + 2\beta^*\hat{n} - \ln N(\beta^*)\right]\right| \beta^*\rangle . \tag{4.3}$$

Here $|\beta^*\rangle$ is a normalization vector of state depending on the arbitrary parameter β^*, which has a physical meaning of the effective inverse temperature for equilibrium excitation system corresponding to (4.2). The value of the parameter β^* will be found later from the condition of the best approximation for PF of the system. The state vector $|\beta^*\rangle$ is written as follows:

$$|\beta^*\rangle = \sqrt{N}\sum_{n=0}^{\infty}\sqrt{g_n}\exp\left[-\beta^* n\right]|n\rangle , \tag{4.4}$$

where the constant N is determined from the normalization of the state vector $|\beta^*\rangle$:

$$N = \left\{\sum_{n=0}^{\infty} g_n \exp\left[-2\beta^* n\right]\right\}^{-1} . \tag{4.5}$$

The operator representation of PF similar to (4.3) can also be obtained with more complex trial vector instead of used one (4.4), which fits the qualitative features of the system in the best way. Such a choice improves the zeroth approximation, however, complicates the construction of the successive approximations. The simple form of the vector (4.4) makes it possible to construct the universal scheme of UAA for arbitrary quantum system and to develop successive approximations [1]. The approximate summation method over the quantum states is based on the cumulant expansion (CE) [4], which is valid for any exponential operator averaged over normalized state vector:

$$\left\langle \exp \hat{A}\right\rangle = \exp\left[\sum_{n=1}^{\infty}\frac{K_n}{n!}\right] , \tag{4.6}$$

where the cumulants K_n are expressed through the moments of the operator $\hat{A}$. The expansion (4.6) is an accurate one, and each cumulant corresponds to the partial summation of the power series. As follows from [4], the cumulants can be found by using recurrent relationships, for example,the first terms in (4.6) are:

$$\begin{aligned} K_1 &= \left\langle \hat{A}\right\rangle , \\ K_2 &= \left\langle \hat{A}^2\right\rangle - \left\langle \hat{A}\right\rangle^2 , \\ K_3 &= \left\langle \hat{A}^3\right\rangle - 3\left\langle \hat{A}\right\rangle\left\langle \hat{A}^2\right\rangle + 2\left\langle \hat{A}\right\rangle^3 . \end{aligned} \tag{4.7}$$

It is worth to note that the use of only first cumulant in (4.6) is equivalent to the inequality of Bogolyubov-Feynman [5].

The cumulant expansion is intensively used in the theory of probability [6] as well as for description of the critical phenomena and phase transition [7], and for investigation of the influence of the random processes on the physical characteristics [8]. The general procedure of the cumulant expansion can be applied to the partition function in operator form (4.3). The successive approximations are further calculated using the cumulants, and here we consider only primary and secondary cumulants:

$$
\begin{aligned}
Z \approx Z(\beta,\beta^*) &= Z_0(\beta,\beta^*)Z_1(\beta,\beta^*) \\
&= \exp\left[\langle A(\hat{n})\rangle + \ln N + \frac{1}{2}\left(\langle A^2(\hat{n})\rangle - \langle A(\hat{n})\rangle^2\right)\right], \\
A(\hat{n}) &= -\beta E(\hat{n}) + 2\beta^*\hat{n}.
\end{aligned} \tag{4.8}
$$

For any fixed number of cumulants in the formula (4.8), the partition function depends on the parameter β^*, which plays a role of the variational parameter for best approximation in each order of CE. For example, the zeroth approximation $Z_0(\beta,\beta^*)$ and first successive correction $Z_1(\beta,\beta^*)$ follow from the formulas [1]:

$$
Z_0 = \exp\left[-\beta\overline{E(\beta^*)} + 2\beta^*\bar{n} - \ln N(\beta^*)\right], \tag{4.9}
$$

$$
Z_1 = \left\{\frac{1}{2}\left[\beta^2(\overline{E^2} - \bar{E}^2) - 4\beta\beta^*(\overline{En} - \bar{E}\bar{n}) + 4\beta^{*2}(\overline{n^2} - \bar{n}^2)\right]\right\}.
$$

In the formulas (4.9), all values are averaged over the trial distribution function, which corresponds to the excitation ensemble (4.2) with degeneracies g_n and effective inverse temperature β^*:

$$
\begin{aligned}
\overline{E(\beta^*)} &= N\sum_{n=0}^{\infty} g_n E_n \exp\left[-2\beta^* n\right], \\
\bar{n}(\beta^*) &= N\sum_{n=0}^{\infty} g_n n \exp\left[-2\beta^* n\right] = \frac{N}{2}\frac{\partial N}{\partial\beta^*}.
\end{aligned} \tag{4.10}
$$

Within the approximation for PF found above, the free energy is expressed as follows:

$$
\begin{aligned}
F(\beta) &= -\frac{1}{\beta}\ln Z(\beta) \approx F^{(0)} + F^{(1)} + \dots \\
&= -\frac{1}{\beta}\left[\ln Z_0(\beta,\beta^*) + \ln Z_1(\beta,\beta^*) + \dots\right].
\end{aligned} \tag{4.11}
$$

The approximation to be found for all above mentioned values and for certain energy spectrum is defined be the variational parameter β^*, which being a function of the inverse temperature β, has to be found from the condition of the best zeroth approximation of CE [1]:

$$\frac{\partial Z_0}{\partial \beta^*} = \beta \frac{\partial \bar{E}}{\partial \beta^*} - 2\bar{n} - 2\beta^* \frac{\partial \bar{n}}{\partial \beta^*} + \frac{1}{N} \frac{\partial N}{\partial \beta^*} = 0. \tag{4.12}$$

For calculation of the moments $\overline{E}$ and $\overline{E^2}$ with distribution function (4.10), the approximation for energy spectrum found by OM can be used, and then the relations (4.9) and (4.12) determine the uniformly suitable approximation for the thermodynamic system with the accuracy up to the second order of OM. However, the equations obtained are still cumbersome and the additional approximation can be applied to simplify the mathematical calculation at the cost of the accuracy of zeroth approximation. The condition

$$\frac{\partial}{\partial \omega} E_n^{(0)} = 0 \tag{4.13}$$

corresponds to the optimal choice of the parameter ω_n for energy level n in zeroth approximation of OM. The convergence of the successive approximations of OM has been shown in [9] to exist for any value of the parameter ω. The partition function and free energy are the integral characteristics in the space of the quantum numbers. Therefore, the parameter ω equal for all quantum numbers can be selected to approximate both functions, and this parameter is determined by the simultaneous optimization of free energy in OM and CE techniques. Using this approach, the moments of the energy spectrum have to account only explicit dependence of the spectrum on quantum number n and the dependence followed from the selection of ω_n from (4.13) can be ignored [1]. This procedure makes the values

$$\begin{aligned} \overline{E(\beta^*, \omega)} &\simeq \sum_{n=0}^{\infty} g_n \left[E_n^{(0)}(\omega) + \Delta E_n(\omega)\right] \exp\left[-2\beta^* n\right], \\ \overline{E^2(\beta^*, \omega)} &\simeq \sum_{n=0}^{\infty} g_n \left[E_n^{(0)}(\omega)\right]^2 \exp\left[-2\beta^* n\right] \end{aligned} \tag{4.14}$$

dependent on both parameters β^* and ω, which are determined from the minimum condition for free energy in zeroth approximation. As a result, a single equation for ω has to be resolved instead of the system of the algebraic equations (4.13) for ω_n:

$$\frac{\partial}{\partial \omega} \sum_{n=0}^{\infty} g_n \left[E_n^{(0)}(\omega)\right] \exp\left[-2\beta^* n\right] = 0. \tag{4.15}$$

The physical meaning of the Eq. (4.15) is the selection of the parameter ω from the condition of the best approximation for energy level with highest population at certain temperate [2]. The results delivered by uniformly suitable approximations obtained below for PF with the use of cumulant expansion are compared with the exact numerical calculations and direct numerical summation over the energy levels found by OM.

4.2 Statistics of Non-interacting Systems with One-Dimensional Energy Spectrum

As a first example, we consider the system consisting of the non-interacting quantum anharmonic oscillators with the Hamiltonian (2.16); the energy spectrum of QAO has been obtained by OM in the Chap. 2. This physical model describes the ideal gas of diatomic molecules with anharmonic interatomic potential or anharmonic atomic oscillations in crystallographic lattice. The accuracy of approximations described in the Sect. 4.1 and applied to this system can be derived as follows. The partition function and free energy of the non-interacting QAO is expressed as:

$$Z(\beta,\lambda)=\sum_{n=0}^{\infty}e^{-\beta E_n(\lambda)};\quad F(\beta,\lambda)=-\frac{1}{\beta}\ln Z(\beta,\lambda), \tag{4.16}$$

where the multiplicity of the degeneration in (4.16) equals to unity $g_n = 1$. First of all, we consider here the calculation of PF and free energy for QAO with 8 energy levels found in [10]. The fixed number of energy levels implements the inaccuracy in partition function in the limit of high temperatures ($\beta \to 0$), where the highly excited quantum states contribute essentially in the sum. In this situation, the asymptotic expansion for partition function of QAO obtained analytically at ($\beta \to 0$) [1, 11]:

$$Z(\beta,\lambda)\simeq 1+e^{-\beta E_0(\lambda)},\quad \beta\gg 1,$$

$$Z(\beta,\lambda)\simeq\frac{e^{-\beta/2}}{1-e^{-\beta}}\left[1+\frac{3\lambda e^{-\beta}}{(1-e^{-\beta})^2}\right],\quad \frac{\lambda}{(1-e^{-\beta})^2}\ll 1,$$

$$Z(\beta,\lambda)\simeq\sum_{n=0}^{\infty}\exp\left[-\beta\lambda^{1/3}b_n\right],\quad \lambda\gg 1, \tag{4.17}$$

where b_n are the coefficients of asymptotic expansion for energy levels of QAO in the limit of strong coupling [10]. The approximate analytical expression for these coefficients in zeroth approximation of OM is given by formula (2.28):

$$b_n\simeq\left(\frac{3}{4}\right)^{4/3}\left[\frac{1+2n+2n^2}{(1+2n)^2}\right]^{1/3}. \tag{4.18}$$

The uniformly available approximation in the second order of OM for this system is provided by the Eqs. (2.23)–(2.24). According to the Sect. 4.1, there are several forms of UAA can be constructed for PF (4.16):

$$Z^{(0)}(\beta,\lambda) = \sum_{n=0}^{\infty} \exp\left[-\beta E_n^{(0)}(\lambda)\right],$$
$$Z^{(1)}(\beta,\lambda) = \sum_{n=0}^{\infty} \exp\left[-\beta\left(E_n^{(0)}(\lambda) + \Delta E_n(\lambda)\right)\right], \tag{4.19}$$
$$Z^{(01)}(\beta,\lambda) = \sum_{n=0}^{\infty} (1-\beta\Delta E_n(\lambda)) \exp\left[-\beta E_n^{(0)}(\lambda)\right],$$

where the energy in zeroth approximation $E_n^{(0)}(\lambda)$ and the second-order correction $\Delta E_n(\lambda)$ are defined by OM formulas (2.21) and (2.49). The former $Z^{(0)}(\beta,\lambda)$ is a zeroth approximation for PF, the latter $Z^{(1)}(\beta,\lambda)$ is determined by the second-order OM approximation for eigenvalues with exact summation over quantum numbers, and the third expression $Z^{(01)}(\beta,\lambda)$ is used in the expansion over $\Delta E_n(\lambda)$. These expressions for PF lead to the approximations for free energy:

$$F^{(0)}(\beta,\lambda) = -\frac{1}{\beta}\ln\left\{\sum_{n=0}^{\infty} \exp\left[-\beta E_n^{(0)}(\lambda)\right]\right\}, \tag{4.20}$$
$$F^{(1)}(\beta,\lambda) = -\frac{1}{\beta}\ln\left\{\sum_{n=0}^{\infty} \exp\left[-\beta\left(E_n^{(0)}(\lambda) + \Delta E_n(\lambda)\right)\right]\right\},$$
$$F^{(01)}(\beta,\lambda) = -\frac{1}{\beta}\ln\left\{\sum_{n=0}^{\infty} (1-\beta\Delta E_n(\lambda)) \exp\left[-\beta E_n^{(0)}(\lambda)\right]\right\}.$$

The results by different approximations for free energy are shown in Figs. 4.1 and 4.2. As follows from the Fig. 4.1, in the domain of small temperatures the accuracy of the calculated by the expansion over $\Delta E_n(\lambda)$ free energy decreases (which makes this approximation invalid at $\beta \sim 1/\Delta E_n$), whereas the accuracy of the result by OM remains constant. In zeroth approximation of OM, the accuracy of UAA is estimated by formula (1.1), $\xi^{(0)} \approx 0.085$. In the second-order approximation and $F^{(1)}(\beta,\lambda)$, the accuracy is $\xi^{(1)} \approx 0.025$.

A similar conclusion about uniformly suitable approximation of the free energy of QAO on the anharmonic parameter λ follows from the Fig. 4.2, where the drawback of the numerical summation over the limited number of levels is evident. An accounting of only 5 instead of 8 levels of the states of QAO worsens the results essentially. Using the direct summation of the levels calculated by OM, the error due to the limitation of the level numbers is considerably less because of 100 levels

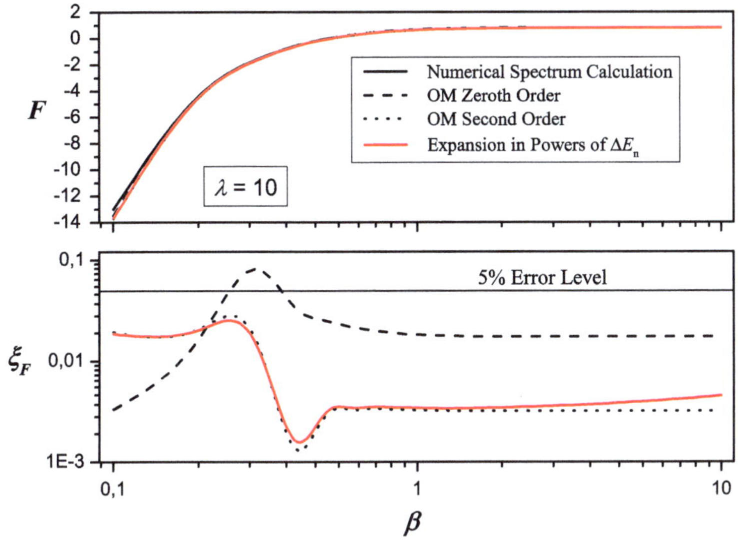

Fig. 4.1 The approximation for free energy F and relative accuracy ξ_F for QAO by using a direct summation of the energy spectrum in OM and by numerical calculations in [10] at $\lambda = 10$

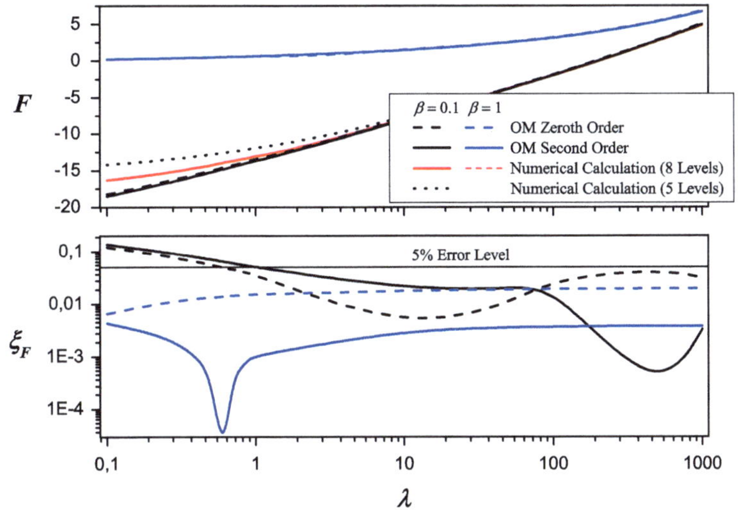

Fig. 4.2 The free energy of QAO and relative accuracy calculated on the basis of direct summation with the use of the energy spectrum simulated by operator method for various inverse temperatures β

are taken into account. This error riches 1 % at $\beta \sim 5/E_{100}^{(0)}(\lambda)$, which corresponds to relatively high temperature even for harmonic oscillator ($\lambda = 0$):

$$\xi \sim O(e^{-\beta E_{100}^{(0)}(\lambda)}) \tag{4.21}$$

The source of this inaccuracy is eliminated when cumulant expansion is used, as will be shown below. Figure 4.2 demonstrates the uniform suitability of the approximation on the inverse temperature β both in zeroth and the first order. The dependence of free energy of QAO on anharmonic parameter λ is similar for $\beta = 0.1$ and $\beta = 1$.

A similar scheme of the approximation is also applicable to the Hamiltonian with polynomial potential (3.1) with the energy spectrum in the zeroth approximation of OM is described by (3.7) and (3.8). For this physical system, the zeroth approximations for free energy and partition function are considered and the summation involves 1000 energy levels:

$$Z^{(0)}(\beta, \lambda) = \sum_{n=0}^{1000} \exp\left[-\beta E_n^{(0)}(\lambda)\right],$$
$$F^{(0)}(\beta, \lambda) = -\frac{1}{\beta} \ln Z^{(0)}(\beta, \lambda). \tag{4.22}$$

In the same way as in the Sect. 3.1 for QAO with polynomial potential, we consider here the particular cases of Hamiltonian of the oscillator of eights order (3.12) and sixth order (3.9) and compare the results for partition function and free energy by numerical calculations [12]. The results for the partition function of the oscillator with polynomial potential are shown in Fig. 4.3, and for free energy in Fig. 4.4.

The results obtained can be used to estimate the precision of the calculations for the thermodynamical characteristics of QAO with polynomial potential in zeroth approximation of OM by comparing the zeroth and the second approximations. In the entire range of the inverse temperature and the parameters of Hamiltonian, the second-order correction improves the values by maximal 7 %, which fits well the estimate for UAA in zeroth approximation $\xi^{(0)} \approx 0.07$.

Another approach for calculation of thermodynamical properties of quantum systems assumes the replacement of exact summation by the approximate one on the basis of cumulant expansion [1]. The advantage of this method is a relatively small number of terms to be summed to rich a required accuracy of the results. As a first step, CE is applied to the system with known energy spectrum, which makes possible to estimate instantly the accuracy of the CE and the UAA based on this technique. As a model for this step, the ensemble of quantum rotators is chosen, which is frequently used for the description of the rotational degrees of freedom

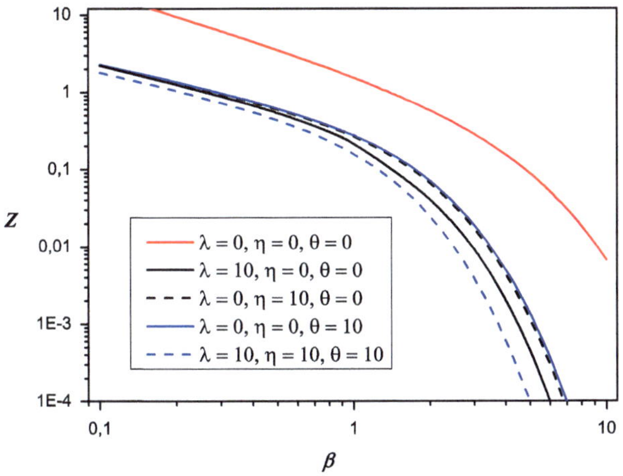

Fig. 4.3 The approximation of the partition function for the oscillator with polynomial potential for various coefficients at the perturbation part of Hamiltonian (3.12). The *red line* corresponds to harmonic oscillator

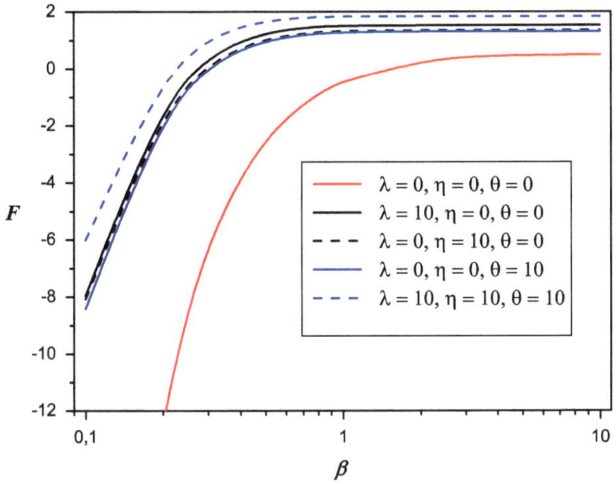

Fig. 4.4 The approximation for free energy of the oscillator with polynomial potential for various coefficients at the perturbation part of the Hamiltonian (3.12). The *red line* corresponds to harmonic oscillator

in molecular gases [13]. The energy levels of this physical system follow from the equation:

$$E_n = \frac{\hbar^2}{2I}(n^2 + n), \tag{4.23}$$

where I is the inertia of the rotator. The multiplicity of degeneracy of the level n is $g_n = 2n + 1$, and with known energy spectrum and degeneracy, the rotational PF Z_r can be calculated as in (4.1):

$$Z_r(x) = \sum_{n=0}^{\infty}(2n+1)\exp\left[-x(n^2+n)\right], \tag{4.24}$$

where $x = \beta\theta_r$ is a dimensionless parameter, $\theta_r = \hbar^2/2I$ is a rotational temperature of the system in the energy units. In the limiting cases of low and high temperatures, where $x \gg 1$ and $x \ll 1$, the asymptotic approximations for PF (4.24) [13] are:

$$Z_r(x) \simeq 1 + 3\exp(-2x) + \ldots, \qquad x \gg 1. \tag{4.25}$$

The use of two terms in the summation Euler formula:

$$\sum_{n=1}^{\infty} f(n+a) \simeq \int_0^{\infty} f(y)dy - \frac{1}{2}f(a) - \frac{1}{12}f'(a) + \ldots, \tag{4.26}$$

the following expansion is obtained:

$$\begin{aligned} Z_r(x) &\simeq 1 + \int_0^{\infty}(2y+1)\exp\left[-x(y^2+y)\right]dy - \frac{1}{2} - \frac{1}{12}(2-x) + \ldots - \\ &= \frac{1}{x} + \frac{1}{3} + \frac{x}{12}, \quad x \ll 1. \end{aligned} \tag{4.27}$$

The expressions (4.25) and (4.27) do not adequately describe the partition function $Z_r(x)$ in the range of intermediate x. However, this PF can be expressed through the Weierstrass function [14], which is too complex for analytical investigations (the application of OM for this PF has been discussed in the Sect. 3.7). The use of cumulant expansion indeed is able to build a uniformly available approximation for $Z_r(x)$ in the entire range of temperatures. All the mean values in the formula (4.10) can be expressed through the moments of trial function of distribution:

$$\begin{aligned} N &= \left[\sum_{n=0}^{\infty}(2n+1)e^{-2\beta^* n}\right]^{-1} = \frac{(1-q)^2}{1+q}, \quad q = e^{-2\beta^*}, \\ \overline{n} &= N\sum_{n=0}^{\infty} n(2n+1)e^{-2\beta^* n} = \frac{q(3+q)}{1-q^2}, \\ \overline{n^2} &= \frac{q(3+8q+q^2)}{(1+q)(1-q)^2}. \end{aligned} \tag{4.28}$$

Substituting the expressions (4.27) into formula (4.9), the following expression for PF in zeroth approximation of CE can be found (the parameter $q \in (0;1)$ is treated as a variational parameter instead of β^*):

$$Z_r^{(0)} = \exp\left[\varphi(x,q)\right], \tag{4.29}$$
$$\varphi(x,q) = -x\frac{6q}{(1-q)^2} - \frac{3q+q^2}{1-q^2}\ln q - \ln\left[\frac{(1-q)^2}{1+q}\right].$$

The formula (4.29) has to be supplemented by the equation for function $q = q(x)$, following from (4.12) when the parameter for the best zeroth approximation is selected:

$$\frac{\partial\varphi}{\partial q} = 0, \qquad x = -\frac{(3+2q+3q^2)(1-q)}{6(1+q)^3}\ln q, \tag{4.30}$$

that results in the following PF in zeroth CE approximation:

$$Z_r^{(0)}(q) = \exp\left[\varphi(q)\right],$$
$$\varphi(q) = q\left[\frac{3+2q+3q^2}{(1-q)(1+q)^2} - (3+q)(1-q^2)\right]\ln q -$$
$$-\ln\left[\frac{(1-q)^2}{1+q}\right]. \tag{4.31}$$

The system of the Eqs. (4.30) and (4.31) defines in a parametric way the function $Z_r^{(0)}(x)$, which is an analytical zeroth approximation for the partition function of the ensemble of rotators. The first order correction for PF is [2]:

$$Z_r^{(1)}(x,q) = \exp\left[\varphi_1(x,q)\right],$$
$$\varphi_1(x,q) = \frac{1}{2}x^2\left[\overline{(n^2+2)^2} - \overline{(n^2+n)}^2\right] - \tag{4.32}$$
$$-2x\left[\overline{n(n^2+n)} - \overline{n}\,\overline{(n^2+n)}\right]\ln q + \left[\overline{n^2} - \overline{n}^2\right]\ln^2 q.$$

Finally, using the Eq. (4.30), the following expression is obtained:

$$Z_r^{(1)}(q) = \exp\left[\varphi_1(q)\right], \tag{4.33}$$
$$\varphi_1(q) = \frac{q^2(15+4q+26q^2+4q^3+15q^4)}{6(1-q)^2(1+q)^6}\ln^2 q.$$

The formulas (4.30), (4.31) and (4.33) determine the uniformly available approximation for rotational partition function. In the limiting cases, the following expressions are valid:

$$x \simeq -\frac{\ln q}{2}, \quad \varphi(q) \simeq 3q, \quad Z_r^{(0)}(x) \simeq 1 + 3\exp(-2x), \quad x \gg 1, \tag{4.34}$$

and

$$x \simeq \frac{(1-q)^2}{6}, \quad \varphi(q) \simeq 1 + \ln 2 - 2\ln(1-q),$$
$$Z_r^{(0)}(x) \simeq \frac{0,906}{x} + 0,303, \quad x \ll 1. \tag{4.35}$$

By comparing these equations with the asymptotic formulas (4.25) and (4.27), the obtained approximation is found to describe correctly the dependence of the partition function on the parameter x in both limiting cases. The general accuracy of the approximation, as defined in (1.1), is determined by the parameter $\xi^{(0)} \simeq 0,1$. In our case, the first-order correction for PF (4.33) improves the accuracy and in the limit of small x:

$$Z_r(x) \simeq Z_r^{(0)}(x) Z_r^{(1)}(x) \simeq \frac{1,063}{x} + 0,348 + \ldots, \quad x \ll 1. \tag{4.36}$$

Figure 4.5 shows the calculation result for free energy of the ensemble of rotators:

$$F(x) = -\frac{1}{\beta} \ln Z(x), \tag{4.37}$$

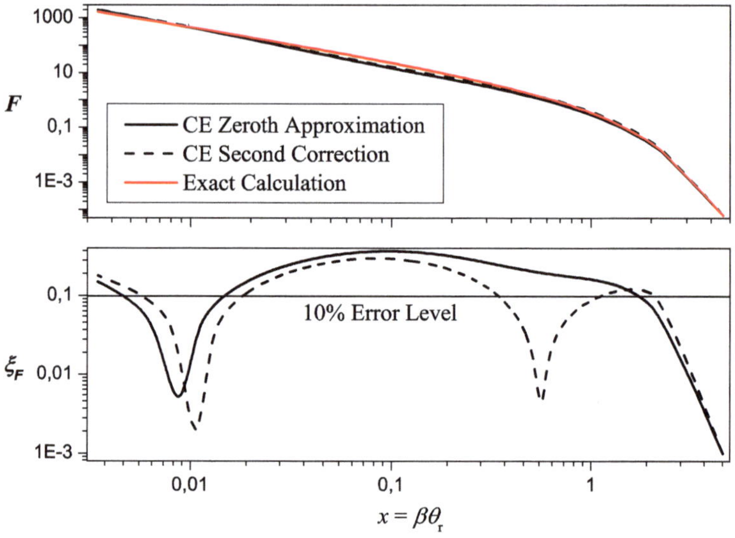

Fig. 4.5 Calculated free energy on the basis of CE for the ensemble of quantum rotators as a dependence on parameter x

based on the direct summation of 1000 energy levels in zeroth and first orders of CE. The acceptable accuracy of zeroth approximation of CE and uniform suitability of the method in entire range of x is evident from the plot.

The poor accuracy at small values of x is explained by the fact, that at high temperatures the entire energy spectrum becomes essential and limitation to only 1000 levels in direct summation causes unavoidable error in the simulation. This error is in fact not a failure of CE, but a failure of direct summation procedure. Applying the CE procedure to QAO with Hamiltonian (2.18) and using the energy of the system calculated by OM (2.22)–(2.23), the CE zeroth approximation is:

$$Z^{(0C)}(\beta,\lambda) = \exp\left[\varphi(\beta,\beta^*,\lambda,\omega)\right], \tag{4.38}$$
$$\varphi(\beta,\beta^*,\lambda,\omega) = -\beta\overline{E_n^{(0)}(\beta^*,\lambda,\omega)} + 2\beta^*\overline{n}(\beta^*) - \ln N,$$

where the normalizing constant N is defined as follows:

$$N = \left[\sum_{n=0}^{\infty}\exp(-2\beta^* n)\right]^{-1} = (1-q), \quad q = e^{-2\beta^*}. \tag{4.39}$$

The mean values in the formula (4.38) are found from:

$$\overline{n} = \frac{q}{1-q}, \quad \overline{n^2} = \frac{q(q+1)}{(1-q)^2}, \tag{4.40}$$
$$\overline{E_n^{(0)}(\beta^*,\lambda,\omega)} = \frac{1}{4\omega}(\omega^2+1)(2\overline{n}+1) + \frac{3\lambda}{4\omega^2}(2\overline{n^2}+2\overline{n}+1),$$
$$\overline{E_n^{(0)}(\beta^*,\lambda,\omega)} = \frac{1}{4\omega}(\omega^2+1)\frac{(1+q)}{(1-q)} + \frac{3\lambda}{4\omega^2}\frac{(1+q)^2}{(1-q)^2}.$$

The expressions (4.39) and (4.40) lead to the following form of the function $\varphi(\beta,\beta^*,\lambda,\omega)$:

$$\varphi = -\beta\left[\frac{1}{4\omega}(\omega^2+1)\frac{(1+q)}{(1-q)} + \frac{3\lambda}{4\omega^2}\frac{(1+q)^2}{(1-q)^2}\right] - $$
$$-\frac{q\ln q}{1-q} - \ln(1-q). \tag{4.41}$$

The variational parameters q,ω are determined from the nullification condition for the derivatives of φ, as follows from (4.12) and (4.15), that results in:

$$\frac{\partial\varphi}{\partial q} = 0 \quad \Rightarrow \quad -\beta\left[\frac{\omega^2+1}{2\omega(1-q)} + \frac{3\lambda(1+q)}{\omega^2(1-q)^2}\right] - \frac{\ln q}{1-q} = 0,$$
$$\frac{\partial\varphi}{\partial\omega} = 0 \quad \Rightarrow \quad \omega^3 - \omega - 6\lambda\frac{1+q}{1-q} = 0. \tag{4.42}$$

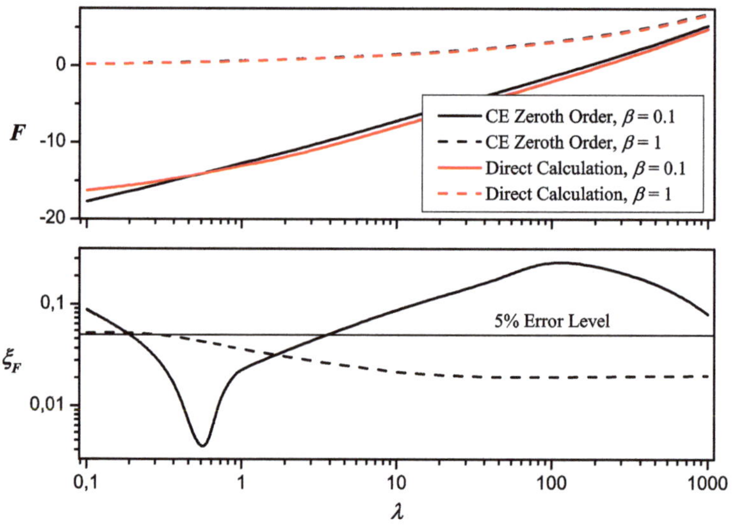

Fig. 4.6 Calculated free energy of QAO on the basis of CE as a dependence on the anharmonic parameter λ

The most convenient technique for the solution of the Eqs. (4.42) is: the functions $\lambda(q, \omega)$ and $\beta(q, \omega)$ are considered as functions of variables q, ω, instead of numerical solution of the Eqs. (4.42). These functions along with the formulas (4.38)–(4.41) define the unknown function $Z^{(0C)}(\beta, \lambda)$ in a parametric form. The results of the described technique applied for calculation of the free energy of QAO are shown in Fig. 4.6.

For direct numerical summation presented above, the results for the energy spectrum of QAO from [10] have been used. For small values of the inverse temperature β, the accuracy of the calculated free energy of QAO decreases, which is explained by unavoidable error caused by the limited number of the accounted energy levels. In the region of high anharmonicity $\lambda \sim 100$, the local decrease of the accuracy is observed, and this fact is explained by a low precision of the calculated energy spectrum. As a prove of this, with the decrease of temperature and increase of parameter β up to unity, the accuracy improves because of the population of levels becomes smaller and excited levels contribute less to free energy. Figure 4.7 shows the dependence of the free energy of QAO ($F^{(0C)} = -(1/\beta) \ln Z^{(0C)}$) on the inverse temperature β.

The local fluctuations of the accuracy in the range of high temperatures are of the same nature as mentioned above, and with the decrease of temperature the accuracy reduces to the level below 2 %. The analysis of the Figs. 4.6 and 4.7 demonstrates the uniform suitability of the approximation for free energy of QAO calculated by cumulant expansion and operator method. The limiting cases with regard to the

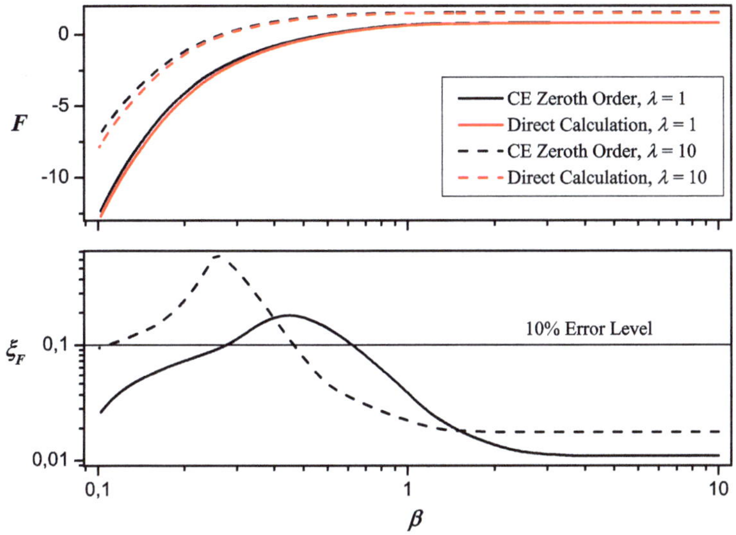

Fig. 4.7 Cumulant expansion for free energy of QAO as a dependence on the inverse temperature β

temperature for this approximation are important, and for low temperatures ($\beta \to \infty$), neglecting the terms inversely proportional to square and higher powers of the temperature β, the approximation is:

$$\omega_0^3 - \omega_0 - 6\lambda = 0, \quad q \simeq e^{-\beta K}, \quad \overline{n} \simeq e^{-\beta K} \ll 1,$$
$$\overline{E^{(0)}} \simeq \frac{1}{4}\left[\omega_0 + \frac{1}{\omega_0} + \frac{3\lambda}{\omega_0^2}\right], \quad Z^{(0C)} \simeq e^{-\beta \overline{E^{(0)}}}, \quad F^{(0C)} \simeq \overline{E^{(0)}},$$
$$K = \frac{1}{2}\left[\omega_0 + \frac{1}{\omega_0}\right] + \frac{3\lambda}{\omega_0^2}. \tag{4.43}$$

In the opposite case of high temperatures ($\beta \to 0$), we obtain under similar assumption:

$$\omega_{\overline{n}} \simeq \sqrt[3]{6\lambda\overline{n}}, \quad \overline{n} \simeq \left(\frac{2}{3\beta}^{3/4}\right)\left(\frac{1}{6\lambda}\right)^{1/4} \gg 1, \quad q \simeq e^{-1/\overline{n}} \to 1,$$
$$\overline{E^{(0)}} \simeq \frac{2}{3\beta}, \quad Z^{(0C)} \simeq \overline{n}, \quad F^{(0C)} \simeq -\frac{1}{\beta}\ln\overline{n}. \tag{4.44}$$

To improve the accuracy of the obtained approximation, the successive approximations can be calculated, and there are exist several options to make the corrections to zeroth approximations. The first one is the use of zeroth approximation of CE with energy levels obtained by OM with the accuracy of the second order at fixed parameter ω: $Z^{(1C)}$ and $F^{(1C)}$. The second option is the account of the second cumulant in the zeroth approximation of OM for energy levels: $Z^{(2C)}$ and $F^{(2C)}$. Finally, the third option is to take into account both the second cumulant and the correction to the energy levels by OM: $Z^{(3C)}$, $F^{(3C)}$. The former approach result in:

$$
\begin{aligned}
Z^{(1C)}(\beta,\lambda) &= \exp\left[\varphi_1(\beta,\beta^*,\lambda,\omega)\right], \\
\varphi_1(\beta,\beta^*,\lambda,\omega) &= -\beta\overline{(E_n^{(0)}(\beta^*,\lambda,\omega)+\Delta E_n(\beta^*,\lambda,\omega))}+ \\
&\quad +2\beta^*\overline{n}(\beta^*)-\ln N,
\end{aligned}
\tag{4.45}
$$

where the mean values are determined by the formulas (4.39), (4.40) and the expression:

$$
\begin{aligned}
&\overline{\Delta E_n(\beta^*,\lambda,\omega)} = \\
&= -\overline{\left[\frac{(n+1)(n+2)\left[\omega(1-\omega^2)+2\lambda(2n+3)\right]^2}{16\omega^2\left[\omega(\omega^2+1)+3\lambda(2n+3)\right]}\right]} - \\
&-\overline{\left[\frac{\lambda^2(n+1)(n+2)(n+3)(n+4)}{32\omega^2\left[\omega(\omega^2+1)+3\lambda(2n+5)\right]}\right]} = \\
&= -N\sum_{n=0}^{\infty}\left\{\left[\frac{(n+1)(n+2)\left[\omega(1-\omega^2)+2\lambda(2n+3)\right]^2}{16\omega^2\left[\omega(\omega^2+1)+3\lambda(2n+3)\right]}\right]+\right. \\
&\left.+\left[\frac{\lambda^2(n+1)(n+2)(n+3)(n+4)}{32\omega^2\left[\omega(\omega^2+1)+3\lambda(2n+5)\right]}\right]\right\}e^{-2\beta^* n} = \\
&= -(1-q)\left\{\frac{1}{H^4(1-q)^4}\times\right. \\
&\times\left[H^3\left(D(1-q)^3+q((1-q)(B+C+q(B-C))+\right.\right. \\
&\left.+A(1+4q+q^2)))\right)- \\
&-H^2(1-q)\left(C(1-q)^2+q(A+B+q(A-B))\right)G+ \\
&\left.+H(1-q)^2\left(B(1-q)+Aq\right)G^2-A(1-q)^3G^3\right]+ \\
&+\left(\frac{T}{G}-\frac{D}{H}+\frac{CG}{H^2}-\frac{BG^2}{H^3}+\frac{AG^3}{H^4}\right){}_2F_1\left(\frac{G}{H},1;\frac{H+G}{H};q\right)+ \\
&+\frac{1}{16H^4(1-q)^4}\times \\
&\times\left[8H^3\left(P(1-q)^3+q((1-q)(L+M+q(L-M))+\right.\right.
\end{aligned}
\tag{4.46}
$$

$$+ K(1 + 4q + q^2)))) -$$
$$-4H^2(1 - q)\left(M(1 - q)^2 + q(K + L + q(K - L))\right) R +$$
$$+ 2H(1 - q)^2 (L(1 - q) + Kq) R^2 - K(1 - q)^3 R^3\big] +$$
$$+\left(\frac{Q}{R} - \frac{P}{2H} + \frac{MR}{4H^2} - \frac{LR^2}{8H^3} + \frac{KR^3}{16H^4}\right) {}_2F_1\left(\frac{R}{2H}, 1; \frac{2H+R}{2H}; q\right)\Bigg\},$$

and the following notations are used:

$$A = 16\lambda^2, \quad B = 8\lambda(12\lambda + \omega(1 - \omega)),$$
$$C = 32\lambda^2 + 24\lambda(6\lambda + \omega(1 - \omega)) + (6\lambda + \omega(1 - \omega))^2,$$
$$D = 16\lambda(6\lambda + \omega(1 - \omega)) + 3(6\lambda + \omega(1 - \omega))^2,$$
$$G = 16\omega^2(\omega(\omega^2 + 1) + 9\lambda), \quad H = 96\lambda\omega^2,$$
$$K = \lambda^2, \quad L = 10\lambda^2, \quad M = 35\lambda^2,$$
$$P = 50\lambda^2, \quad Q = 24, \quad R = 32\omega^2(\omega(\omega^2 + 1) + 15\lambda),$$
$$T = 2(6\lambda + \omega(1 - \omega))^2, \quad q = e^{-2\beta^*},$$

and ${}_2F_1(a, b; c; z)$ is hypergeometric function [15].

The second approach assumes the accounting of the second cumulant in the formulas (4.8) and (4.9). The mean values in (4.9) with substituted from (2.22) and (2.24) energy spectrum of QAO are expressed as:

$$\begin{aligned}\overline{(E^{(0)})^2} &= \frac{8\bar{n}^2 + 8\bar{n} + 1}{16}\left(\omega_{\bar{n}} + \frac{1}{\omega_{\bar{n}}}\right)^2 + \\ &+ \frac{9\lambda^2}{16\omega_{\bar{n}}^4}\left(96\bar{n}^4 + 192\bar{n}^3 + 120\bar{n}^2 + 24\bar{n} + 1\right) + \\ &+ \frac{3\lambda}{8\omega_{\bar{n}}^2}\left(\omega_{\bar{n}} + \frac{1}{\omega_{\bar{n}}}\right)\left(24\bar{n}^3 + 36\bar{n}^2 + 14\bar{n} + 1\right), \\ \overline{E^{(0)}n} &= \frac{\bar{n}(4\bar{n}+3)}{4}\left(\omega_{\bar{n}} + \frac{1}{\omega_{\bar{n}}}\right) + \frac{3\lambda}{4\omega_{\bar{n}}^2}\bar{n}\left(12\bar{n}^2 + 16\bar{n} + 5\right).\end{aligned} \tag{4.47}$$

which leads to the following expression for the partition function of QAO:

$$Z^{(2C)}(\beta, \lambda) = \exp\left[\varphi_2(\beta, \beta^*, \lambda, \omega)\right],$$
$$\varphi_2(\beta, \beta^*, \lambda, \omega) = -\beta(\overline{E_n^{(0)}(\beta^*, \lambda, \omega))} + 2\beta^*\bar{n}(\beta^*) - \ln N +$$

$$+\frac{1}{2}\left[\beta^2(\overline{(E_n^{(0)})^2}-\overline{E_n^{(0)}}^2)-4\beta\beta^*(\overline{E_n^{(0)}n}-\overline{E_n^{(0)}}\bar{n})+\right.$$

$$\left.+4\beta^{*2}(\overline{n^2}-\bar{n}^2)\right]. \tag{4.48}$$

Finally, in the third approach for calculation of PF accounting both second cumulant and second correction for energy by OM, the partition function has a form:

$$Z^{(3C)}(\beta,\lambda)=\exp\left[\varphi_3(\beta,\beta^*,\lambda,\omega)\right],$$

$$\varphi_3(\beta,\beta^*,\lambda,\omega)=-\beta\overline{(E_n^{(2)}(\beta^*,\lambda,\omega))}+2\beta^*\overline{n}(\beta^*)-\ln N+$$

$$+\frac{1}{2}\left[\beta^2(\overline{(E_n^{(2)})^2}-\overline{E_n^{(2)}}^2)-4\beta\beta^*(\overline{E_n^{(2)}n}-\overline{E_n^{(2)}}\bar{n})+\right.$$

$$\left.+4\beta^{*2}(\overline{n^2}-\bar{n}^2)\right], \tag{4.49}$$

where the second approximation of OM for energy is:

$$E_n^{(2)}=E_n^{(0)}+\Delta E_n. \tag{4.50}$$

Figure 4.8 shows the comparison of various second-order approximations as a dependence on the anharmonic parameter λ at different inverse temperatures β.

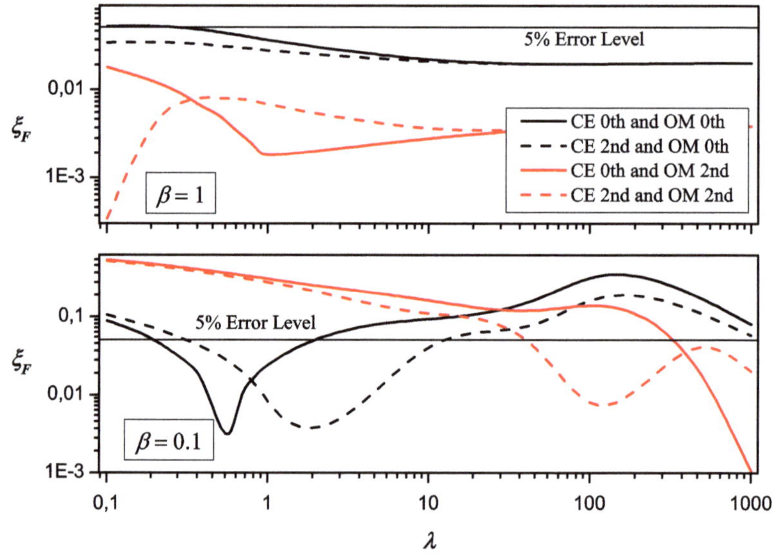

Fig. 4.8 The comparison of the accuracy of second-order approximations for free energy of QAO at different values of the anharmonic constant λ

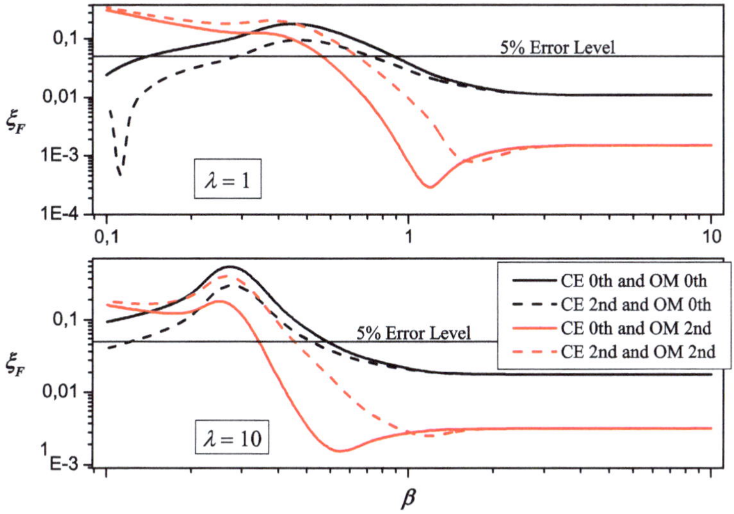

Fig. 4.9 Comparison of the accuracy for various approximations of the second order for free energy of QAO as a dependence on the inverse temperature β

The accounting of the second-order OM correction is a most essential factor for improvement of accuracy. A large errors at $\beta = 0.1$ are explained by the failures of numerical summations. This becomes evident at small values of λ for $\beta = 0.1$, when the number of excitations exceeds essentially 8 taken from [10]. At the same time, Fig. 4.8 demonstrate a uniform suitability for all approximations based on operator method in dependence on the anharmonicity parameter λ.

Figure 4.9 compares the second-order approximations as a dependence on the inverse temperature β. Similar to previous case, the second-order correction in OM plays crucial role in the achievement of high accuracy. All the approaches provide the UAA for thermodynamical characteristics and the best accuracy was supplied by zeroth approximation of CE and second approximation of OM.

For any value of the parameter λ and inverse temperature, the following inequality is fulfilled:

$$\left| \frac{F^{(1)} - F^{(0)}}{F^{(0)}} \right| \ll 1, \qquad (4.51)$$

which corresponds to the definition of UAA (1.1). The higher corrections improve the accuracy of the approximation, which satisfies the requirement (1.2) for the convergence of the successive approximations in quantum statistics. The non-perturbative approach based on combination of OM and CE is effective for even more complex potentials described by Hamiltonian (3.1). The zeroth approximation for energy is provided by (3.7) and (3.8). The expression for PF in zeroth CE approximation follows from the formulas (4.9). To perform an averaging of energy in (4.9), the mean values of operator of particle number have to be calculated:

$$\begin{aligned} \overline{n} &= \overline{n}, \\ \overline{n^2} &= \overline{n}(2\overline{n}+1), \\ \overline{n^3} &= \overline{n}(6\overline{n}^2+6\overline{n}+1), \\ &\vdots \\ \overline{n^p} &= \frac{1}{\overline{n}+1}\mathrm{Li}_{-p}\left(\frac{\overline{n}}{\overline{n}+1}\right), \end{aligned} \tag{4.52}$$

where $\mathrm{Li}_n(z)$ is a polylogarithmic function [15],

$$\mathrm{Li}_n(z) = \sum_{j=1}^{\infty} \frac{z^j}{j^n}. \tag{4.53}$$

Taking into account the relationships (4.52) and (4.53), the mean value for energy (3.7) is:

$$\begin{aligned} \overline{E^{(0)}(n,\omega_n)} &= \frac{2\overline{n}+1}{4}\left(\omega_{\bar{n}}+\frac{1}{\omega_{\bar{n}}}\right)+\frac{1}{\bar{n}+1}\sum_{m=2}^{Q}\frac{\lambda_m(2m)!}{(2\omega_{\bar{n}})^m}\times \\ &\times\left[\sum_{k=0}^{m-1}\frac{1}{2^k k!((m-k)!)^2}\times\right. \\ &\times\left.\sum_{p=0}^{m-k-1}(-1)^p\zeta_p\mathrm{Li}_{-(m-k-p)}\left(\frac{\overline{n}}{\overline{n}+1}\right)+\frac{1}{2^m m!}\right], \end{aligned} \tag{4.54}$$

where ζ_p is a sum of all possible products of natural numbers from 1 to $m-k-1$ in amount of p multiplicands:

$$\begin{aligned} \zeta_0 &= 1, \\ \zeta_1 &= 1+2+\ldots+(m-k-1), \\ \zeta_2 &= 1\cdot 2+1\cdot 3+\ldots+1\cdot(m-k-1)+ \\ &\quad +2\cdot 3+2\cdot 4+\ldots+2\cdot(m-k-1)+ \\ &\quad +\ldots+(m-k-2)\cdot(m-k-1), \\ &\vdots \end{aligned} \tag{4.55}$$

$$\zeta_{m-k-1} = 1 \cdot 2 \cdot 3 \cdot \ldots \cdot (m-k-1),$$

$$\zeta_p = \underbrace{\sum_{i_1=1}^{m-k-p} i_1 \sum_{i_2=i_1+1}^{m-k-p+1} i_2 \sum_{i_3=i_2+1}^{m-k-p+2} i_3 \cdot \ldots \cdot \sum_{i_p=i_{p-1}+1}^{m-k-1} i_p}_{p}.$$

which leads to the expressions for PF and free energy in zeroth CE approximation:

$$Z^{(0)} = \exp\left[-\beta \overline{E^{(0)}(n, \omega_n)} + (\bar{n}+1)\ln(\bar{n}+1) - \bar{n}\ln\bar{n}\right],$$
$$F^{(0)} = -\frac{1}{\beta}\ln Z^{(0)}. \tag{4.56}$$

The condition (4.12) takes the following form if the formula (4.54) is taken into account:

$$\beta\left\{\frac{1}{2}\left(\omega_{\bar{n}} + \frac{1}{\omega_{\bar{n}}}\right) + \frac{1}{\bar{n}(1+\bar{n})^2}\sum_{m=2}^{Q}\frac{\lambda_m}{(2\omega_{\bar{n}})^m}\times\right.$$
$$\times\sum_{k=0}^{m-1}\frac{(2m)!}{2^k k!((m-k)!)^2}\sum_{p=0}^{m-k-1}(-1)^p\zeta_p\times \tag{4.57}$$
$$\left.\times\left[\mathrm{Li}_{-(m-k-p+1)}\left(\frac{\overline{n}}{\overline{n}+1}\right) - \bar{n}\mathrm{Li}_{-(m-k-p)}\left(\frac{\overline{n}}{\overline{n}+1}\right)\right]\right\} =$$
$$= \ln\frac{\overline{n}+1}{\overline{n}},$$

where the following expression is used for the derivative of the polylogarithmic function [15]:

$$\frac{d\mathrm{Li}_n(z)}{dz} = \frac{1}{z}\mathrm{Li}_{n-1}(z), \tag{4.58}$$

and the mean number of the excitations $\bar{n}$ is an variational parameter. For the limiting low temperature case ($\beta \to \infty$), and by neglecting the terms of the second order, the approximation becomes:

$$\omega_0^{Q+1} - \omega_0^{Q-1} - \sum_{m=2}^{Q}\frac{m\lambda_m(2m-1)!!\omega_0^{Q-m}}{2^{m-2}} = 0,$$
$$\overline{E^{(0)}} \simeq \frac{1}{4}\left(\omega_0 + \frac{1}{\omega_0}\right) + \sum_{m=2}^{Q}\frac{\lambda_m(2m)!}{2^{2m}m!\omega_0^m}, \tag{4.59}$$
$$Z^{(0)} \simeq e^{-\beta\overline{E^{(0)}}}, \quad F^{(0)} \simeq \overline{E^{(0)}}.$$

In the limit of high temperatures ($\beta \to 0$) and by neglecting the terms of the second order:

$$
\begin{aligned}
\omega_{\bar{n}}^{Q+1} &\simeq \frac{(2Q)!}{Q!(Q-1)!2^{Q-1}}\lambda_Q \bar{n}^{Q-1},\\
\overline{E_n^{(0)}} &\simeq \frac{\omega_{\bar{n}}\bar{n}}{2} + \frac{(2Q)!\lambda_Q}{Q!2^Q\omega_{\bar{n}}^Q}\bar{n}^Q = \frac{1}{\beta}\frac{1+(Q-1)!}{1+Q!},\\
\bar{n} &\simeq \left[\frac{1}{\beta(1+Q!)}\right]^{\frac{Q+1}{2Q}}\left[\frac{(Q-1)!Q!2^{2Q}}{(2Q)!\lambda_Q}\right]^{\frac{1}{2Q}} \gg 1, \qquad (4.60)\\
Z^{(0)} &\simeq \bar{n}, \quad F^{(0)} \simeq -\frac{1}{\beta}\ln \bar{n}.
\end{aligned}
$$

To calculate the second-order approximation, the second cumulant in the formula (4.9) has to be computed, which involves the mean values $\overline{(E_n^{(0)})^2}$, $\overline{E_n^{(0)}n}$, and for energy spectrum of QAO with polynomial potential they are:

$$
\begin{aligned}
\overline{(E_n^{(0)})^2} &= \frac{8\bar{n}^2+8\bar{n}+1}{16}\left(\omega_{\bar{n}}+\frac{1}{\omega_{\bar{n}}}\right)^2 +\\
&+\frac{1}{2}\left(\omega_{\bar{n}}+\frac{1}{\omega_{\bar{n}}}\right)\frac{1}{\bar{n}+1}\sum_{m=2}^{Q}\frac{\lambda_m(2m)!}{(2\omega_{\bar{n}})^m}\sum_{k=0}^{m-1}\frac{1}{2^k k!((m-k)!)^2}\times\\
&\times\left[2\sum_{p=0}^{m-k-1}(-1)^p\zeta_p \mathrm{Li}_{-(m-k-p+1)}\left(\frac{\bar{n}}{\bar{n}+1}\right)+\right. \qquad (4.61)\\
&\left.+\sum_{p=0}^{m-k-1}(-1)^p\zeta_p \mathrm{Li}_{-(m-k-p)}\left(\frac{\bar{n}}{\bar{n}+1}\right)+\frac{2\bar{n}+1}{2^m m!}\right]+\\
&+\frac{1}{\bar{n}+1}\sum_{m=2}^{Q}\sum_{l=2}^{Q}\frac{\lambda_m\lambda_l}{(2\omega_{\bar{n}})^{m+l}}\left[\frac{(2m)!(2l)!}{2^{m+l}m!l!}+\right.\\
&+\frac{1}{2^l l!}\sum_{k=0}^{m-1}\frac{(2m)!}{2^k k!((m-k)!)^2}\sum_{p=0}^{m-k-1}(-1)^p\zeta_p \mathrm{Li}_{-(m-k-p)}\left(\frac{\bar{n}}{\bar{n}+1}\right)+\\
&+\frac{1}{2^m m!}\sum_{j=0}^{l-1}\frac{(2l)!}{2^j j!((l-j)!)^2}\sum_{q=0}^{l-j-1}(-1)^q\zeta_q \mathrm{Li}_{-(l-j-q)}\left(\frac{\bar{n}}{\bar{n}+1}\right)+
\end{aligned}
$$

$$+\sum_{k=0}^{m-1}\sum_{j=0}^{l-1}\frac{(2m)!(2l)!}{2^{k+j}k!j!((m-k)!)^2((l-j)!)^2}\times$$

$$\times\sum_{p=0}^{m-k-1}\sum_{q=0}^{l-j-1}(-1)^{p+q}\zeta_p\zeta_q\mathrm{Li}_{-(m+l-k-j-p-q)}\left(\frac{\bar{n}}{\bar{n}+1}\right)\Bigg].$$

$$\overline{E_n^{(0)}n}=\frac{\bar{n}(4\bar{n}+3)}{4}\left(\omega_{\bar{n}}+\frac{1}{\omega_{\bar{n}}}\right)+\frac{1}{\bar{n}+1}\sum_{m=2}^{Q}\frac{\lambda_m(2m)!}{(2\omega_{\bar{n}})^m}\times$$

$$\times\left[\sum_{k=0}^{m-1}\frac{1}{2^k k!((m-k)!)^2}\times\right.$$

$$\left.\times\sum_{p=0}^{m-k-1}(-1)^p\zeta_p\mathrm{Li}_{-(m-k-p+1)}\left(\frac{\bar{n}}{\bar{n}+1}\right)+\frac{\bar{n}}{2^m m!}\right], \tag{4.62}$$

Basing on the derived above analytical expressions, the particular cases of the calculation of thermodynamical parameters are demonstrated below. The uniformly suitable CE approximation for QAO of the sixth order with Hamiltonian (3.9) and energy spectrum (3.10), (3.11) can be written in explicit form using the relation (4.54):

$$\overline{E^{(0)}}=\frac{2\bar{n}+1}{4}\left(\omega_{\bar{n}}+\frac{1}{\omega_{\bar{n}}}\right)+\frac{3\lambda}{4\omega_{\bar{n}}^2}(4\bar{n}^2+4\bar{n}+1)+$$

$$+\frac{5\eta}{8\omega_{\bar{n}}^3}(24\bar{n}^3+36\bar{n}^2+18\bar{n}+3), \tag{4.63}$$

that simplifies the expression (4.57):

$$\beta\left[\frac{1}{2}\left(\omega_{\bar{n}}+\frac{1}{\omega_{\bar{n}}}\right)+\frac{3\lambda}{\omega_{\bar{n}}^2}(2\bar{n}+1)+\frac{45\eta}{4\omega_{\bar{n}}^3}(4\bar{n}^2+4\bar{n}+1)\right]=\ln\frac{\bar{n}+1}{\bar{n}}. \tag{4.64}$$

By calculating the mean number of excitation $\bar{n}$ for each inverse temperature β from formula (4.64) and substituting it into (4.63), the expressions for PF and free energy of the QAO of sixth order are obtained in zeroth CE approximation. The correction of the second order is derived in a similar way by simplifying the expressions (4.62) and using the energy spectrum.

$$\overline{E_n^{(0)}n}=\frac{\bar{n}(4\bar{n}+3)}{4}\left(\omega_{\bar{n}}+\frac{1}{\omega_{\bar{n}}}\right)+\frac{3\lambda}{4\omega_{\bar{n}}^2}\bar{n}(12\bar{n}^2+16\bar{n}+5)+$$

$$+\frac{15\eta}{8\omega_{\bar{n}}^3}\bar{n}(32\bar{n}^3+60\bar{n}^2+36\bar{n}+3), \tag{4.65}$$

$$
\begin{aligned}
\overline{(E_n^{(0)})^2} = & \frac{8\bar{n}^2+8\bar{n}+1}{16}\left(\omega_{\bar{n}}+\frac{1}{\omega_{\bar{n}}}\right)^2 + \\
& + \frac{3\lambda}{8\omega_{\bar{n}}^2}\left(\omega_{\bar{n}}+\frac{1}{\omega_{\bar{n}}}\right)(24\bar{n}^3+36\bar{n}^2+14\bar{n}+1) + \\
& + \frac{9\lambda^2}{16\omega_{\bar{n}}^4}(96\bar{n}^4+192\bar{n}^3+120\bar{n}^2+24\bar{n}+1) + \\
& + \frac{15\eta}{16\omega_{\bar{n}}^3}\left(\omega_{\bar{n}}+\frac{1}{\omega_{\bar{n}}}\right)(64\bar{n}^4+128\bar{n}^3+84\bar{n}^2+22\bar{n}+1) + \\
& + \frac{45\lambda\eta}{\omega_{\bar{n}}^5}(320\bar{n}^5+960\bar{n}^4+704\bar{n}^3+256\bar{n}^2+34\bar{n}+1) + \\
& + \frac{225\eta^2}{64\omega_{\bar{n}}^6}(1280\bar{n}^6+3840\bar{n}^5+4320\bar{n}^4+2240\bar{n}^3 + \\
& + 528\bar{n}^2+48\bar{n}+1).
\end{aligned}
$$

The results are presented in the Fig. 4.10. The results obtained by CE zeroth approximation are very similar to ones after direct summation, see Fig. 4.4. Figure 4.11 shows the accuracy of both methods. A relatively large error in the domain of high temperatures is related to the inaccuracy of direct summation method in this range. The accounting of the second cumulant reduces the error of the calculation of free energy. The analogous simulations can be performed for QAO of eighth order with Hamiltonian (3.12) and energy spectrum (3.13) and (3.14). The expressions for mean value of energy (4.55) and for mean number

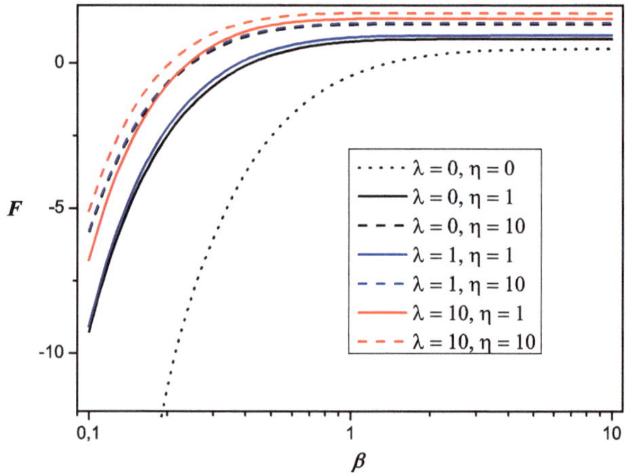

Fig. 4.10 Zeroth approximation of free energy for QAO of sixth order at different parameters of Hamiltonian as a dependence on inverse temperature β

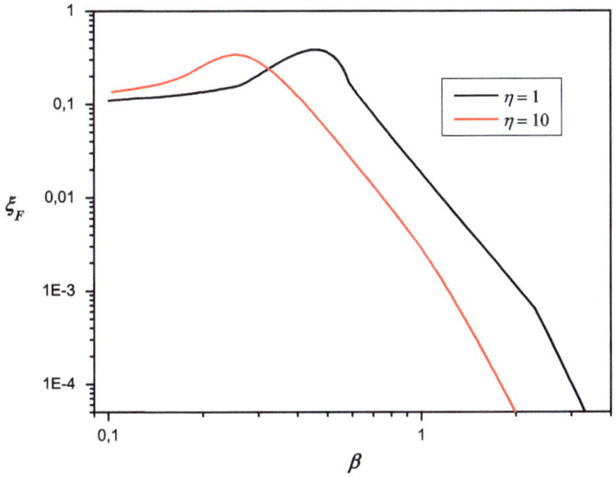

Fig. 4.11 The error for calculated by OM free energy of QAO of sixth order at $\lambda = 0$ as a dependence on inverse temperature β

of excitations (4.57) in this case are written as:

$$\overline{E^{(0)}} = \frac{2\bar{n}+1}{4}\left(\omega_{\bar{n}} + \frac{1}{\omega_{\bar{n}}}\right) + \frac{3\lambda}{4\omega_{\bar{n}}^2}(4\bar{n}^2 + 4\bar{n} + 1) +$$
$$+\frac{5\eta}{8\omega_{\bar{n}}^3}(24\bar{n}^3 + 36\bar{n}^2 + 18\bar{n} + 3) + \tag{4.66}$$
$$+\frac{105\theta}{16\omega_{\bar{n}}^4}(16\bar{n}^4 + 32\bar{n}^3 + 24\bar{n}^2 + 8\bar{n} + 1)$$

and

$$\beta\left[\frac{1}{2}\left(\omega_{\bar{n}} + \frac{1}{\omega_{\bar{n}}}\right) + \frac{3\lambda}{\omega_{\bar{n}}^2}(2\bar{n}+1) + \frac{45\eta}{4\omega_{\bar{n}}^3}(4\bar{n}^2 + 4\bar{n} + 1) +\right.$$
$$\left. + \frac{105\theta}{2\omega_{\bar{n}}^4}(8\bar{n}^3 + 12\bar{n}^2 + 6\bar{n} + 1)\right] = \ln\frac{\bar{n}+1}{\bar{n}}. \tag{4.67}$$

The mean values required for the second cumulant are:

$$\overline{(E_n^{(0)})^2} = \frac{8\bar{n}^2 + 8\bar{n} + 1}{16}\left(\omega_{\bar{n}} + \frac{1}{\omega_{\bar{n}}}\right)^2 +$$
$$\frac{3\lambda}{8\omega_{\bar{n}}^2}\left(\omega_{\bar{n}} + \frac{1}{\omega_{\bar{n}}}\right)(24\bar{n}^3 + 36\bar{n}^2 + 14\bar{n} + 1) +$$

$$+\frac{9\lambda^2}{16\omega_{\bar{n}}^4}(96\bar{n}^4+192\bar{n}^3+120\bar{n}^2+24\bar{n}+1)+$$

$$+\frac{15\eta}{16\omega_{\bar{n}}^3}\left(\omega_{\bar{n}}+\frac{1}{\omega_{\bar{n}}}\right)(64\bar{n}^4+128\bar{n}^3+84\bar{n}^2+22\bar{n}+1)+$$

$$+\frac{45\lambda\eta}{\omega_{\bar{n}}^5}(320\bar{n}^5+960\bar{n}^4+704\bar{n}^3+256\bar{n}^2+34\bar{n}+1)+$$

$$+\frac{225\eta^2}{64\omega_{\bar{n}}^6}(1280\bar{n}^6+3840\bar{n}^5+4320\bar{n}^4+2240\bar{n}^3+528\bar{n}^2+48\bar{n}+1)+$$

$$+\frac{105\theta}{64\omega_{\bar{n}}^4}\left(\omega_{\bar{n}}+\frac{1}{\omega_{\bar{n}}}\right)(80\bar{n}^5+208\bar{n}^4+200\bar{n}^3+88\bar{n}^2+17\bar{n}+1)+$$

$$+\frac{105\lambda\theta}{64\omega_{\bar{n}}^6}(2880\bar{n}^6+8640\bar{n}^5+9840\bar{n}^4+5160\bar{n}^3+1328\bar{n}^2+125\bar{n}+3)+$$

$$+\frac{1575\eta\theta}{128\omega_{\bar{n}}^7}(4480\bar{n}^7+15680\bar{n}^6+21600\bar{n}^5+14800\bar{n}^4+$$

$$+5240\bar{n}^3+900\bar{n}^2+62\bar{n}+1)+$$

$$+\frac{11025\theta^2}{256\omega_{\bar{n}}^8}(17920\bar{n}^8+71680\bar{n}^7+116480\bar{n}^6+$$

$$+98560\bar{n}^5+46240\bar{n}^4+11840\bar{n}^3+1520\bar{n}^2+80\bar{n}+1).$$

$$\overline{E_n^{(0)}n}=\frac{\bar{n}(4\bar{n}+3)}{4}\left(\omega_{\bar{n}}+\frac{1}{\omega_{\bar{n}}}\right)+\frac{3\lambda}{4\omega_{\bar{n}}^2}\bar{n}(12\bar{n}^2+16\bar{n}+5)+$$

$$+\frac{15\eta}{8\omega_{\bar{n}}^3}\bar{n}(32\bar{n}^3+60\bar{n}^2+36\bar{n}+3)+$$

$$+\frac{105\theta}{16\omega_{\bar{n}}^4}\bar{n}(80\bar{n}^4+192\bar{n}^3+168\bar{n}^2+64\bar{n}+9);$$

and the results for free energy of QAO of eighth order for various Hamiltonian's parameters as a dependence on the inverse temperature β are presented in Fig. 4.12.

Figure 4.13 shows the relative error of the approximation for free energy of the QAO of eighth order by CE with respect to the direct summation method. As in case of QAO of sixth order, the large errors are seen at high temperatures, which is typical for direct summation method.

Figure 4.14 shows the results calculated on the basis of approximations for free energy of sixth and eighth order QAO by CE, direct summation of energy levels computed by OM and exact ones [12]. For general comparison, the results of the thermodynamical perturbation theory are also shown. The direct summation on the

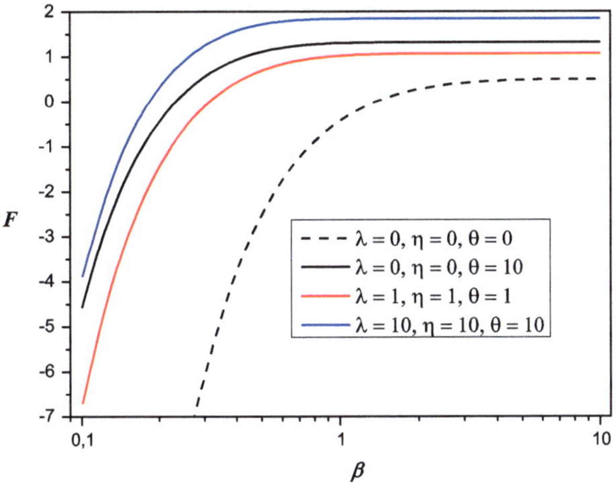

Fig. 4.12 Zeroth approximation for free energy of QAO of eighth order at varied Hamiltonian's parameters as a dependence on the inverse temperature β

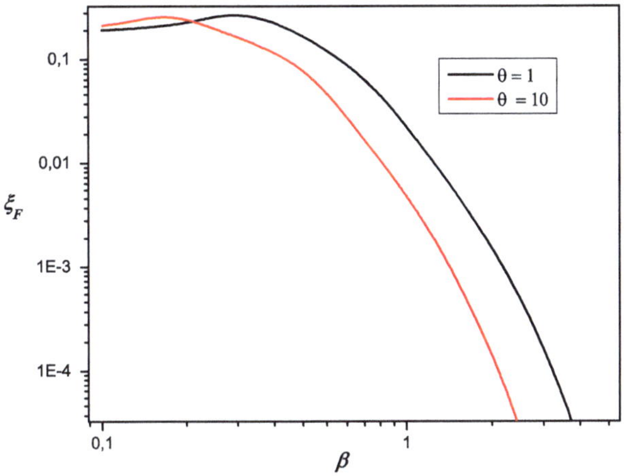

Fig. 4.13 The calculation error of free energy for QAO of eighth order by operator method ($\lambda = 0$, $\eta = 0$) with respect to the results of direct summation as a dependence on the inverse temperature β

basis of OM and CE approximation give a good accuracy within entire range of inverse temperatures β. A certain discrepancy is observed between CE method and direct summation, which is explained by the loss of accuracy in the summation over the limited number of levels. Indeed at small inverse temperature β the level population becomes essential $n \to \infty$, and for direct summation the number of accounted levels is limited apriori, which implements an error at high temperatures.

In this part of the section, we pay a special attention to the uniformly available approximation of the thermodynamical characteristics of quantum systems with asymmetric potential and Hamiltonian (3.15). The cumulant expansion is used to

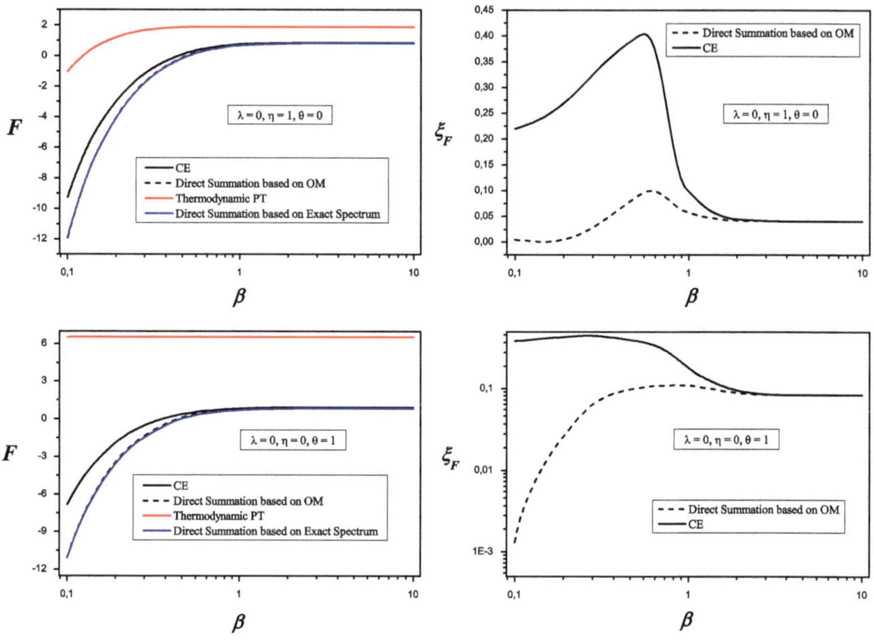

Fig. 4.14 The approximation for free energy of QAO of sixth and eighth order based on CE, direct summation by OM and exact levels as well as the errors of these approximations in comparison with the direct summation on the basis of exact energy spectrum

build an UAA for this physical system. According to the formulas (4.8) and (4.9), the zeroth approximation by cumulant expansion for partition function of the system with Hamiltonian (2.10) and energy spectrum (3.17)–(3.18) is chosen as:

$$
\begin{aligned}
&Z^{(0)}(\beta) = \exp\left[\varphi(\beta, \beta^*, \gamma, \lambda, \omega, u)\right], \\
&\varphi(\beta, \beta^*, \gamma, \lambda, \omega, u) = -\beta \left\{ \frac{2\bar{n}+1}{4}\left(\omega_{\bar{n}} + \frac{1}{\omega_{\bar{n}}}\right) + \frac{u_{\bar{n}}^2}{2} + \right. \\
&+\gamma\left[u_{\bar{n}}^3 + \frac{3u_{\bar{n}}}{2\omega_{\bar{n}}}(2\bar{n}+1)\right] + \\
&\left. + \lambda\left[u_{\bar{n}}^4 + \frac{3u_{\bar{n}}^2}{\omega_{\bar{n}}}(2\bar{n}+1) + \frac{3}{4\omega_{\bar{n}}^2}(4\bar{n}^2 + 4\bar{n} + 1)\right]\right\} + \\
&+(\bar{n}+1)\ln(\bar{n}+1) - \bar{n}\ln\bar{n},
\end{aligned}
\tag{4.68}
$$

where $\bar{n} = \bar{n}(\beta)$ is an average number of excitations at certain temperature. The values $\bar{n}$, $\omega_{\bar{n}}$, $u_{\bar{n}}$ are determined from the best fit of approximation in zeroth order of CE:

$$\frac{\partial \varphi}{\partial \bar{n}} = 0 \quad \Rightarrow$$
$$\beta \left\{ \frac{1}{2}\left(\omega_{\bar{n}} + \frac{1}{\omega_{\bar{n}}}\right) + \frac{3\gamma u_{\bar{n}}}{\omega_{\bar{n}}} + 3\lambda \left[\frac{2u_{\bar{n}}^2}{\omega_{\bar{n}}} + \frac{1}{\omega_{\bar{n}}^2}(2\bar{n}+1)\right]\right\} =$$
$$= \ln\left(\frac{\bar{n}+1}{\bar{n}}\right), \tag{4.69}$$
$$\frac{\partial \varphi}{\partial u_{\bar{n}}} = 0 \quad \Rightarrow$$
$$u_{\bar{n}}^3 + \frac{3\gamma}{4\lambda}u_{\bar{n}}^2 + \frac{\omega_{\bar{n}} + 6\lambda(2\bar{n}+1)}{4\lambda\omega_{\bar{n}}}u_{\bar{n}} + \frac{3\gamma(2\bar{n}+1)}{8\lambda\omega_{\bar{n}}} = 0,$$
$$\frac{\partial \varphi}{\partial \omega_{\bar{n}}} = 0 \quad \Rightarrow$$
$$\omega_{\bar{n}}^3 - [1 + 6u_{\bar{n}}(\gamma + \lambda u_{\bar{n}})]\,\omega_{\bar{n}} - \frac{6\lambda(4\bar{n}^2 + 4\bar{n} + 1)}{2\bar{n}+1} = 0.$$

In a similar way as former applications of CE to quantum systems, the following algorithm is optimal for the solution of the Eq. (4.69). First step is to find the inverse temperature β, which corresponds to the selected number of excitations $\bar{n}$. A next stage is to compute the parameters $u_{\bar{n}}$ and $\omega_{\bar{n}}$ using the known parameters of Hamiltonian and average number of excitations, and further to calculate the value of PF and free energy. The Hamiltonian's parameters have to satisfy the above described requirement (3.21). This criteria can be slightly violated, as in the case of QAO with cubic potential, unless at least one energy level appears in the region of the second local minimum. Because of the missing data for the energy spectrum and the thermodynamical characteristics of the system with Hamiltonian (3.37), we compare the obtained results with the ones by direct summation with the use of the approximate energy spectrum obtained in OM zeroth order. The comparison with the results of thermodynamical perturbation theory is also provided, which are knowingly non-uniformly suitable and demonstrate good results in the region of very small values of Hamiltonian's parameters. Figure 4.15 shows the approximation results for free energy of QAO with cubic term in the potential.

Figure 4.15 shows that the approximation by CE improves the results obtained by direct summation in the entire range of Hamiltonian's parameters and is an uniformly available approximation. The relative error of this approximation is shown in Fig. 4.16.

The curves demonstrate an error less than 10 % in the entire range, which riches this maximum only in the range of high temperatures, where direct summation

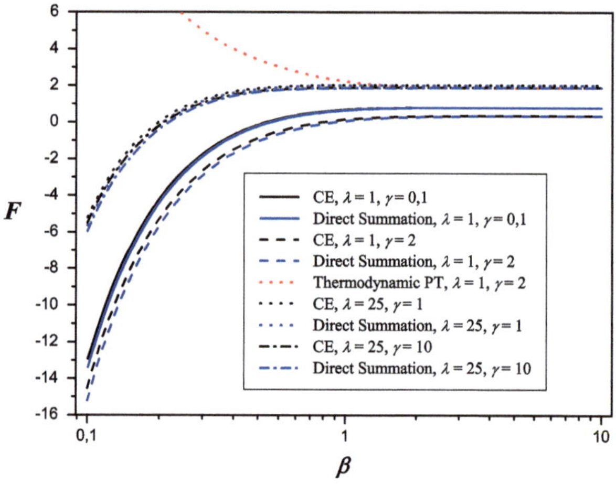

Fig. 4.15 The comparison of different approximations (zeroth order CE, direct summation of energy spectrum by zeroth order OM, thermodynamical perturbation theory) for free energy of QAO with cubic term in the potential as a dependence on the inverse temperature β for various combinations of Hamiltonian's parameters. The *red dots* correspond to TPT

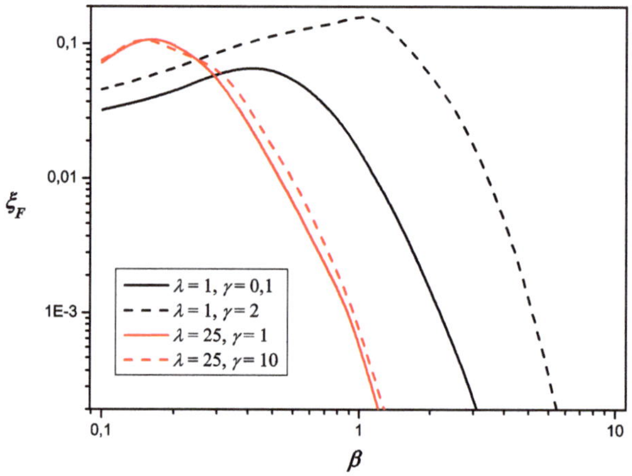

Fig. 4.16 The errors of the calculation of free energy for QAO with cubic term in the potential by OM as a dependence on the inverse temperature β for various combinations of Hamiltonian's parameters

apriori fails, and at low temperatures the error tends to zero. In opposite to the range of high temperatures, where the number of energy levels is unlimited ($n \to \infty$), the systems with limited number of excited levels (e.g., oscillator Morse) is described well by the direct summation technique. The procedure of summation over all quantum states of the system using approximation energy spectrum is applied below for the oscillator with Hamiltonian (3.24). The exact spectrum of this system is given by (3.25), and the approximate one by OM zeroth approximation by (3.31) and (3.34). Both exact and approximate spectra have a limited number of energy levels:

$$n_{max} = \left[\frac{\sqrt{2D_c}}{\alpha} - \frac{1}{2} \right], \tag{4.70}$$

where brackets mean the integer numbers. As for exact PF and free energy, the following expressions are chosen:

$$Z_{ex}(\beta) = \sum_{n=0}^{n_{max}} \exp\left[-\beta E_n\right],$$
$$F_{ex}(\beta) = -\frac{1}{\beta} \ln Z_{ex}(\beta), \tag{4.71}$$

with exact spectrum E_n from (3.25). As for zeroth approximation of operator method, the following relationships are used:

$$Z^{(0)}(\beta) = \sum_{n=0}^{n_{max}} \exp\left[-\beta E_n^{(0)}\right],$$
$$F^{(0)}(\beta) = -\frac{1}{\beta} \ln Z^{(0)}(\beta), \tag{4.72}$$

where $E_n^{(0)}$ is a zeroth approximation of OM for energy from (2.25) and (2.26). Figure 4.17 shows the results of the approximation for free energy of Morse

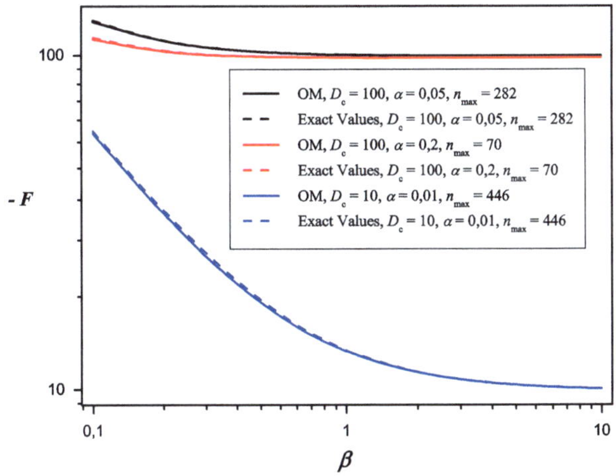

Fig. 4.17 The approximations for free energy (with negative sign) of Morse oscillator as a dependence on the inverse temperature β for different combinations of the parameters of Hamiltonian. The direct summation of both exact energy spectrum and approximate spectrum calculated by zeroth order OM are depicted

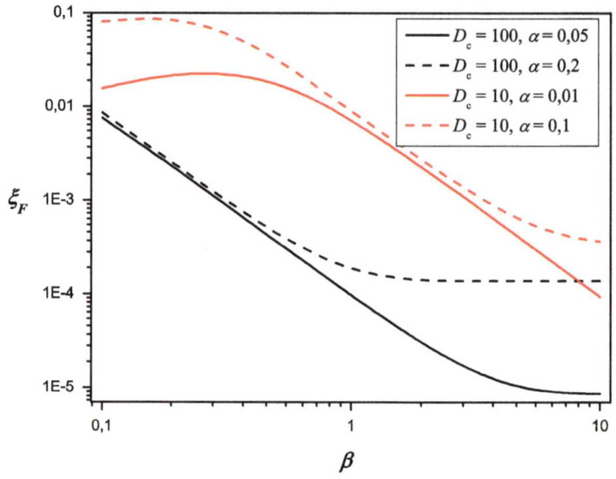

Fig. 4.18 The errors of calculation of free energy of Morse oscillator in direct summation method using approximate energy spectrum by zeroth order OM in relation to exact spectrum. The error plot is depicted in dependence on the inverse temperature β for different combinations of Hamiltonian's parameters

oscillator as a dependence on the inverse temperature β for various combinations of Hamiltonian's parameters. To confirm the uniform suitability of the approximation, the calculation accuracy for free energy of Morse oscillator is shown in Fig. 4.18 for direct summation method with exact spectrum replaced by the approximate values from OM.

The demonstrated results display the uniform suitability of the proposed approximation scheme: for any set of the parameters of the Hamiltonian the error is not exceeding 8 %.

4.3 Coupled Quantum Anharmonic Oscillators (CQAO)

In the preceding sections, we analyzed the uniformly suitable approximations for energy spectra and thermodynamical characteristics for the systems with a single degree of freedom. However, to generalize the application area of the non-perturbative methods proposed in this monograph, the systems with multiple degrees of freedom have to be considered, too. As a typical one, the system of asymmetric CQAO is discussed, described by the Hamiltonian:

$$\hat{H} = \frac{1}{2}\hat{p}_1^2 + \frac{1}{2}\hat{p}_2^2 + \frac{1}{2}(\hat{x}_1^2 + \Omega^2\hat{x}_2^2) + \lambda\hat{x}_1\hat{x}_2 + A\hat{x}_1^4 + B\hat{x}_2^4 + C\hat{x}_1^2\hat{x}_2^2, \quad (4.73)$$

where A, B, C are dimensionless parameters ($A > 0, B > 0, C > -2\sqrt{AB}$). There are both quadratic and linear interactions of oscillators under consideration, because of latter plays an essential role in the modeling of molecular potentials and potentials of crystallographic lattice [16]. The energy spectrum of this system in zeroth approximation of OM is found in [17] (see also Sect. 5.1) and is written as:

$$
\begin{aligned}
E^{(0)}_{n_1 n_2}(\omega_1, \omega_2, \alpha) = & \frac{1}{4}\left(\omega_1 + \frac{u^2 + \Omega^2 v^2}{\omega_1}\right)(2n_1 + 1) + \\
& + \frac{1}{4}\left(\omega_2 + \frac{v^2 + \Omega^2 u^2}{\omega_2}\right)(2n_2 + 1) + \\
& + \frac{\lambda u v}{4\omega_1\omega_2}\left[\omega_1(2n_2 + 1) - \omega_2(2n_1 + 1)\right] + \\
& + \frac{3}{4\omega_1^2}(2n_1^2 + 2n_1 + 1)(Au^4 + Bv^4 + Cu^2v^2) + \\
& + \frac{3}{4\omega_2^2}(2n_2^2 + 2n_2 + 1)(Av^4 + Bu^4 + Cu^2v^2) + \\
& + \frac{1}{4\omega_1\omega_2}(2n_1 + 1)(2n_2 + 1) \times \\
& \times \left[2u^2v^2(3A + 3B - 2C) + C(u^4 + v^4)\right],
\end{aligned}
\tag{4.74}
$$

where $u = \cos\alpha$, $v = \sin\alpha$, and α is a parameter. The non-trivial dependence of the energy on quantum numbers is governed by parameters $\omega_1, \omega_2, \alpha$, which are found from the algebraic equations:

$$
\begin{aligned}
& \omega_1^3 - \omega_1\left\{u^2 + \Omega^2 v^2 + 2\lambda u v + \frac{4}{\omega_2}\left(n_2 + \frac{1}{2}\right) \times \right. \\
& \left. \times\left[(3A + 3B - 2C)u^2v^2 + \frac{1}{2}C(u^4 + v^4)\right]\right\} - \\
& - \frac{6(2n_1^2 + 2n_1 + 1)}{2n_1 + 1}(Au^4 + Bv^4 + Cu^2v^2) = 0,
\end{aligned}
\tag{4.75}
$$

$$
\begin{aligned}
& \omega_2^3 - \omega_2\left\{v^2 + \Omega^2 u^2 + 2\lambda u v + \frac{4}{\omega_1}\left(n_1 + \frac{1}{2}\right) \times \right. \\
& \left. \times\left[(3A + 3B - 2C)u^2v^2 + \frac{1}{2}C(u^4 + v^4)\right]\right\} - \\
& - \frac{6(2n_2^2 + 2n_2 + 1)}{2n_2 + 1}(Av^4 + Bu^4 + Cu^2v^2) = 0,
\end{aligned}
\tag{4.76}
$$

$$\left[\omega_1\left(n_2+\frac{1}{2}\right)-\omega_2\left(n_1+\frac{1}{2}\right)\right](1-\Omega^2+2\lambda\,\mathrm{ctg}\,2\alpha)+$$
$$+\frac{3\omega_2}{\omega_1}(2n_1^2+2n_1+1)\times$$
$$\times\left[B-(A+B)\cos^2\alpha+\frac{1}{2}C\cos 2\alpha\right]+ \tag{4.77}$$
$$+\frac{3\omega_1}{\omega_2}(2n_2^2+2n_2+1)\times$$
$$\times\left[A-(A+B)\cos^2\alpha+\frac{1}{2}C\cos 2\alpha\right]+$$
$$+12\left(n_1+\frac{1}{2}\right)\left(n_2+\frac{1}{2}\right)(A+B-C)\cos 2\alpha=0.$$

These formulas have to be modified to take into account the additional degeneration of the energy levels caused by the transposed symmetry of the system in case of identical oscillators ($A = B, \Omega = 1$). The ignorance of this fact leads to the pseudo-singularities in high order corrections, which are typical for any modification of the perturbation theory for energy spectrum close to the degeneracy [18]. This problem is bypassed in OM by the selection of proper linear combinations [17], which leads to the following expression for the energy levels of the system of CQAO:

$$E^{\pm}_{n_1n_2}=\frac{1}{2}\left(E^{(0)}_{n_1n_2}+E^{(0)}_{n_2n_1}\right)\pm \tag{4.78}$$
$$\pm\sqrt{\frac{1}{4}\left(E^{(0)}_{n_1n_2}-E^{(0)}_{n_2n_1}\right)^2+\left|\langle n_1,n_2|\,\hat{V}\,|n_2,n_1\rangle\right|^2},$$

where matrix elements:

$$\langle n_1,n_2|\,\hat{V}\,|n_2,n_1\rangle=\frac{C}{4\omega_1\omega_2}\left[(n_1+1)(n_1+2)\delta_{n_1,n_2-2}+\right.$$
$$\left.+(n_2+1)(n_2+2)\delta_{n_1,n_2+2}\right]. \tag{4.79}$$

In the case of the systems with multiple degrees of freedom like CQAO, the relationships of cumulant expansion technique are modified and the partition function is written as:

$$Z(\beta)=\sum_{n_1}\sum_{n_2}g_{n_1n_2}e^{-\beta E_{n_1n_2}}, \tag{4.80}$$

where the multiplicity of degeneration $g_{n_1 n_2}$ for CQAO equals unity. The partition function can be represented in operator form:

$$Z(\beta) = \langle \alpha^*, \beta^* | \exp\left[-\beta E(\hat{n}_1, \hat{n}_2) + 2\beta^* \hat{n}_1 + 2\alpha^* \hat{n}_2 - \right. \\ \left. - \ln N(\alpha^*, \beta^*)\right] |\alpha^*, \beta^*\rangle . \tag{4.81}$$

According to definition, the trial vector $|\alpha^*, \beta^*\rangle$ has a following form:

$$|\alpha^*, \beta^*\rangle = \sqrt{N} \sum_{n_1=0}^{\infty} \sum_{n_2=0}^{\infty} \sqrt{g_{n_1 n_2}} \exp\left[-(\beta^* n_1 + \alpha^* n_2)\right] |n_1, n_2\rangle , \tag{4.82}$$

where the constant N is defined from the condition for normalization of vector $|\alpha^*, \beta^*\rangle$:

$$N = \left\{ \sum_{n_1=0}^{\infty} \sum_{n_2=0}^{\infty} g_{n_1 n_2} \exp\left[-(2\beta^* n_1 + 2\alpha^* n_2)\right] \right\}^{-1} , \tag{4.83}$$

or in the case of interacting CQAO:

$$\begin{aligned} N &= \left[1 - e^{-2\alpha^*}\right]\left[1 - e^{-2\beta^*}\right] = (1-q)(1-p), \\ q &= e^{-2\alpha^*}, \\ p &= e^{-2\beta^*}. \end{aligned} \tag{4.84}$$

As a result, in the zeroth order of cumulant expansion the partition function of CQAO is:

$$\begin{aligned} Z^{(0)}(\beta) &= \exp\left[\varphi(\beta, \alpha^*, \beta^*)\right], \\ \varphi(\beta, \alpha^*, \beta^*) &= -\beta \overline{E^{(0)}_{n_1 n_2}(\omega_1, \omega_2, \alpha)} + 2\beta^* \bar{n}_1(\beta^*) + \\ &\quad + 2\alpha^* \bar{n}_2(\alpha^*) - \ln N. \end{aligned} \tag{4.85}$$

The mean values in the formula (4.85) are written as:

$$\bar{n}_1 = \frac{q}{1-q}, \quad \bar{n}_2 = \frac{p}{1-p},$$

$$\overline{n_1^2} = \frac{q(q+1)}{(1-q)^2}, \quad \overline{n_2^2} = \frac{p(p+1)}{(1-p)^2}, \quad \overline{n_1 n_2} = \frac{qp}{(1-q)(1-p)},$$

$$\overline{E^{(0)}_{n_1 n_2}(\omega_1, \omega_2, \alpha)} = \frac{1}{4}\left(\omega_1 + \frac{u^2 + \Omega^2 v^2}{\omega_1}\right)\frac{(1+q)}{(1-q)} +$$

$$
\begin{aligned}
&+\frac{1}{4}\left(\omega_2+\frac{v^2+\Omega^2u^2}{\omega_2}\right)\frac{(1+p)}{(1-p)}+ \\
&+\frac{\lambda uv}{4\omega_1\omega_2}\left[\omega_1\frac{(1+q)}{(1-q)}-\omega_2\frac{(1+p)}{(1-p)}\right]+ \\
&+\frac{3}{4\omega_1^2}\frac{(1+q)^2}{(1-q)^2}(Au^4+Bv^4+Cu^2v^2)+ \\
&+\frac{3}{4\omega_2^2}\frac{(1+p)^2}{(1-p)^2}(Av^4+Bu^4+Cu^2v^2)+ \\
&+\frac{1}{4\omega_1\omega_2}\frac{(1+q)}{(1-q)}\frac{(1+p)}{(1-p)}\times \\
&\times\left[2u^2v^2(3A+3B-2C)+C(u^4+v^4)\right].
\end{aligned}
\tag{4.86}
$$

The expressions (4.86) result in the following explicit form for the function $\varphi(\beta,\alpha^*,\beta^*)$:

$$
\begin{aligned}
\varphi(\beta,\alpha^*,\beta^*) = -\beta\Bigg[&\frac{1}{4}\left(\omega_1+\frac{u^2+\Omega^2v^2}{\omega_1}\right)\frac{(1+q)}{(1-q)}+ \\
&+\frac{1}{4}\left(\omega_2+\frac{v^2+\Omega^2u^2}{\omega_2}\right)\frac{(1+p)}{(1-p)}+\frac{\lambda uv}{4\omega_1\omega_2}\left[\omega_1\frac{(1+q)}{(1-q)}-\omega_2\frac{(1+p)}{(1-p)}\right]+ \\
&+\frac{3}{4\omega_1^2}\frac{(1+q)^2}{(1-q)^2}(Au^4+Bv^4+Cu^2v^2)+ \\
&+\frac{3}{4\omega_2^2}\frac{(1+p)^2}{(1-p)^2}(Av^4+Bu^4+Cu^2v^2)+ \\
&+\frac{1}{4\omega_1\omega_2}\frac{(1+q)}{(1-q)}\frac{(1+p)}{(1-p)}\times\left[2u^2v^2(3A+3B-2C)+C(u^4+v^4)\right]\Bigg]- \\
&-\frac{q\ln q}{1-q}-\frac{p\ln p}{1-p}-\ln(1-q)-\ln(1-p).
\end{aligned}
\tag{4.87}
$$

The variational parameters q, p, ω_1, ω_2, and α are determined from the nullification of corresponding derivatives of φ, which yields the following equations:

$$
\begin{aligned}
&\frac{\partial\varphi}{\partial q}=0 \quad\Rightarrow \\
&\beta\left[\frac{1}{2(1-q)}\left(\omega_1+\frac{u^2+\Omega^2v^2}{\omega_1}\right)+\frac{\lambda uv}{2\omega_2(1-q)}+\right. \\
&+\frac{3(1+q)}{\omega_1^2(1-q)^2}(Au^4+Bv^4+Cu^2v^2)+
\end{aligned}
\tag{4.88}
$$

$$
\begin{aligned}
&+\frac{(1+p)}{2\omega_1\omega_2(1-q)^2(1-p)}\left[2u^2v^2(3A+3B-2C)+C(u^4+v^4)\right]\Bigg]+\\
&+\frac{\ln q}{1-q}=0,
\end{aligned}
$$

$$
\begin{aligned}
&\frac{\partial\varphi}{\partial p}=0 \quad \Rightarrow\\
&\beta\left[\frac{1}{2(1-p)}\left(\omega_2+\frac{v^2+\Omega^2u^2}{\omega_2}\right)-\frac{\lambda uv}{2\omega_1(1-p)}+\right.\\
&+\frac{3(1+p)}{\omega_1^2(1-p)^2}(Av^4+Bu^4+Cu^2v^2)+ \qquad (4.89)\\
&+\frac{(1+q)}{2\omega_1\omega_2(1-p)^2(1-q)}\left[2u^2v^2(3A+3B-2C)+C(u^4+v^4)\right]\Bigg]+\\
&+\frac{\ln p}{1-p}=0,
\end{aligned}
$$

$$
\begin{aligned}
&\frac{\partial\varphi}{\partial\omega_1}=0 \quad \Rightarrow\\
&\omega_1^3-\omega_1\left[(u^2+\Omega^2v^2)\frac{(1+q)}{(1-q)}+\frac{(1+p)}{(1-p)}\times\right. \qquad (4.90)\\
&\left.\times\left(\frac{(1-q)}{\omega_2(1+q)}\left[2u^2v^2(3A+3B-2C)+C(u^4+v^4)\right]-\lambda uv\right)\right]-\\
&-6\frac{(1+q)^2}{(1-q)^2}(Au^4+Bv^4+Cu^2v^2)=0,
\end{aligned}
$$

$$
\begin{aligned}
&\frac{\partial\varphi}{\partial\omega_2}=0 \quad \Rightarrow\\
&\omega_2^3-\omega_2\left[(v^2+\Omega^2u^2)\frac{(1+p)}{(1-p)}+\frac{(1+q)}{(1-q)}\times\right. \qquad (4.91)\\
&\left.\times\left(\frac{(1-p)}{\omega_1(1+p)}\left[2u^2v^2(3A+3B-2C)+C(u^4+v^4)\right]+\lambda uv\right)\right]-\\
&-6\frac{(1+p)^2}{(1-p)^2}(Av^4+Bu^4+Cu^2v^2)=0,
\end{aligned}
$$

$$
\begin{aligned}
&\frac{\partial\varphi}{\partial\alpha}=0 \quad \Rightarrow\\
&\frac{1}{4}\left[\omega_1-\frac{1-2\Omega^2uv}{\omega_1}\right]\frac{(1+q)}{(1-q)}+
\end{aligned}
$$

$$
\begin{aligned}
&+\frac{1}{4}\left[\omega_2+\frac{1-2\Omega^2uv}{\omega_2}\right]\frac{(1+p)}{(1-p)}+\\
&+\frac{\lambda(u^2-v^2)}{\omega_1\omega_2}\left[\omega_1\frac{(1+q)}{(1-q)}-\omega_2\frac{(1+p)}{(1-p)}\right]+\\
&+\frac{1}{\omega_1\omega_2}\frac{(1+q)(1+p)}{(1-q)(1-p)}uv\times\\
&\times\left[\frac{3\omega_2}{4\omega_1}\frac{(1+q)}{(1-q)}(4Bv^2-4Au^2+C(u^2-v^2))+\right.\\
&+\frac{3\omega_1}{4\omega_2}\frac{(1+p)}{(1-p)}(4Av^2-4Bu^2+C(u^2-v^2))+\\
&\left.+(3A+3B-4C)(u^2-v^2)\right].
\end{aligned}
\tag{4.92}
$$

The equations obtained along with the expressions (4.85)–(4.87) define a uniformly available approximation for the partition function of CQAO. The convenient technique for the solution of these equations is to treat the functions $\Omega = \Omega(q,p,\omega_1,\omega_2,\alpha)$, $\lambda = \lambda(q,p,\omega_1,\omega_2,\alpha)$, $A = A(q,p,\omega_1,\omega_2,\alpha)$, $B = B(q,p,\omega_1,\omega_2,\alpha)$, $C = C(q,p,\omega_1,\omega_2,\alpha)$ as functions of the variables $q,p,\omega_1,\omega_2,\alpha$, instead of numerical solution of the Eqs. (4.88)–(4.90). These functions along with the formulas (4.83)–(4.87) define the functions in question in a parametric form.

4.4 Density Matrix

Alongside with the partition function, the statistical operator (density matrix) plays an important role in the description of the quantum systems. Indeed the known density matrix permits to calculate a mean value of any physical parameter and the distribution of the probability density for any physical characteristics of the system. The density matrix is a most general form of the quantum mechanical description due to its capability to interpret the mixed states of the system, whereas the wave function deals with the pure states only [18]. The widely used methods for the approximation of the density matrix are mainly related to the formulation of the quantum mechanics by path integrals (see, for example, [11] and citations therein). However, as mentioned in the Chap. 1, this approach is not universal because of the calculations of energy spectrum and thermodynamical characteristics involves heavy mathematical operations with path integrals even for simple Hamiltonians.

Here we introduce the algorithm for the construction of UAA for the density matrix within the framework of the cumulant expansion [19]. All the formulas are derived for zeroth order of the approximate methods, which are of the importance for analytical analysis of physical systems. Let us assume the system is described by the Hamiltonian $\hat{H}$, and has the energy spectrum $E_n(\omega)$ in zeroth approximation of the operator method. The wave functions of this system $< x|n >= \Psi_n(x)$ are known: for example, in OM zeroth approximation they are chosen as wave functions of the harmonic oscillator with arbitrary frequency ω [18]:

$$\Psi_n(x) = \sqrt{\frac{\sqrt{\omega}}{2^n n!}} e^{-\omega x^2} \mathrm{H}_n(x\sqrt{\omega}), \tag{4.93}$$

where $\mathrm{H}_n(z)$ are the orthogonal polynomials of Hermit [15]. Then the density matrix is approximated by the following function [19]:

$$\rho(x, x', \beta) = \sum_{n=0}^{\infty} e^{-\beta E_n(\omega)} \Psi_n(x) \Psi_n^*(x'), \tag{4.94}$$

based on the approximate energy spectrum $E_n(\omega)$.

According to the algorithm of cumulant expansion, a new normalized basis is selected which depends on the additional variational parameter β^*, meaning the effective inverse temperature. The value of this parameter will be found later from the condition of best fit of approximation:

$$|\beta^*, x\rangle = \frac{1}{\sqrt{A(x)}} \sum_{n=0}^{\infty} e^{-\beta^* n} \Psi_n(x)\, |n\rangle\,. \tag{4.95}$$

The function $A(x)$ is determined from the normalization condition for the basis (4.95):

$$\langle x, \beta^* |\, \beta^*, x\rangle = \frac{1}{A} \sum_{n=0}^{\infty} e^{-2\beta^* n}\, |\Psi_n(x)|^2 = \frac{1}{A} \rho_0(x, 2\beta^*) = 1. \tag{4.96}$$

The sum $\rho_0(x, 2\beta^*)$ in the (4.96) is represented by the diagonal elements of the density matrix of harmonic oscillator corresponding to the frequency β^* [18], and the normalizing function is then expressed as:

$$A(x) = \rho_0(x, 2\beta^*). \tag{4.97}$$

In the same way, the normalizing function for a new basis related to different coordinates can be implemented:

$$\begin{aligned} \langle x', \beta^* | \beta^*, x\rangle &= \frac{1}{A(x,x')} \sum_{n=0}^{\infty} e^{-2\beta^* n} \Psi_n^*(x') \Psi_n(x) = \\ &= \frac{1}{A(x,x')} \rho_0(x, x', 2\beta^*) = 1, \quad A(x,x') = \rho_0(x, x', 2\beta^*), \end{aligned} \tag{4.98}$$

where the function $\rho_0(x, x^{'}, 2\beta^*)$ is determined by non-diagonal elements of the density matrix of harmonic oscillator [18]. As a result, the density matrix (4.94) is identically written in operator form:

$$\rho(x, x^{'}, \beta) = \rho_0(x, x^{'}, 2\beta^*) \left\langle x^{'}, \beta^* \left| \exp\left[-\beta \hat{H}(\hat{n}) + \hat{n} \ln 2\beta^* \right] \right| x, \beta^* \right\rangle, \qquad (4.99)$$

and in zeroth order of CE:

$$\rho(x, x^{'}, \beta) \simeq \rho_0(x, x^{'}, 2\beta^*) \exp\left[-\beta \left\langle x^{'}, \beta^* \left| \hat{H}(\hat{n}) \right| x, \beta^* \right\rangle + \right.$$
$$\left. + \ln 2\beta^* \left\langle x^{'}, \beta^* \left| \hat{n} \right| x, \beta^* \right\rangle \right]. \qquad (4.100)$$

All the values in exponent can be expressed through the density matrix of the harmonic oscillator $\rho_0(x, x^{'}, 2\beta^*)$, for example:

$$\begin{aligned} \langle x^{'}, \beta^* | \hat{n} | x, \beta^* \rangle &= \frac{1}{2\rho_0} \frac{\partial}{\partial \beta^*} \rho_0(x, x^{'}, 2\beta^*), \\ \langle x^{'}, \beta^* | \hat{n}^2 | x, \beta^* \rangle &= \frac{1}{4\rho_0} \frac{\partial^2}{\partial (\beta^*)^2} \rho_0(x, x^{'}, 2\beta^*), \qquad (4.101) \\ &\vdots \\ \langle x^{'}, \beta^* | \hat{n}^p | x, \beta^* \rangle &= \frac{1}{2^p \rho_0} \frac{\partial^p}{\partial (\beta^*)^p} \rho_0(x, x^{'}, 2\beta^*). \end{aligned}$$

The variational parameters β^*, ω have to be derived from the best fit of the approximation for density matrix in zeroth order of CE, which means the solution of the following equations:

$$\frac{\partial \rho}{\partial \omega} = \frac{\partial \rho}{\partial \beta^*} = 0 \quad \Rightarrow \quad \beta_0^*, \omega_0. \qquad (4.102)$$

The relationships (4.100)–(3.100) establish the uniformly suitable approximation for the density matrix of the quantum system with known energy spectrum. The density matrix used in the expressions above is written in explicit form as [18]:

$$\rho_0(x, x^{'}, \beta) = \sqrt{\frac{\omega}{\pi} \operatorname{th} \frac{\beta\omega}{2}} \exp\left[-\frac{\omega(x + x^{'})^2}{4} \operatorname{th} \frac{\beta\omega}{2} - \right.$$
$$\left. - \frac{\omega(x - x^{'})^2}{4} \operatorname{cth} \frac{\beta\omega}{2} \right]. \qquad (4.103)$$

The use of this expression for calculation of mean values in the formulas (4.101) results in the construction of UAA for the density matrix of quantum system in analytical form. The formulas obtained are essentially simplified for the diagonal elements of the density matrix. Assuming $x = x'$ in (4.100)–(4.103), the result is [19]:

$$\rho_0(x,\beta) = \sqrt{\frac{\omega}{\pi}\,\mathrm{th}\,\frac{\beta\omega}{2}}\,\exp\left[-\omega x^2\,\mathrm{th}\,\frac{\beta\omega}{2}\right],$$

$$\rho(x,\beta) \simeq \rho_0(x,2\beta^*)\exp\Big[-\beta\,\langle x,\beta^*|\,\hat{H}(\hat{n})\,|x,\beta^*\rangle + \\ + \ln 2\beta^*\,\langle x,\beta^*|\,\hat{n}\,|x,\beta^*\rangle\Big]. \tag{4.104}$$

The proposed above algorithm for the approximation of the density matrix has several advantages in comparison with other known methods: (i) the generalization of this technique for the systems with arbitrary number of the degrees of freedom is easy and straightforward; (ii) the variational basis is composed of the arbitrary set of normalized state vectors; (iii) the approach is well fitted to the systems with special Hamiltonians, for example, Coulomb's system or spin system. In the case of QAO, the system is described by Hamiltonian (2.18) and has the energy spectrum defined by Eqs. (2.21) and (2.23) in zeroth approximation of OM. The maximal order of the operator of the number of excitations for spectrum (2.21) equals two, therefore the explicit expressions for the mean values (4.101) with the use of the density matrix of harmonic oscillator (4.103) are:

$$\tilde{n} = \frac{\omega}{2}\left\{\frac{1}{\mathrm{sh}(2\beta^*\omega)} - \frac{\omega}{4}\left[\frac{(x+x')^2}{\mathrm{ch}^2\,\beta^*\omega} - \frac{(x-x')^2}{\mathrm{sh}^2\,\beta^*\omega}\right]\right\},$$

$$\widetilde{n^2} = \frac{\omega^2}{4}\left\{-\frac{1}{\mathrm{sh}^2(2\beta^*\omega)} - \frac{1}{\mathrm{ch}^2\,\beta^*\omega} + \\ + \frac{\omega}{2\,\mathrm{sh}(2\beta^*\omega)}\left[\frac{(x+x')^2}{\mathrm{ch}^2\,\beta^*\omega} - \frac{(x-x')^2}{\mathrm{sh}^2\,\beta^*\omega}\right] + \\ + \frac{\omega^2}{16}\left[\frac{(x+x')^2}{\mathrm{ch}^2\,\beta^*\omega} - \frac{(x-x')^2}{\mathrm{sh}^2\,\beta^*\omega}\right]^2 + \\ + \frac{\omega^3}{2}\left[\frac{(x+x')^2\,\mathrm{th}\,\beta^*\omega}{\mathrm{ch}^2\,\beta^*\omega} - \frac{(x-x')^2\,\mathrm{cth}\,\beta^*\omega}{\mathrm{sh}^2\,\beta^*\omega}\right]\right\}. \tag{4.105}$$

The equations obtained give an analytical approximation for density matrix of QAO with Hamiltonian (2.18) in CE zeroth approximation.

$$\rho(x,x',\beta) \simeq \sqrt{\frac{\omega}{\pi}\,\text{th}\,\beta^*\omega}\exp\left\{-\frac{\omega(x+x')^2}{4}\,\text{th}\,\beta^*\omega - \right.$$
$$-\frac{\omega(x-x')^2}{4}\,\text{cth}\,\beta^*\omega - \beta\left[\frac{2\tilde{n}+1}{4}\left[\omega+\frac{1}{\omega}(1+2\mu)\right]+\right.$$
$$\left.\left.+\frac{3\lambda}{4\omega^2}\left(2\widetilde{n^2}+2\tilde{n}+1\right)\right]+\tilde{n}\ln 2\beta^*\right\}. \tag{4.106}$$

In accordance with the variational relationships for optimization of parameters β^* and ω, the following equations have to be resolved:

$$\frac{\omega}{2\sqrt{\text{th}\,\beta^*\omega}\,\text{ch}^2\,\beta^*\omega}+\sqrt{\text{th}\,\beta^*\omega}\left\{-\frac{\omega^2}{4}\left[\frac{(x+x')^2}{\text{ch}\,\beta^*\omega}-\frac{(x-x')^2}{\text{sh}\,\beta^*\omega}\right]-\right.$$
$$-\beta\left[\frac{1}{2}\left[\omega+\frac{1}{\omega}(1+2\mu)\right]\frac{\partial\tilde{n}}{\partial\beta^*}+\frac{3\lambda}{2\omega^2}\left[\frac{\partial\widetilde{n^2}}{\partial\beta^*}+\frac{\partial\tilde{n}}{\partial\beta^*}\right]\right]+$$
$$\left.+\frac{\partial\tilde{n}}{\partial\beta^*}\ln 2\beta^*+\frac{2\tilde{n}}{\beta^*}\right\}=0,$$
$$\frac{\text{th}\,\beta^*\omega\,\text{ch}^2\,\beta^*\omega+\beta^*\omega}{2\sqrt{\omega\,\text{th}\,\beta^*\omega}\,\text{ch}^2\,\beta^*\omega}+\sqrt{\omega\,\text{th}\,\beta^*\omega}\times$$
$$\times\quad\left\{-\frac{1}{4}\left[(x+x')^2\,\text{th}\,\beta^*\omega+(x-x')^2\,\text{cth}\,\beta^*\omega\right]-\right. \tag{4.107}$$
$$-\frac{\beta^*\omega}{4}\left[\frac{(x+x')^2}{\text{ch}^2\,\beta^*\omega}-\frac{(x-x')^2}{\text{sh}^2\,\beta^*\omega}\right]-$$
$$\beta\left[\frac{2\tilde{n}+1}{4}\left(1-\frac{1}{\omega^2}(1+2\mu)\right)+\right.$$
$$+\frac{1}{2}\frac{\partial\tilde{n}}{\partial\omega}\left(\omega+\frac{1}{\omega}(1+2\mu)\right)-\frac{3\lambda}{8\omega^3}\left(2\widetilde{n^2}+2\tilde{n}+1\right)+$$
$$\left.\left.+\frac{3\lambda}{2\omega^2}\left(\frac{\partial\widetilde{n^2}}{\partial\omega}+\frac{\partial\tilde{n}}{\partial\omega}\right)\right]+\frac{\partial\tilde{n}}{\partial\omega}\ln 2\beta^*\right\}=0,$$

where the derivatives of the mean values are:

$$\frac{\partial\tilde{n}}{\partial\beta^*}=\frac{\omega^2}{2}\left\{-\frac{2\,\text{cth}(2\beta^*\omega)}{\text{sh}(2\beta\omega)}+\frac{\omega^2}{2}\left[\frac{(x+x')^2\,\text{th}\,\beta^*\omega}{\text{ch}^2\,\beta^*\omega}-\right.\right.$$
$$\left.\left.-\frac{(x-x')^2\,\text{cth}\,\beta^*\omega}{\text{sh}^2\,\beta^*\omega}\right]\right\}, \tag{4.108}$$

$$\frac{\partial\widetilde{n^2}}{\partial\beta^*}=\frac{\omega^2}{4}\left\{\frac{4\omega\,\mathrm{cth}(2\beta^*\omega)}{\mathrm{sh}^2(2\beta^*\omega)}+\frac{2\omega\,\mathrm{th}\,\beta^*\omega}{\mathrm{ch}^2\,\beta^*\omega}-\right.$$
$$-\frac{\omega^2\,\mathrm{cth}(2\beta^*\omega)}{\mathrm{sh}(2\beta^*\omega)}\left[\frac{(x+x')^2}{\mathrm{ch}^2\,\beta^*\omega}-\frac{(x-x')^2}{\mathrm{sh}^2\,\beta^*\omega}\right]-$$
$$-\frac{\omega^2}{\mathrm{sh}(2\beta^*\omega)}\left[\frac{(x+x')^2\,\mathrm{th}\,\beta^*\omega}{\mathrm{ch}^2\,\beta^*\omega}-\frac{(x-x')^2\,\mathrm{cth}\,\beta^*\omega}{\mathrm{sh}^2\,\beta^*\omega}\right]-$$
$$-\frac{\omega^3}{4}\left[\frac{(x+x')^2}{\mathrm{ch}^2\,\beta^*\omega}-\frac{(x-x')^2}{\mathrm{sh}^2\,\beta^*\omega}\right]\times \tag{4.109}$$
$$\times\left[\frac{(x+x')^2\,\mathrm{th}\,\beta^*\omega}{\mathrm{ch}^2\,\beta^*\omega}-\frac{(x-x')^2\,\mathrm{cth}\,\beta^*\omega}{\mathrm{sh}^2\,\beta^*\omega}\right]+$$
$$+\frac{\omega^4}{2}\left[\frac{(x+x')^2\left(1-2\,\mathrm{sh}^2\,\beta^*\omega\right)}{\mathrm{ch}^4\,\beta^*\omega}+\frac{(x-x')^2\left(1+2\,\mathrm{ch}^2\,\beta^*\omega\right)}{\mathrm{sh}^4\,\beta^*\omega}\right]$$

and the differentiation over the parameter ω results in:

$$\frac{\partial\tilde{n}}{\partial\omega}=\frac{1}{2}\left\{\frac{1}{\mathrm{sh}(2\beta^*\omega)}-\frac{\omega}{4}\left[\frac{(x+x')^2}{\mathrm{ch}^2\,\beta^*\omega}-\frac{(x-x')^2}{\mathrm{sh}^2\,\beta^*\omega}\right]\right\}+$$
$$+\frac{\omega}{2}\left\{-\frac{2\beta^*\,\mathrm{cth}(2\beta^*\omega)}{\mathrm{sh}(2\beta^*\omega)}-\frac{1}{4}\left[\frac{(x+x')^2}{\mathrm{ch}^2\,\beta^*\omega}-\frac{(x-x')^2}{\mathrm{sh}^2\,\beta^*\omega}\right]+\right.$$
$$\left.+\frac{\beta^*\omega}{2}\left[\frac{(x+x')^2\,\mathrm{th}\,\beta^*\omega}{\mathrm{ch}^2\,\beta^*\omega}-\frac{(x-x')^2\,\mathrm{cth}\,\beta^*\omega}{\mathrm{sh}^2\,\beta^*\omega}\right]\right\}, \tag{4.110}$$

$$\frac{\partial\widetilde{n^2}}{\partial\omega}=\frac{\omega}{2}\left\{-\frac{1}{\mathrm{sh}^2(2\beta^*\omega)}-\frac{1}{\mathrm{ch}^2\,\beta^*\omega}+\right.$$
$$+\frac{\omega}{2\,\mathrm{sh}(2\beta^*\omega)}\left[\frac{(x+x')^2}{\mathrm{ch}^2\,\beta^*\omega}-\frac{(x-x')^2}{\mathrm{sh}^2\,\beta^*\omega}\right]+$$
$$+\frac{\omega^2}{16}\left[\frac{(x+x')^2}{\mathrm{ch}^2\,\beta^*\omega}-\frac{(x-x')^2}{\mathrm{sh}^2\,\beta^*\omega}\right]^2+$$
$$\left.+\frac{\omega^3}{2}\left[\frac{(x+x')^2\,\mathrm{th}\,\beta^*\omega}{\mathrm{ch}^2\,\beta^*\omega}-\frac{(x-x')^2\,\mathrm{cth}\,\beta^*\omega}{\mathrm{sh}^2\,\beta^*\omega}\right]\right\}+$$
$$+\frac{\omega^2}{4}\left\{\frac{4\beta^*\,\mathrm{cth}(2\beta^*\omega)}{\mathrm{sh}^2(2\beta^*\omega)}+\frac{2\beta^*\,\mathrm{th}\,\beta^*\omega}{\mathrm{ch}^2\,\beta^*\omega}+\right.$$

$$
\begin{aligned}
&+\frac{\operatorname{sh}(2\beta^*\omega)-\omega\operatorname{ch}(2\beta^*\omega)}{2\operatorname{sh}^2(2\beta^*\omega)}\left[\frac{(x+x')^2}{\operatorname{ch}^2\beta^*\omega}-\frac{(x-x')^2}{\operatorname{sh}^2\beta^*\omega}\right]-\\
&-\frac{\beta^*\omega}{\operatorname{sh}(2\beta^*\omega)}\left[\frac{(x+x')^2\operatorname{th}\beta^*\omega}{\operatorname{ch}^2\beta^*\omega}-\frac{(x-x')^2\operatorname{cth}\beta^*\omega}{\operatorname{sh}^2\beta^*\omega}\right]+\\
&+\frac{\omega}{8}\left[\frac{(x+x')^2}{\operatorname{ch}^2\beta^*\omega}-\frac{(x-x')^2}{\operatorname{sh}^2\beta^*\omega}\right]^2-\\
&-\frac{\beta^*\omega^2}{4}\left[\frac{(x+x')^2}{\operatorname{ch}^2\beta^*\omega}-\frac{(x-x')^2}{\operatorname{sh}^2\beta^*\omega}\right]\\
&\times\left[\frac{(x+x')^2\operatorname{th}\beta^*\omega}{\operatorname{ch}^2\beta^*\omega}-\frac{(x-x')^2\operatorname{cth}\beta^*\omega}{\operatorname{sh}^2\beta^*\omega}\right]+\\
&+\frac{\beta^*\omega^3}{2}\left[\frac{(x+x')^2\left(1-2\operatorname{sh}^2\beta^*\omega\right)}{\operatorname{ch}^4\beta^*\omega}+\right.\\
&\left.\left.+\frac{(x-x')^2\left(1+2\operatorname{ch}^2\beta^*\omega\right)}{\operatorname{sh}^4\beta^*\omega}\right]\right\}.
\end{aligned}
\tag{4.111}
$$

The relations (4.105)–(4.110) define the UAA for full density matrix of QAO. The form of these equations is simplified essentially if only the diagonal elements of the density matrix is considered, i.e. the probability of the distribution of QAO over the coordinates:

$$
\begin{aligned}
\tilde{n} &= \frac{\omega}{2}\left\{\frac{1}{\operatorname{sh}(2\beta^*\omega)}-\frac{\omega x^2}{\operatorname{ch}^2\beta^*\omega}\right\},\\
\widetilde{n^2} &= \frac{\omega^2}{4}\left\{-\frac{1}{\operatorname{sh}^2(2\beta^*\omega)}-\frac{1}{\operatorname{ch}^2\beta^*\omega}+\frac{2\omega x^2}{\operatorname{sh}(2\beta^*\omega)\operatorname{ch}^2\beta^*\omega}+\right.\\
&\left.+\frac{\omega^2x^4}{\operatorname{ch}^4\beta^*\omega}+\frac{2\omega^3x^2\operatorname{th}\beta^*\omega}{\operatorname{ch}^2\beta^*\omega}\right\}.
\end{aligned}
\tag{4.112}
$$

$$
\begin{aligned}
\rho(x,\beta) &\simeq \sqrt{\frac{\omega}{\pi}\operatorname{th}\beta^*\omega}\exp\left\{-\omega x^2\operatorname{th}\beta^*\omega-\right.\\
&-\beta\left[\frac{2\tilde{n}+1}{4}\left[\omega+\frac{1}{\omega}(1+2\mu)\right]+\frac{3\lambda}{4\omega^2}\left(2\widetilde{n^2}+2\tilde{n}+1\right)\right]+\\
&\left.+\tilde{n}\ln 2\beta^*\right\},
\end{aligned}
\tag{4.113}
$$

where the mean values are from formulas (4.112), and then the variational conditions (4.107)–(4.110) are transformed into:

$$
\frac{\omega}{2\sqrt{\operatorname{th}\beta^*\omega}\operatorname{ch}^2\beta^*\omega} + \sqrt{\operatorname{th}\beta^*\omega}\left\{-\frac{\omega^2x^2}{\operatorname{ch}\beta^*\omega} - \right.
$$

$$
-\beta\left[\frac{1}{2}\left[\omega + \frac{1}{\omega}(1+2\mu)\right]\frac{\partial\tilde{n}}{\partial\beta^*} + \frac{3\lambda}{2\omega^2}\left[\frac{\partial\widetilde{n^2}}{\partial\beta^*} + \frac{\partial\tilde{n}}{\partial\beta^*}\right]\right] +
$$

$$
\left. + \frac{\partial\tilde{n}}{\partial\beta^*}\ln 2\beta^* + \frac{2\tilde{n}}{\beta^*}\right\} = 0, \tag{4.114}
$$

$$
\frac{\operatorname{th}\beta^*\omega\operatorname{ch}^2\beta^*\omega + \beta^*\omega}{2\sqrt{\omega\operatorname{th}\beta^*\omega}\operatorname{ch}^2\beta^*\omega} + \sqrt{\omega\operatorname{th}\beta^*\omega}\left\{- x^2\operatorname{th}\beta^*\omega - \frac{\beta^*\omega x^2}{\operatorname{ch}^2\beta^*\omega} - \right.
$$

$$
-\beta\left[\frac{2\tilde{n}+1}{4}\left(1 - \frac{1}{\omega^2}(1+2\mu)\right) + \frac{1}{2}\frac{\partial\tilde{n}}{\partial\omega}\left(\omega + \frac{1}{\omega}(1+2\mu)\right) - \right.
$$

$$
\left.\left. - \frac{3\lambda}{8\omega^3}\left(2\widetilde{n^2}+2\tilde{n}+1\right) + \frac{3\lambda}{2\omega^2}\left(\frac{\partial\widetilde{n^2}}{\partial\omega} + \frac{\partial\tilde{n}}{\partial\omega}\right)\right] + \frac{\partial\tilde{n}}{\partial\omega}\ln 2\beta^*\right\} = 0,
$$

with the derivatives:

$$
\frac{\partial\tilde{n}}{\partial\beta^*} = \omega^2\left\{-\frac{\operatorname{cth}(2\beta^*\omega)}{\operatorname{sh}(2\beta^*\omega)} + \frac{\omega^2x^2\operatorname{th}\beta^*\omega}{\operatorname{ch}^2\beta^*\omega}\right\},
$$

$$
\frac{\partial\widetilde{n^2}}{\partial\beta^*} = \frac{\omega^2}{2}\left\{\frac{2\omega\operatorname{cth}(2\beta^*\omega)}{\operatorname{sh}^2(2\beta^*\omega)} + \frac{\omega\operatorname{th}\beta^*\omega}{\operatorname{ch}^2\beta^*\omega} - \right.
$$

$$
- \frac{2\omega^2x^2\left(\operatorname{cth}(2\beta^*\omega) + \operatorname{th}(2\beta^*\omega)\right)}{\operatorname{sh}(2\beta^*\omega)\operatorname{ch}^2\beta^*\omega} -
$$

$$
\left. - \frac{2\omega^3x^4\operatorname{th}\beta^*\omega}{\operatorname{ch}^4\beta^*\omega} + \frac{\omega^2x^2\left(1 - 2\operatorname{sh}^2\beta^*\omega\right)}{\operatorname{ch}^4\beta^*\omega}\right\},
$$

$$
\frac{\partial\tilde{n}}{\partial\omega} = \frac{1}{2}\left\{\frac{1}{\operatorname{sh}(2\beta^*\omega)} - \frac{\omega x^2}{\operatorname{ch}^2\beta^*\omega}\right\} + \frac{\omega}{2}\left\{-\frac{2\beta^*\operatorname{cth}(2\beta^*\omega)}{\operatorname{sh}(2\beta^*\omega)} - \right.
$$

$$
\left. - \frac{x^2}{\operatorname{ch}^2\beta^*\omega} + \frac{2\beta^*\omega x^2\operatorname{th}\beta^*\omega}{\operatorname{ch}^2\beta^*\omega}\right\}, \tag{4.115}
$$

$$
\frac{\partial\widetilde{n^2}}{\partial\omega} = \frac{\omega}{2}\left\{-\frac{1}{\operatorname{sh}^2(2\beta^*\omega)} - \frac{1}{\operatorname{ch}^2\beta^*\omega} + \frac{2\omega x^2}{\operatorname{sh}(2\beta^*\omega)\operatorname{ch}^2\beta^*\omega} + \right.
$$

$$
\left. + \frac{\omega^2x^4}{\operatorname{ch}^4\beta^*\omega} + \frac{2\omega^3x^2\operatorname{th}\beta^*\omega}{\operatorname{ch}^2\beta^*\omega}\right\} + \frac{\omega^2}{4}\left\{\frac{4\beta^*\operatorname{cth}(2\beta^*\omega)}{\operatorname{sh}^2(2\beta^*\omega)} + \right.
$$

$$+\frac{2\beta^* \operatorname{th} \beta^* \omega}{\operatorname{ch}^2 \beta^* \omega}+\frac{2\,[\operatorname{sh}(2\beta^* \omega)-\omega \operatorname{ch}(2\beta^* \omega)]\,x^2}{\operatorname{sh}^2(\beta^* \omega) \operatorname{ch}^2 \beta^* \omega}-\frac{2\beta^* \omega x^2}{\operatorname{ch}^4 \beta^* \omega}+$$

$$\left.+\frac{2\omega x^4}{\operatorname{ch}^4 \beta^* \omega}-\frac{4\beta^* \omega^2 x^4 \operatorname{th} \beta^* \omega}{\operatorname{ch}^4 \beta^* \omega}+\frac{2\beta^* \omega^3 x^2\left(1-2 \operatorname{sh}^2 \beta^* \omega\right)}{\operatorname{ch}^4 \beta^* \omega}\right\}.$$

To evaluate the accuracy of the approximation obtained, the numerical results for the probability density of QAO for certain values of the anharmonic parameter λ are compared with the results received in [11] on the basis of the direct numerical solution of the Schrödinger equation using Feynman formulation of the quantum mechanics. Figure 4.19 shows the calculations of the probability density of QAO at $\lambda = 40$ for different inverse temperatures β.

To illustrate the universal character of the proposed approach, the density matrix of the Morse oscillator with the Hamiltonian (3.24) and exact energy spectrum (3.25) possessing a limited number of the discrete levels is considered hereafter. Using the density matrix of the harmonic oscillator as approximation,

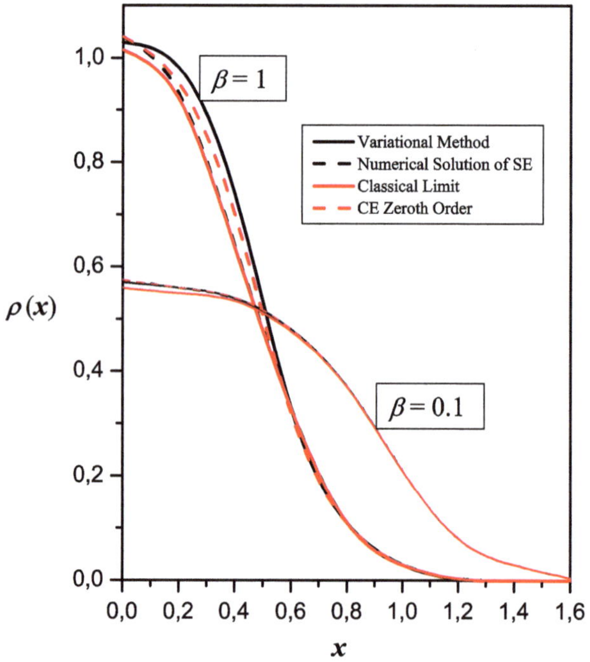

Fig. 4.19 The comparison of different approximations for probability density of QAO at $\lambda = 40$ for several values of the inverse temperature β

the density matrix of the coupled states of the system in zeroth order of cumulant expansion can be obtained. Indeed starting from the expressions (4.100)–(4.103), we arrive at:

$$\rho(x,x',\beta) \simeq \sqrt{\frac{\omega}{\pi}\,\mathrm{th}\,\beta^*\omega}\,\exp\left\{-\frac{\omega(x+x')^2}{4}\,\mathrm{th}\,\beta^*\omega-\right.$$
$$-\frac{\omega(x-x')^2}{4}\,\mathrm{cth}\,\beta^*\omega+\beta D_c\left[1-\frac{\sqrt{2}\alpha}{\sqrt{D_c}}\left(\tilde{n}+\frac{1}{2}\right)+\right.$$
$$\left.\left.+\frac{\alpha^2}{2D_c}\left(\widetilde{n^2}+\tilde{n}+\frac{1}{4}\right)\right]+\tilde{n}\ln 2\beta^*\right\}, \tag{4.116}$$

where the mean values follow from the relations (4.101).

The expressions for the zeroth approximation of the density matrix has to be complemented with the variational equations to determine the parameters ω and β^*. Following the standard CE algorithm, we obtain:

$$\frac{\omega}{2\sqrt{\mathrm{th}\,\beta^*\omega}\,\mathrm{ch}^2\,\beta^*\omega}+\sqrt{\mathrm{th}\,\beta^*\omega}\left\{-\frac{\omega^2}{4}\left[\frac{(x+x')^2}{\mathrm{ch}\,\beta^*\omega}-\frac{(x-x')^2}{\mathrm{sh}\,\beta^*\omega}\right]+\right.$$
$$+\beta D_c\left[-\frac{\sqrt{2}\alpha}{\sqrt{D_c}}\frac{\partial\tilde{n}}{\partial\beta^*}+\frac{\alpha^2}{2D_c}\left(\frac{\partial\widetilde{n^2}}{\partial\beta^*}+\frac{\partial\tilde{n}}{\partial\beta^*}\right)\right]+$$
$$\left.+\frac{\partial\tilde{n}}{\partial\beta^*}\ln 2\beta^*+\frac{2\tilde{n}}{\beta^*}\right\}=0, \tag{4.117}$$

$$\frac{\mathrm{th}\,\beta^*\omega\,\mathrm{ch}^2\,\beta^*\omega+\beta^*\omega}{2\sqrt{\omega\,\mathrm{th}\,\beta^*\omega}\,\mathrm{ch}^2\,\beta^*\omega}+\sqrt{\omega\,\mathrm{th}\,\beta^*\omega}$$
$$\times\left\{-\frac{1}{4}\left[(x+x')^2\,\mathrm{th}\,\beta^*\omega+(x-x')^2\,\mathrm{cth}\,\beta^*\omega\right]-\right.$$
$$-\frac{\beta^*\omega}{4}\left[\frac{(x+x')^2}{\mathrm{ch}^2\,\beta^*\omega}-\frac{(x-x')^2}{\mathrm{sh}^2\,\beta^*\omega}\right]+$$
$$\left.+\beta D_c\left[-\frac{\sqrt{2}\alpha}{\sqrt{D_c}}\frac{\partial\tilde{n}}{\partial\omega}+\frac{\alpha^2}{2D_c}\left(\frac{\partial\widetilde{n^2}}{\partial\omega}+\frac{\partial\tilde{n}}{\partial\omega}\right)\right]+\frac{\partial\tilde{n}}{\partial\omega}\ln 2\beta^*\right\}=0,$$

where the derivatives of the mean values are obtained again from the Eqs. (4.108)–(4.110), as in case of QAO. The expressions (4.116)–(4.117) and (4.108)–(4.110) determine the uniformly suitable approximation for the density matrix of the Morse oscillator. For the probability density, the Eq. (4.116) is transformed to:

$$\rho(x,\beta) \simeq \sqrt{\frac{\omega}{\pi}\,\mathrm{th}\,\beta^*\omega}\,\exp\Bigg\{ -\omega x^2\,\mathrm{th}\,\beta^*\omega + \\ +\beta D_c\left[1-\frac{\sqrt{2}\alpha}{\sqrt{D_c}}\left(\tilde{n}+\frac{1}{2}\right)+\frac{\alpha^2}{2D_c}\left(\widetilde{n^2}+\tilde{n}+\frac{1}{4}\right)\right]+ \\ +\tilde{n}\ln 2\beta^*\Bigg\}, \tag{4.118}$$

and the variational equations are transformed into the form:

$$\frac{\omega}{2\sqrt{\mathrm{th}\,\beta^*\omega}\,\mathrm{ch}^2\,\beta^*\omega}+\sqrt{\mathrm{th}\,\beta^*\omega}\Bigg\{-\frac{\omega^2x^2}{\mathrm{ch}\,\beta^*\omega}+ \\ +\beta D_c\left[-\frac{\sqrt{2}\alpha}{\sqrt{D_c}}\frac{\partial\tilde{n}}{\partial\beta^*}+\frac{\alpha^2}{2D_c}\left(\frac{\partial\widetilde{n^2}}{\partial\beta^*}+\frac{\partial\tilde{n}}{\partial\beta^*}\right)\right]+ \\ +\frac{\partial\tilde{n}}{\partial\beta^*}\ln 2\beta^*+\frac{2\tilde{n}}{\beta^*}\Bigg\}=0, \tag{4.119}$$

$$\frac{\mathrm{th}\,\beta^*\omega\,\mathrm{ch}^2\,\beta^*\omega+\beta^*\omega}{2\sqrt{\omega\,\mathrm{th}\,\beta^*\omega}\,\mathrm{ch}^2\,\beta^*\omega}+\sqrt{\omega\,\mathrm{th}\,\beta^*\omega}\Bigg\{-x^2\,\mathrm{th}\,\beta^*-\frac{\beta^*\omega x^2}{\mathrm{ch}^2\,\beta^*\omega}+ \\ +\beta D_c\left[-\frac{\sqrt{2}\alpha}{\sqrt{D_c}}\frac{\partial\tilde{n}}{\partial\omega}+\frac{\alpha^2}{2D_c}\left(\frac{\partial\widetilde{n^2}}{\partial\omega}+\frac{\partial\tilde{n}}{\partial\omega}\right)\right]+\frac{\partial\tilde{n}}{\partial\omega}\ln 2\beta^*\Bigg\}=0. \tag{4.120}$$

Finally, the derivatives of the mean functions are defined by the Eqs. (4.115).

4.5 Calculation of Physical Characteristics

As has been demonstrated in the previous section, the use of OM and CE result in a correct description of the mixed quantum states, and the analytical expressions for mean values of the observed physical parameters can be obtained within the quantum statistics. In this section, we provide further examples how the above presented technique can be used for calculation of physical characteristics, and we start with the coefficient of thermal expansion: the parameter defined by the mean value of the operator of atomic coordinates. For symmetric potential, this value equals zero, and therefore we consider here the QAO with the cubic term in the potential as in Sect. 3.2. The mean value of the coordinate operator of non-interacting two-atomic molecules with the Hamiltonian (3.15) and for the interatomic potential with a canonic distribution on the oscillating degrees of freedom of the molecule is written as:

$$\bar{x}(\beta) = \frac{1}{Z} \sum_{n=0}^{\infty} u_n \exp\left[-\beta E_n\right]. \tag{4.121}$$

This quantity defines the average distance between the atoms and the average size of the molecule as a function of the temperature. In the description of real gases, this value is used to take into account the additional dependence of the gas kinetic coefficients on the temperature. These coefficients take a part in the transfer equations due to their proportionality to $\bar{x}(\beta)^2$ [20]. Moreover, adopting the Van-der-Waals law for real gases [13], the average size of the molecule also determines the critical parameters P_0, V_0, T_0. In particular, using the explicit expressions for the critical parameters [13], the corrections for these parameters due to asymmetry of the potential are expressed through $\bar{x}(\beta)$ in the following way:

$$\frac{\Delta T_0}{T_0} \simeq -\frac{\bar{x}(\beta_0)}{R_0}, \quad \frac{\Delta V_0}{V_0} \simeq \frac{\bar{x}(\beta_0)}{R_0}, \quad \frac{\Delta P_0}{P_0} \simeq -2\frac{\bar{x}(\beta_0)}{R_0}. \tag{4.122}$$

Another application of the approximate methods for calculation of physical thermodynamical characteristics is a microscopical evaluation of the coefficient of the linear expansion of crystals. According to crystallographical chemistry [21], the distance between neighbor atoms inside the real crystals is almost equal to the equilibrium distance between the atoms in two-atomic molecule composed of the same atoms. Because of the distance between the atoms in crystal makes the linear dimension of the elementary crystallographic unit cell, the parameter $\bar{x}(\beta)$ defines well the coefficient of the thermal expansion α of solid state at temperatures far from the melting point, when the expansion is governed by linear law.

In general case, it is difficult to derive the analytical expression for the coefficient of linear expansion for Hamiltonian (3.15) because of essentially non-linear dependence of the parameters u_n and ω_n on quantum number n. The situation, however, is simplified in the limiting case of low temperatures ($\beta \to \infty$), where the linear law of expansion is valid. Indeed, simplifying the Eqs. (4.68), (4.69), and (4.121) and neglecting the terms of the second order, we obtain:

$$\begin{aligned}
&\beta \to \infty, \quad \bar{n} \ll 1, \\
&\omega_0^3 - \left[1 + 6u_0(\gamma + \lambda u_0)\right] \omega_0 - 6\lambda = 0, \\
&u_0^3 + \frac{3\gamma}{4\lambda} u_0^2 + \frac{\omega_0 + 6\lambda}{4\lambda\omega_0} u_0 + \frac{3\gamma}{8\lambda\omega_0} = 0, \\
&\overline{E} \simeq \frac{1}{4}\left(\omega_0 + \frac{1}{\omega_0}\right) + \frac{u_0^2}{2} + \gamma u_0 \left[u_0^2 + \frac{3}{2\omega_0}\right] +
\end{aligned} \tag{4.123}$$

$$
\begin{aligned}
&+\lambda\left[u_0^4+\frac{3u_0^2}{\omega_0}+\frac{3}{4\omega_0^2}\right],\\
&\bar{n}\simeq e^{-\beta K},\quad K=\frac{1}{2}\left(\omega_0+\frac{1}{\omega_0}\right)+\frac{3\gamma u_0}{\omega_0}+\frac{3\lambda}{\omega_0}\left[2u_0^2+\frac{1}{\omega_0}\right],\\
&\bar{x}\simeq u_0,
\end{aligned}
$$

Because for the energy ground state the parameters u_0, ω_0 take on a non-zero constant value determined by the Eqs. (4.123), they can be represented as:

$$
\begin{aligned}
u_n &= u_0\left(1+\frac{\delta}{\beta}\right),\\
\omega_n &= \omega_0\left(1+\frac{\mu}{\beta}\right),
\end{aligned}
\tag{4.124}
$$

where the terms of the second order are shown in brackets. Substituting these parameters in the Eqs. (4.69), and using (4.123), the following equations are obtained:

$$
\begin{aligned}
&\frac{\delta}{\beta}\left[3u_0^3+\frac{3\gamma u_0^2}{2\lambda}+\frac{u_0}{4\lambda}+\frac{3}{2\omega_0}\right]-\frac{\mu}{\beta}\left[\frac{3u_0}{2\omega_0}+\frac{3\gamma}{8\lambda\omega_0}\right]+\\
&+\bar{n}\left[\frac{3\lambda u_0}{\omega_0}+\frac{3\gamma}{4\lambda\omega_0}\right],\\
&\frac{\mu}{\beta}\left[\omega_0(\omega_0^2-1)\right]-\frac{\delta}{\beta}\left[6u_0\omega_0(\gamma+2\lambda u_0)\right]-12\lambda\bar{n}=0.
\end{aligned}
\tag{4.125}
$$

The coefficients are then found from the solution of the linear equations above:

$$
\begin{aligned}
\frac{\delta}{\beta} &= \frac{CE-GB}{AE-BD}\bar{n},\\
\frac{\mu}{\beta} &= \frac{CD-GA}{AE-BD}\bar{n},
\end{aligned}
\tag{4.126}
$$

where the following notations are introduced for constant quantities:

$$
\begin{aligned}
A &= 3u_0^3+\frac{3\gamma u_0^2}{2\lambda}+\frac{u_0}{4\lambda}+\frac{3}{2\omega_0},\\
B &= \frac{3u_0}{2\omega_0}+\frac{3\gamma}{8\lambda\omega_0},\\
C &= \frac{3\lambda u_0}{\omega_0}+\frac{3\gamma}{4\lambda\omega_0},
\end{aligned}
\tag{4.127}
$$

$$D = 6u_0\omega_0(\gamma + 2\lambda u_0),$$
$$E = \omega_0(\omega_0^2 - 1), \qquad G = 12\lambda.$$

Using the expression for the mean number of the excited states from (4.123) and notations (4.124), the explicit functional of the mean operator of coordinate depending on the Hamiltonian's parameters and inverse temperature is:

$$\bar{x} = u_0\left[1 + \frac{CE - GB}{AE - BD}e^{-\beta K}\right], \tag{4.128}$$

and the coefficients of the thermal expansion depending on Hamiltonian's parameters and inverse temperature are:

$$\alpha(\beta) = \beta\frac{CE - GB}{AE - BD}e^{-\beta K}, \tag{4.129}$$

where the constant coefficients are taken from formulas (4.123) and (4.127).

In an analogous way, the coefficient of the thermal expansion can be calculated for arbitrary temperature. The algorithm is as follows: (i) for a certain temperature T_0, the mean number of excitations $\bar{n}$ is computed using (4.69); (ii) using the second and third equations from (4.69), the parameters $u_{\bar{n}}$ and $\omega_{\bar{n}}$ are calculated, which correspond to the mean number of excitations; (iii) in a small vicinity of the temperature T_0, make the parameter u to vary linearly with the change of temperature:

$$u(T) = u(T_0)\left(1 + \alpha(T_0)(T - T_0)\right), \tag{4.130}$$

(iv) the variational equations (4.69) are expanded into series over $\alpha(T)(T - T_0)$ in the vicinity of T_0 with the accuracy of the second order; (v) the coefficient of thermal expansion is calculated from the linear equations system obtained in the step (iv). This local determination of the coefficient of thermal expansion makes it possible to numerically simulate it at any temperature and for any parameters of Hamiltonian, i.e. to construct a uniformly suitable approximation.

A next example of the calculation of physical characteristics of quantum systems is a determination of the mean value of the energy of the system with known Hamiltonian, using QAO [2]. For this purpose, we use the numerical values for the QAO spectrum $E_n^{(num)}$ [10], and the values obtained in the lower order of the perturbation theory $E_n^{(PT)}$ and operator method $E_n^{(OM)}$ as well as the cumulant expansion. Thus, the following approximations are considered for this observed physical parameter:

$$\overline{E^{(num)}} = \frac{1}{Z_A^{(nm)}}\sum_n E_n^{(nm)}\exp\left[-\beta E_n^{(nm)}\right],$$

$$\overline{E^{(OM)}} = \frac{1}{Z_A^{(0)}}\sum_n E_n^{(0)}\exp\left[-\beta E_n^{(0)}\right],$$

$$\overline{E^{(PT)}} = \frac{1}{Z_A^{(TB)}} \sum_n E_n^{(TB)} \exp\left[-\beta E_n^{(TB)}\right],$$

$$\overline{E^{(CE)}} = -\frac{1}{Z^{(KP)}} \frac{\partial}{\partial \beta} Z^{(KP)}(\beta). \tag{4.131}$$

The numerical results of these approximations are presented in the Fig. 4.20, which demonstrates the uniform suitability of CE for mean energy values of the QAO ensemble.

Finally, we calculate in this section the partition function of the ideal gas consisting on two-atoms molecules possessing the limited number n of the coupled states:

$$Z = \sum_{k=0}^{n} \left\langle \Psi_k \left| \exp(-\beta \hat{H}) \right| \Psi_k \right\rangle, \tag{4.132}$$

where $|\Psi_k\rangle$ are the eigenvectors of the Hamiltonian of the molecule $\hat{H}$:

$$\hat{H} |\Psi_k\rangle = E_k |\Psi_k\rangle, \tag{4.133}$$

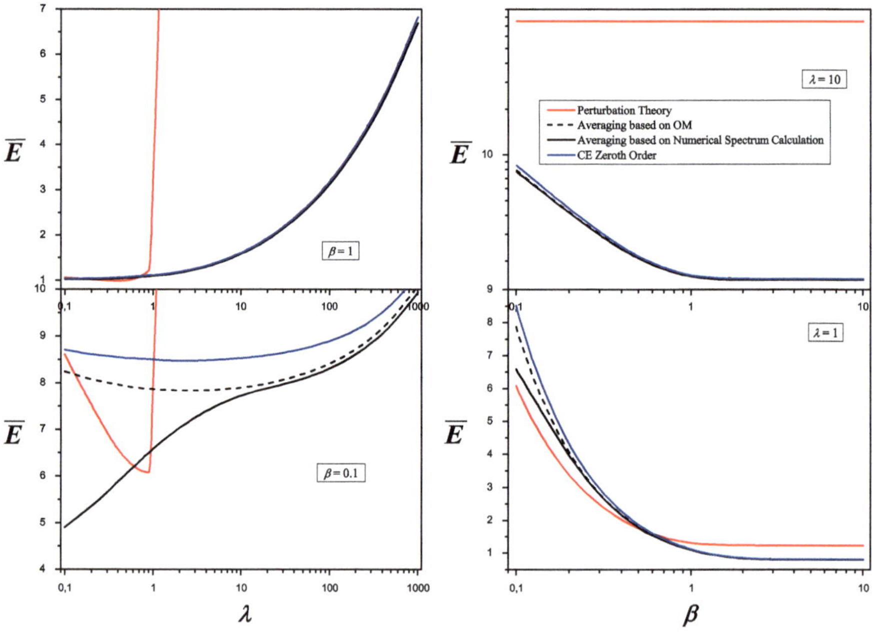

Fig. 4.20 A mean energy value of QAO depending on the anharmonic parameter λ for different inverse temperatures β (left panel) and as a dependence on the inverse temperature β for different values of the anharmonic parameter λ (right panel)

and $\beta = 1/kT$ is the inverse temperature. Introducing the normalized state vector:

$$|v\rangle = \sqrt{C}\sum_{k=0}^{n}\sqrt{v_k}\,|\Psi_k\rangle\,, \qquad 0 \le v \le 1$$

$$\langle v|\, v\rangle = 1, \qquad C(v) = \frac{1-v}{1-v^{n+1}}, \tag{4.134}$$

the PF is represented in the form:

$$Z(\beta) = \left\langle v \left| e^{-\hat{R}} \right| v \right\rangle,$$

$$\hat{R} = \beta\hat{H} + \hat{k}\ln v + \ln C(v), \tag{4.135}$$

where the operator $\hat{k}$ is defined by the equation:

$$\hat{k}\,|\Psi_k\rangle = k\,|\Psi_k\rangle\,.$$

To approximately calculate the partition function, the zeroth order CE is used to estimate the average in the formula (4.132):

$$Z \simeq \exp\left[-\varphi(\beta, v)\right], \tag{4.136}$$

$$\varphi(\beta, v) = \beta C\sum_{k=0}^{n} v^k E_k + \left[\frac{v}{1-v} - \frac{(n+1)v^{n+1}}{1-v^{n+1}}\right]\ln v -$$

$$- \ln\frac{1-v}{1-v^{n+1}}.$$

The accuracy of the estimate (4.136) depends on two factors: (i) accuracy of the eigenvalues of Hamiltonian, (ii) optimal choice of the parameter v. Here we discuss only the influence of the second factor for the system with a finite number of the levels as in the case of real molecules. Therefore, the Morse model is used for the interatomic potential, which delivers the exact eigenvalues E_n with respect to the anharmonicity and the upper limitation of the spectrum:

$$E_n = \left[\kappa\sqrt{\frac{2D}{m}} - \frac{\kappa^2}{2m}\left(n+\frac{1}{2}\right)\right]\left(n+\frac{1}{2}\right), \tag{4.137}$$

where D and κ are the parameters of the Morse potential ($\hbar = c = 1$):

$$U(x) = D\left(1-e^{-\kappa x}\right)^2,$$

and the coordinate $x = R - R_0$ is defined with the center in the equilibrium point R_0 of the atoms with the mass m. The function $\varphi(\beta, v)$ in (4.136) is calculated analytically:

$$\begin{aligned}\varphi(\beta, v) = {} & \beta\kappa\sqrt{\frac{2D}{m}}\left[\frac{v}{1-v} - \frac{(n+1)v^{n+1}}{1-v^{n+1}} + \frac{1}{2}\right] - \\ & - \beta\frac{\kappa^2}{2m}\left[\frac{2v}{(1-v)^2} - \frac{(1+v)}{(1-v)}\frac{(n+1)v^{n+1}}{(1-v^{n+1})} - \right. \\ & \left. - \frac{(n+1)^2 v^{n+1}}{1-v^{n+1}} + \frac{1}{4}\right] + \\ & + \left[\frac{v}{1-v} - \frac{(n+1)v^{n+1}}{1-v^{n+1}}\right]\ln v - \ln\frac{1-v}{1-v^{n+1}},\end{aligned} \tag{4.138}$$

and the parameter v follows from the extremum condition for the right side of the equation:

$$\frac{\partial}{\partial v}\varphi(\beta, v) = 0. \tag{4.139}$$

Equations (4.138) and (4.139) specify the parametric approximation of the function $Z(\beta)$. However, the real computations show the equivalence up to $\sim 99\,\%$ of the solutions for these equations at $n \geq 5$, which applicable to real molecules, to the solutions of the equations in the limit $n \to \infty$, which are considerably easier to solve:

$$\begin{aligned}\varphi(\beta, v) = {} & \frac{1}{2}\beta\kappa\sqrt{\frac{2D}{m}}\frac{(1+v)}{(1-v)} - \beta\frac{\kappa^2}{2m}\left[\frac{2v}{(1-v)^2} + \frac{1}{4}\right] - \\ & - \frac{v}{1-v}\ln v - \ln(1-v),\end{aligned} \tag{4.140}$$

$$\frac{1}{\beta}\ln v = \frac{\kappa^2}{m}\frac{(1+v)}{(1-v)} - \kappa\sqrt{\frac{2D}{m}}.$$

The approximations derived above can also be used for UAA of other thermodynamical characteristics of gas, for example, an average energy of the molecule $(n \gg 1)$:

$$\overline{E} = \frac{\partial\varphi}{\partial\beta} = \frac{\kappa}{2}\sqrt{\frac{2D}{m}}\frac{(1+v)}{(1-v)} - \frac{\kappa^2}{2m}\left[\frac{2v}{(1-v)^2} + \frac{1}{4}\right], \tag{4.141}$$

and for heat capacity:

$$C_V(\beta) = \left[\kappa\sqrt{\frac{2D}{m}} - \frac{\kappa^2}{m}\frac{(1+v)}{(1-v)}\right]\frac{v\ln v}{\left[2v\frac{\kappa^2}{m} - \frac{1-v^2}{\beta}\right]}, \tag{4.142}$$

where the parameter v is calculated from the second equation in (4.140).

The presented in this chapter methods can be used for calculation of the atomic deviations from the equilibrium state $\bar{x}$, which is caused by the anharmonicity of the potential and is not reduced to the derivative of PF:

$$\bar{x} = \frac{1}{Z}\sum_{k=0}^{n} x_k e^{-\beta E_k}, \tag{4.143}$$

where x_k is a diagonal matrix element of the coordinate operator. To estimate such a sum quantity, the use of the ansatz $|v_k\rangle$ is very beneficial, which has more general form than (4.134):

$$|v_k\rangle = \sqrt{C_k}\sum_{k=0}^{n}\sqrt{x_k v^k}\,|\Psi_k\rangle\,. \tag{4.144}$$

For Morse potential this yields:

$$x_k = \frac{1}{\kappa}\left[\Psi\left(2\frac{\sqrt{2mD}}{\kappa}\right) - \Psi\left(2\frac{\sqrt{2mD}}{\kappa} - k\right)\right],$$

with $\Psi(z)$ as a logarithmic derivative of Γ-function [15]. Because for the real molecules the condition is satisfied:

$$\frac{\sqrt{2mD}}{\kappa} \gg 1$$

the expression for x_k is substantially simplified:

$$x_k \simeq \frac{\kappa}{2\sqrt{2mD}}\left(k + \frac{1}{2}\right),$$

and produces the following estimate for $\bar{x}$:

$$\bar{x} = \frac{1}{4\sqrt{2mD}}\frac{(1+v)}{(1-v)}. \tag{4.145}$$

Table 4.1 compares the thermodynamical characteristics of the ideal gas of molecules N_2, calculated by zeroth order OM and CE, and exact values

Table 4.1 Thermodynamical characteristics of the ideal gas of molecules N_2. $m = 7.004$ a.e.m., $\kappa = 4.28 \times 10^7$ cm^{-1}, $D = 79890$ cm^{-1}, $n = 56$, $\omega_0 = \frac{\kappa}{2\pi C}\sqrt{\frac{2D}{m}} = 2358$ cm^{-1}, $\omega_0 x_0 = \frac{\hbar\kappa^2}{4\pi m C} = 14.14$ cm^{-1}, $T_0 = \frac{hC\omega_0}{k} = 3391$ K

	T, K				
	1000	2000	5000	10000	20000
$v(n \to \infty)$	0.035	0.189	0.521	0.732	0.868
Z	0.191	0.529	1.474	3.053	6.243
Z_{ex}	0.191	0.529	1.474	3.053	6.431
Z_h	0.190	0.525	1.447	2.935	5.891
$E - 0.5$	0.035	0.228	1.059	2.608	5.977
$E_{ex} - 0.5$	0.035	0.228	1.059	2.607	5.987
$E_h - 0.5$	0.035	0.225	1.031	2.477	5.410
C_V	0.424	0.807	1.002	1.081	1.253
$C_{V,ex}$	0.414	0.807	1.004	1.091	1.254
$C_{V,h}$	0.362	0.791	0.963	0.990	0.997
$\bar{x} \cdot 10^3$ Å	9.287	12.696	27.610	57.462	135.120

(subscript *ex*) and frequently used approximate values computed with harmonic interatomic potential (subscript *h*). The parameters of the Morse potential are taken from [22].

References

1. I.D. Feranchuk, A.A. Ivanov, J. Phys. A: Math. Gen. **37**, 9841 (2004)
2. A.A. Ivanov, I.D. Feranchuk, Proc. Natl. Acad. Sci. Belarus, Ser. Phys. Math. Sci (in Russian) **4**, 64 (2004)
3. I.D. Feranchuk, A.A. Ivanov, *Fluctuating Path and Fields*, vol. 1 (World Scientific, Singapore, 2001), p. 329.
4. H. Cramer, *Mathematical Methods of Statistics* (Princeton University Press, Princeton, 1999)
5. R.P. Feynman, A.R. Hibbs, *Quantum Mechanics and Path Integrals* (McGraw-Hill, New York, 1965)
6. A.N. Malakhov, *Cumulant Analysis of Random NonGaussian Processes (in Russian)* (Nauka, Moscow, 1978)
7. C. Domh, J.L. Lebowitz, *Phase Transitions and Crytical Phenomena* (Academic Press, New York, 1991)
8. Y.V. Nazamov, *Quantum Noise in Mesoscopic Physics* (NATO Science Series, Dordrecht, 2003)
9. I.D. Feranchuk, L.I. Komarov, I.V. Nechipor, A.P. Ulyanenkov, Ann. Phys. (N Y) **238**, 370 (1995)
10. F.T. Hioe, D. MacMillen, E.W. Montroll, Phys. Rep. **335**, 307 (1978)
11. H. Kleinert, *Path Integrals in Quantum Mechanics, Statistics, Polymer Physics, and Financial Markets*, 5th edn. (World Scientific, Singapore, 2009)
12. H. Meissner, E.O. Steinborn, Phys. Rev. A **56**, 1189 (1997)
13. L.D. Landau, E.M. Lifshitz, *Statistical Physics* (Nauka, Moscow, 1976)

14. I.S. Gradshtein, I.M. Ryzhik, *Tables of Integrals, Sums, Series and Products* (FIZMATGIZ, Moscow, 1963)
15. M. Abramowitz, I.A. Stegun, *Handbook of Mathematical Functions* (Dover, New York, 1964)
16. I.D. Feranchuk, V.S. Kuz'min, A.P. Ulyanenkov, Chem. Phys. **157**, 61 (1991)
17. I.D. Feranchuk, A.L. Tolstik, J. Phys. A Math. Gen. **32**, 2115 (1999)
18. L.D. Landau, E.M. Lifshitz, *Quantum Mechanics* (Fizmatgiz, Moscow, 2004)
19. I.D. Feranchuk, A.A. Ivanov, Nonlinear Phenom. Complex Syst. **7**, 377 (2004)
20. L.D. Landau, E.M. Lifshitz, *Physical Kinetics* (Nauka, Moscow, 1979)
21. B.G. Bokii, *Crystallochemie (in Russian)* (Nauka, Moscow, 1971)
22. K.P. Huber, G. Gerzberg, *Molecular Spectra and Molecular Structure. V.4 Constants of Diatomic Molecules* (Van Rostrand Reinhold, New York, 1979)

Chapter 5
Quantum Systems with Several Degrees of Freedom

The multi-dimensional physical systems with low number of the degrees of freedom take a special place in the development of new analytical and approximate methods for theoretical physics. The approaches used for the systems with the large number of the degrees of freedom and applied to the statistical physics and to the theory of the quantum field are not useable for multi-dimensional systems. For the latter, the alternative approaches are used, which split the variables in Schrödinger equation. The Born-Oppenheimer [1] technique, called also the adiabatic approximation, is one of the important members of these methods, which separates the variables for fast and slow sub-systems. Another member of this class of methods is one-particle approximation (Hartree or Hartree–Fock methods), where the wave function of the physical system is approximated by the product of the wave functions of single particles [2] .

For many physical phenomena, the non-adiabatic and correlational multi-particle effects are essential and have to be taken into account already in zeroth approximation. These effects may play a crucial role in the correct classification of the energy levels or in qualitative analysis of the systems near the degeneracy points of eigenvalues with different quantum numbers. Therefore, the methods are demanded, which deliver the approximate solutions of Schrödinger equation in the entire range parameters without assumption of negligibility for correlations and non-adiabaticity.

Operator method finds an analytical and uniformly suitable solution for collective energy levels and classifications for any quantum numbers and Hamiltonian parameters. The similar results after the standard perturbation theory, adiabatic expansion or method of the strong coupling interaction can be obtained as limiting cases from the algebraic formulas for OM zeroth approximation. Moreover, the successive approximations of OM improve the results and converge to the exact values for the energy and the level width.

This chapter illustrates the application of OM for the description of the systems with several degrees of freedom: (1) two coupled quantum anharmonic oscillators (CQAO), and (2) non-relativistic quantum theory of atom with two electrons.

I. Feranchuk et al., *Non-perturbative Description of Quantum Systems*,
Lecture Notes in Physics 894, DOI 10.1007/978-3-319-13006-4_5

The model of CQAO possesses numerous properties, which are typical for multi-dimensional systems and therefore is frequently used to testify the non-perturbative methods [3, 4]. The theory of two-electrons atom is a canonic problem of the quantum mechanics [5], and OM is shown to construct a new form of the perturbation theory accounting the correlational corrections.

5.1 Analytical Approximation for the Energy Levels of CQAO

The system of coupled harmonic oscillator (CHO) has been used in [6] to analyze the applicability of adiabatic and one-particle approaches to the description of multi-particle systems. The dimensionless form of the Hamiltonian for CHO is written as:

$$\hat{H} = \frac{1}{2}\hat{p}_x^2 + \frac{1}{2M}\hat{p}_y^2 + \frac{1}{2}\left(x^2 + y^2\right) + \lambda\, xy, \tag{5.1}$$

where M is a parameter for the ratio between the oscillators masses; λ is dimensionless parameter of the interaction. The classic trajectories of this system are described by Lissajous curves, which in quantum case are related to the non-trivial dependence of energy levels on the parameters of Hamiltonian. The exact eigenvalues follow from the formula:

$$E_{nm} = \nu_1 \left(n + \frac{1}{2}\right) + \nu_2 \left(m + \frac{1}{2}\right), \tag{5.2}$$

where the frequencies of normal oscillations $\nu_{1,2}$ are:

$$\nu_{1,2}^2 = \frac{1}{2M}(1 + M \pm \sqrt{(1 - M)^2 + 4\lambda^2 M}). \tag{5.3}$$

As follows from the formula (5.2), the energy, being an analytical function in complex plane of the parameters λ and M, has the singularities (branching points). The series over these parameters have a limited convergence radius, and this fact is a reason of the restriction for the application of the perturbation theory. Due to the same reason, the one-particle (Hartree) approximation is not also applicable to interpret the coupling of the oscillators. The wave function of zeroth approximation in this case is selected as a product of the one-particle wave functions (the symmetrization at $M = 1$ is obsolete for this analysis):

$$\Psi_{OA}(x, y) = \varphi_n(x)\chi_m(y);$$

$$\int \varphi_n(x)\varphi_m(x)\, dx = \int \chi_n(x)\chi_m(x)\, dx = \delta_{mn}. \tag{5.4}$$

The system of approximate equations for one-particle functions follows from the exact Schrödinger equation:

$$\left\{\frac{1}{2}\hat{p}_x^2 + \frac{1}{2}x^2 + \lambda\, x y_{mm} - \epsilon_n\right\} \varphi_n(x) = 0;$$
$$\left\{\frac{1}{2M}\hat{p}_y^2 + \frac{1}{2}y^2 + \lambda y x_{nn} - \epsilon_m\right\} \varphi_m(y) = 0. \tag{5.5}$$

and this equations system corresponds to the uncoupled harmonic oscillators with shifted equilibrium positions:

$$\bar{x} = -\lambda y_{mm}; \qquad \bar{y} = -\lambda x_{nn}.$$

The energy spectrum of the system in this approximation is:

$$E_{nm}^{(OA)} = n + \frac{1}{2} + \frac{1}{\sqrt{M}}\left(m + \frac{1}{2}\right) - \frac{1}{2}\lambda^2(y_{mm}^2 + x_{nn}^2) - \lambda x_{nn} y_{mm}, \tag{5.6}$$

and represents the primary terms of the power series over the parameter λ. The self-consistency of $\bar{x}, \bar{y}$ leads to:

$$\bar{x} = -\lambda \bar{y}, \qquad \bar{y} = -\lambda \bar{x}, \tag{5.7}$$

and for arbitrary λ we arrive at:

$$\bar{x} = \bar{y} = 0. \tag{5.8}$$

Thus, the zeroth order of one-particle approximation E_{mn}^{OA} differs essentially from the exact values. The correlational effects can be accounted in the successive orders, however, the one-particle approximation does not provide the uniformly suitable approximation for energy E in the entire range of the parameters λ and M. The adiabatic approximation for the Schrödinger equation with Hamiltonian (5.1) leads to a similar problem [6]. This case corresponds to the condition $M \gg 1$, and the operator $\hat{p}_y^2$ can be omitted in zeroth approximation. As a result, the adiabatic terms $\epsilon_n(y)$ are defined by the energy levels of the Hamiltonian's part depending on the fast variable x:

$$\epsilon_n(y) = n + \frac{1}{2} - \frac{1}{2}\lambda^2 y^2. \tag{5.9}$$

These quantities play a role of the potential energy in Schrödinger equation for slow oscillator y in the successive order of the adiabatic approximation. The resulting energy spectrum with the accuracy up to the second order on $\frac{1}{\sqrt{M}}$ is:

$$E_{nm}^{(AA)} = n + \frac{1}{2} + \sqrt{\frac{1-\lambda^2}{M}}(m + \frac{1}{2}). \tag{5.10}$$

The comparison of this expression with the formula (5.2) demonstrates the non-uniform suitability of the adiabatic approximation for this simple system. In opposite, the OM zeroth approximation provides the uniformly suitable expression. According to routine OM algorithm described in Chap. 3, the set of normalized state vectors is introduced, which depend on two variables and expressed through the creation a^+, b^+ and annihilation a, b operators:

$$|nm> = \frac{(a^+)^n}{\sqrt{n!}} \frac{(b^+)^m}{\sqrt{m!}} |00>; \; a|00> = b|00> = 0. \tag{5.11}$$

Similarly to one-dimensional case (2.2), this selection of the basis corresponds to the transition from the coordinate representation to the one of particle number by using the canonic transformations:

$$\begin{aligned}
x &= x' \cos\alpha + y' \sin\alpha, \qquad y = y' \cos\alpha - x' \sin\alpha, \\
\hat{p}_x &= \hat{p}'_x \cos\alpha + \hat{p}'_y \sin\alpha, \quad \hat{p}_y = -\hat{p}'_x \sin\alpha + \hat{p}'_y \cos\alpha, \\
x' &= \frac{a + a^+}{\sqrt{2\omega_1}}, \qquad y' = \frac{b + b^+}{\sqrt{2\omega_2}}, \\
\hat{p}'_x &= -i(a - a^+)\sqrt{\frac{\omega_1}{2}}, \quad \hat{p}'_y = -i(b - b^+)\sqrt{\frac{\omega_2}{2}}.
\end{aligned} \tag{5.12}$$

The variational parameters $\omega_{1,2}$ for coordinates in the formulas (5.12) are introduced in a similar way to one-dimensional case. The parameter α for transformation of the operator's phases is related to the additional degree of freedom for two-dimensional systems. The transformation (5.12) in the considered case is equivalent to the transformation of Bogolyubov-Tyablikov [7], and is shown later to be applicable in OM for complex systems. The Hamiltonian (5.1) in a new representation is:

$$\begin{aligned}
\hat{H} = \frac{1}{4}\Big\{ &(a^2 + a^{+2})\left(\frac{1}{\omega_1} - \omega_1\right) + (b^2 + b^{+2})\left(\frac{1}{\omega_2} - \omega_2\right) + \\
&+(2a^+a + 1)\left(\frac{1}{\omega_1} + \omega_1\right) + (2b^+b + 1)\left(\frac{1}{\omega_2} + \omega_2\right)\Big\} + \\
&+\frac{\lambda}{2}\Big\{ \frac{\cos 2\alpha}{\sqrt{\omega_1\omega_2}}(ab + a^+b^+ + a^+b + ab^+) + \\
&\frac{\sin 2\alpha}{2}\Big[-\frac{1}{\omega_1}(a^2 + a^{+2} + 2a^+a + 1) + \\
&\frac{1}{\omega_2}(b^2 + b^{+2} + 2b^+b + 1)\Big]\Big\} +
\end{aligned}$$

$$\frac{1}{4}\left(1-\frac{1}{M}\right)[(a^{+}-a)^{2}\omega_{1}\sin^{2}\alpha+$$
$$(b^{+}-b)^{2}\omega_{2}\cos^{2}\alpha-(a^{+}-a)(b^{+}-b)\sqrt{\omega_{1}\omega_{2}}\sin 2\alpha]. \quad (5.13)$$

The zeroth approximation of the operator method corresponds to the part of Hamiltonian, which commutates with the operator of the excitation numbers $a^{+}a = \hat{n}$ and $b^{+}b = \hat{m}$:

$$\hat{H}_{0}=\frac{1}{4}\left[\left(\frac{1}{\omega_{1}}+\omega_{1}\right)(2\hat{n}+1)+\left(\frac{1}{\omega_{2}}+\omega_{2}\right)(2\hat{m}+1)\right]+$$
$$\frac{\lambda\sin 2\alpha}{4}\left[\frac{1}{\omega_{2}}(2\hat{m}+1)-\frac{1}{\omega_{1}}(2\hat{n}+1)\right]+$$
$$+\frac{1}{4}\left(\frac{1}{M}-1\right)[\omega_{1}(2\hat{n}+1)\sin^{2}\alpha+\omega_{2}(2\hat{m}+1)\cos^{2}\alpha]. \quad (5.14)$$

As shown for one-dimensional case in Sect. 2.1, the series over the operator $\hat{V}=\hat{H}-\hat{H}_{0}$ converges in the entire range of parameters λ, M and $\alpha, \omega_{1,2}$. However, the best zeroth approximation of OM corresponds to the independence condition of eigenvalues on the choice of the representation:

$$\frac{\partial E_{mn}}{\partial\omega_{1}}=\frac{\partial E_{mn}}{\partial\omega_{2}}=\frac{\partial E_{mn}}{\partial\alpha}=0, \quad (5.15)$$

where the energy $E_{mn}(\omega_{1},\omega_{2},\alpha)$ is determined from the formula (5.14) with substitution of operators by numbers m, n. Two equations (5.15) allow to find $\omega_{1,2}(\alpha)$ and exclude them from the expression for energy:

$$E_{mn}^{(OM)}(\alpha)=\nu_{1}(\alpha)\left(n+\frac{1}{2}\right)+\nu_{2}(\alpha)\left(m+\frac{1}{2}\right);$$
$$\nu_{1,2}^{2}=\frac{1}{2M}(1\pm\lambda\sin 2\alpha)[(M+1)\mp(M-1)\cos 2\alpha)]. \quad (5.16)$$

The eigenvalue (5.15) has an extremum, when the transformation parameter α equals to:

$$z_{1,2}=\frac{1}{2M\lambda}[(M-1)\pm\sqrt{(M-1)^{2}+4M\lambda^{2}}], \quad (5.17)$$

where $z = \tan\alpha$. The substitution of the quantities (5.17) into Eq. (5.14) demonstrates the good agreement of the energy levels $E_{mn}^{(OM)}$ with the exact values (5.2) in zeroth approximation of OM. The quadratic form of the Hamiltonian (5.1) emphasizes this coincidence, however, the uniform suitability of the eigenvalues calculated by operator method remains unchanged for complex systems, too. We consider here the general form of the asymmetric Hamiltonian CQAO [8]:

$$\hat{H} = \frac{1}{2}\hat{p}_x^2 + \frac{1}{2}\hat{p}_y^2 + \frac{1}{2}(x^2 + \Omega^2 y^2) + \lambda xy + Ax^4 + By^4 + Cx^2y^2. \tag{5.18}$$

Here A, B, C are the dimensionless parameters ($A, B > 0; C > -2\sqrt{AB}$), and additionally to the quadratic interaction between oscillators, the linear component is introduced, which is essential for the modeling of several molecular and atomic potentials [9]. Following OM algorithm, the zeroth approximation is constructed by transition to the secondary quantization in Hamiltonian (5.18) using the transformation (5.12) and by separation of zeroth order operator $\hat{H}_0$ as a part of $\hat{H}$, which commutates with the operators of particle numbers $\hat{n}_a$ and $\hat{n}_b$. As a result, the operator $\hat{H}_0$ is written as:

$$\begin{aligned} \hat{H}_0 = \frac{1}{2}\left(\omega_1 + \frac{u^2 + \Omega^2 v^2}{\omega_1}\right)\left(\hat{n}_a + \frac{1}{2}\right) + \\ \frac{1}{2}\left(\omega_2 + \frac{v^2 + \Omega^2 u^2}{\omega_2}\right)\left(\hat{n}_b + \frac{1}{2}\right) \\ + \frac{\lambda uv}{\omega_1\omega_2}\left[\omega_1\left(\hat{n}_b + \frac{1}{2}\right) - \omega_2\left(\hat{n}_a + \frac{1}{2}\right)\right] + \\ + \frac{3}{2\omega_1^2}\left(\hat{n}_a^2 + \hat{n}_a + \frac{1}{2}\right)(Au^4 + Bv^4 + Cu^2v^2) + \\ + \frac{3}{2\omega_2^2}\left(\hat{n}_b^2 + \hat{n}_b + \frac{1}{2}\right)(Av^4 + Bu^4 + Cu^2v^2) + \\ \frac{1}{\omega_1\omega_2}\left(\hat{n}_a + \frac{1}{2}\right)\left(\hat{n}_b + \frac{1}{2}\right)[2u^2v^2(3A + 3B - 2C) + \\ C(u^4 + v^4)], \end{aligned} \tag{5.19}$$

where

$$u = \cos\alpha; \qquad v = \sin\alpha,$$

and the oscillator frequencies are considered as variational parameters. The eigenfunctions of the operator (5.19) are identical to the eigenvectors of the operators of particle numbers:

$$\hat{n}_a|N, M >= N|N, M >; \qquad \hat{n}_b|N, M >= M|N, M >, \tag{5.20}$$

where the eigenvalues $E_{NM}^{(0)}(\omega_1, \omega_2, \alpha)$ are found from formula (5.19) by substitution of quantum numbers N, M instead of corresponding operators. However, these eigenvalues play an intermediate role and lead to OM zeroth approximation only in case, if the parameters ω_1, ω_2 and α are calculated from the condition of extremum of the energy levels (5.19) for arbitrary quantum state. After this procedure, the eigenvalues become the cumbersome functions of quantum numbers and parameters of Hamiltonian, and namely this fact pre-defines the uniform suitability of OM zeroth approximation. The analytical approximation for energy levels of CQAO is then found as follows:

$$\begin{aligned} E_{NM}^{(OM)} &= \frac{1}{4}\left(3\omega_1 + \frac{u^2+\Omega^2 v^2}{\omega_1}\right)\left(N+\frac{1}{2}\right) + \\ &\frac{1}{4}\left(3\omega_2 + \frac{v^2+\Omega^2 u^2}{\omega_2}\right)\left(M+\frac{1}{2}\right) + \\ &+\frac{\lambda uv}{2\omega_1\omega_2}\left[\omega_1\left(M+\frac{1}{2}\right) - \omega_2\left(N+\frac{1}{2}\right)\right]. \end{aligned} \tag{5.21}$$

The non-trivial dependence of energy on quantum numbers is defined by the parameters $\omega_1, \omega_2, \alpha$, which are the solutions of the following algebraic equations:

$$\begin{aligned} &\omega_1^3 - \omega_1\left\{u^2 + \Omega^2 v^2 - 2\lambda uv + \right. \\ &\left.\frac{4}{\omega_2}\left(M+\frac{1}{2}\right)\left[(3A+3B-2C)u^2v^2 + \frac{C}{2}\left(u^4+v^4\right)\right]\right\} - \\ &\frac{6(2N^2+2N+1)}{2N+1}(Au^4 + Bv^4 + Cu^2v^2) = 0. \end{aligned} \tag{5.22}$$

$$\begin{aligned} &\omega_2^3 - \omega_2\left\{v^2 + \Omega^2 u^2 + 2\lambda uv + \right. \\ &\left.\frac{4}{\omega_1}\left(N+\frac{1}{2}\right)\left[(3A+3B-2C)u^2v^2 + \frac{C}{2}\left(u^4+v^4\right)\right]\right\} - \\ &\frac{6(2M^2+2M+1)}{2M+1}(Av^4 + Bu^4 + Cu^2v^2) = 0. \end{aligned} \tag{5.23}$$

$$\begin{aligned} &\left[\omega_1\left(M+\frac{1}{2}\right) - \omega_2\left(N+\frac{1}{2}\right)\right](1-\Omega^2+2\lambda\cot 2\alpha) + \\ &\frac{3\omega_2}{\omega_1}(2N^2+2N+1)[B-(A+B)\cos^2\alpha + \frac{C}{2}\cos 2\alpha] + \\ &\frac{3\omega_1}{\omega_2}(2M^2+2M+1)[A-(A+B)\cos^2\alpha + \frac{C}{2}\cos 2\alpha] + \\ &12\left(N+\frac{1}{2}\right)(M+\frac{1}{2})(A+B-C)\cos 2\alpha = 0. \end{aligned} \tag{5.24}$$

In case of identical oscillators ($A = B, \Omega = 1$), the Hamiltonian (5.18) obtains an additional motion integral, which is due to the transpositional symmetry. According to the rules of Sect. 2.8, the exact motion integrals have to be accounted in OM zeroth approximation to correctly describe the condition of degeneracy for energy spectrum. Partially this degeneracy is removed due to the transformation by the coordinate rotation, which is governed by the parameter α and leads to the term in formula (5.22), which is asymmetric with respect to the transposition $N \leftrightarrow M$. This term is proportional to the parameter of linear coupling λ, and the degeneracy remains in the particular case of even degrees of the coordinate operators of oscillators [8]. The known algorithm for the solution of this problem consists of the construction of proper linear combination of degenerated wave functions and further creation of the perturbation operator on the basis of these functions [10]. This approach has been already considered for OM description of the two-level system in quantum field (Sect. 3.5). The modification of OM zeroth approximation within above approach means that the wave functions of collective states

$$|\Psi_{NM}^{(0)}>= C_1|N, M> + C_2|M, N>; \qquad N \neq M, \tag{5.25}$$

have to be used for calculation of the energy levels instead of the eigenfunctions of the operators $\hat{n}_a, \hat{n}_b$. The coefficients $C_{1,2}$ for arbitrary set of the quantum numbers become the additional variational parameters. Then the degeneracy is removed due to the terms in the perturbation operator $\hat{V} = \hat{H} - \hat{H}_0$, which are proportional to the quantities:

$$\hat{V}_{ab} = a^{+2}b^2 + a^2 b^{+2}, \tag{5.26}$$

commutating with the operator of a full particle number:

$$\hat{n} = \hat{n}_a + \hat{n}_b,$$

and mixing the states, which differ by the order of the quantum numbers N and M. The condition of the existence of non-zero coefficients $C_{1,2}$ lead to the following expression for energy levels of CQAO instead of formula (5.21):

$$E_{NM}^{\pm} = \frac{1}{2}(E_{NM}^{(OM)} + E_{MN}^{(OM)}) \pm \sqrt{\frac{1}{4}(E_{NM}^{(OM)} - E_{MN}^{(OM)})^2 + |<N, M|\hat{V}_{ab}|M, N>|^2} \tag{5.27}$$

with the matrix elements:

$$<N, M|\hat{V}_{ab}|M, N>= \frac{C}{4\omega_1\omega_2}\Big[(N+1)(N+2)\delta_{N,M-2} + (M+1)(M+2)\delta_{N,M+2}\Big], \tag{5.28}$$

where $\delta_{N,M}$ is a Kronecker symbol. The transformation parameters $\omega_1, \omega_2, \alpha$ are determined by the Eqs. (5.22)–(5.24).

The expressions (5.21)–(5.24), (5.27) are the principal result of this section. They deliver the algebraic representation for energy levels of CQAO, which is accurate in the entire range of Hamiltonian's parameters and quantum numbers, and thus is the uniformly suitable approximation. These formulas complement essentially the asymptotic and numerical methods when the real physical systems are qualitatively modeled by CQAO.

5.2 Comparison with Known Analytical and Numerical Results

The uniformly suitable approximation provided by formulas (5.21)–(5.24) and (5.27) means that this result has to agree with the analytical results by asymptotic series in all limiting cases. Moreover, the energy $E_{NM}^{(OM)}$ calculated by OM zeroth approximation is worth to compare with the numerically calculated one $E_{NM}^{(acc)}$ for intermediate values of parameters. Last but not least, the operator method calculates the corrections of the second order, estimates the accuracy of zeroth and successive approximations according to the definition (1.1).

The standard asymptotic series are meaningful in a narrow range of parameters, whereas OM zeroth approximation is an analytical function of the parameters in their entire range. By expanding these functions into power series, the coefficients of this series are not equal to ones of asymptotic series, and the difference between both is of the same order as the accuracy of OM zeroth approximation within the intermediate range of parameters.

(1) *Small anharmonicity and low excitation level* ($AN; CM \ll 1$, and for simplicity sake $\Omega = 1, B = \lambda = 0$).

The perturbation series with the accuracy up to the second order of A and C results in:

$$E_{NM}^{(PT)} \simeq N + M + 1 + \frac{3}{4}(2N^2 + 2N + 1)A +$$
$$\frac{1}{4}[(2N + 1)(2M + 1) \pm (N^2 + 3N + 2)\delta_{N,M-2} \pm$$
$$(M^2 + 3M + 2)\delta_{N,M+2}]C -$$
$$\frac{1}{8}(34N^3 + 51N^2 + 59N + 21)A^2 -$$
$$\frac{3}{4}(2M + 1)(2N^2 + 2N + 1)AC -$$

$$\frac{3}{16}[3\,NM(N+M+2)+\frac{3}{2}(N^2+M^2)+ \\ \frac{5}{2}(N+M)+1]C^2. \qquad (5.29)$$

The applicability of the perturbation theory is limited both by small values of the anharmonicity constants ($A, C \ll 1$) and low excitation levels. To analyze the OM zeroth approximation (5.27) in the above mentioned limit, the approximate solutions of the Eqs. (5.22)–(5.24) have to be found:

$$\omega_1 \simeq 1+\frac{p_1}{2}-\frac{p_1^2}{8}-\frac{p_1 p_2}{4}-\frac{p_1 q}{2}+\frac{q}{2}-\frac{3q^2}{8};$$

$$\omega_2 \simeq 1+\frac{p_2}{2}-\frac{p_2^2}{8}-\frac{p_1 p_2}{4}-\frac{p_2 q}{4};\alpha=0;$$

$$p_1=(2M+1)C;\; p_2=(2N+1)C;\; q=6A\frac{2N^2+2N+1}{2N+1}. \qquad (5.30)$$

By substituting these solutions into formula (5.27), the obtained expansion for $E_{NM}^{(\pm)}$ in the case of small anharmonicity has a similar form as $E_{NM}^{(PT)}$ in the expression (5.29), however, with different coefficients in the terms of the second order:

$$-\frac{9}{4}\frac{(2N^2+2N+1)^2}{2N+1}A^2;$$

$$-\frac{C^2}{8}[4NM(N+M+2)+2(N^2+M^2)+3N+3M+1]. \qquad (5.31)$$

(2) *Strong anharmonicity and quasi-classic approximation.*

The equivalent simplification of OM formulas can be performed both in the limit of *strong coupling* ($A, B, C \gg 1$) and in *quasi-classic approximation* ($N, M \gg 1$). The asymptotic expressions in the limit of strong coupling remains very cumbersome, and therefore we present these formulas here for the particular case only $A = B; \Omega = 1; \lambda = 0$. Starting from the equation for OM parameters:

$$\alpha=0;\quad u=1;\quad v=0;$$

$$\omega_1 \simeq \left[Cz(2M+1)+6A\frac{2N^2+2N+1}{2N+1}\right]^{1/3};$$

$$\omega_2 \simeq \left[\frac{C}{z}(2N+1)+6A\frac{2M^2+2M+1}{2M+1}\right]^{1/3}, \qquad (5.32)$$

where z is a positive solution of the following algebraic equation:

$$z^3 6A\frac{2M^2+2M+1}{2M+1} + z^2 C(2N+1) -$$
$$zC(2M+1) - 6A\frac{2N^2+2N+1}{2N+1} = 0, \quad (5.33)$$

the following approximate formula for energy is obtained:

$$E^{\pm)}_{NM} \simeq \frac{3\omega_1}{8}(2N+1) + \frac{3\omega_2}{8}(2M+1) \pm \frac{C}{4\omega_1\omega_2}$$
$$\{(N+1)(N+2)\delta_{N,M-2} + (M+1)(M+2)\delta_{N,M+2}\}. \quad (5.34)$$

In the case of non-interacting systems ($C = 0$), the energy for each oscillator is:

$$\epsilon_N \simeq \frac{3}{8}[6A(2N+1)^2(2N^2+2N+1)]^{1/3},$$

and as shown in (2.26) these quantities make a good approximation for principal term in asymptotic expansion of energy of one-dimensional anharmonic oscillator in the limit of strong coupling. There are no exact asymptotic formulas for the energy of interacting oscillators exist in the limit of strong coupling, however, a comparison of the calculations by formula (5.34) with numerical ones proves the small error (~3–5 %) for the coefficients of the series, which determine the function dependence of the energy levels on the parameters of Hamiltonian.

(3) *Adiabatic approximation* ($\Omega \ll 1$).

The Hamiltonian (5.18) is simplified to find exact asymptotic series and compare them with the results after OM. Assuming $A = B = \lambda = 0$; $\Omega y = Y$, the following Hamiltonian is considered:

$$\hat{H}_{ad} = \frac{1}{2}\hat{p}_x^2 + \frac{\Omega^2}{2}\hat{p}_Y^2 + \frac{1}{2}(x^2+Y^2) + \frac{C}{\Omega^2}x^2Y^2. \quad (5.35)$$

The adiabatic terms of fast oscillator (variable x) define the zeroth order of Born-Oppenheimer approximation:

$$\epsilon_N(Y) = \sqrt{1+2\frac{C}{\Omega^2}Y^2}\left(N+\frac{1}{2}\right) \simeq \frac{\sqrt{2C}}{\Omega}|Y|\left(N+\frac{1}{2}\right).$$

The oscillating spectrum of the slow oscillator (Y) is obtained in quasi-classic approximation, where the principal term of the asymptotic approximation is:

$$E^{(AD)}_{NM} \simeq (\frac{3\pi}{4})^{2/3}\left[C\left(N+\frac{1}{2}\right)^2\left(M+\frac{1}{2}\right)^2\right]^{1/3}. \quad (5.36)$$

A similar expansion follows from zeroth approximation of OM by using corresponding limiting transition in formulas (5.21)–(5.24), which results in:

$$E_{NM}^{(OM)} \simeq \left(\frac{6\sqrt{3}}{4}\right)^{2/3}\left[C\left(N+\frac{1}{2}\right)^2\left(M+\frac{1}{2}\right)^2\right]^{1/3}. \tag{5.37}$$

in good agreement with the expression (5.36).

(4) *Numerical simulations.*

Table 5.1 compares the analytical results by operator method with ones obtained by numerical computing in [3, 8] for intermediate region of Hamiltonian parameters and quantum numbers. In this table, the values $E^{(0)}$ are calculated using analytical formulas of OM zeroth approximation, and $E^{(A)}$ are simulated numerically in [3, 8]. Moreover, the quantities $E^{(2)} = E^{(0)} + E^{(2)}$ are displayed containing the successive correction of operator method $E^{(2)}$, which is calculated by using perturbation theory of the second order over the operator $\hat{V} = \hat{H} - \hat{H}_0$ and with operators $\hat{H}$ and $\hat{H}_0$, from formulas (5.18) and (5.19). More details about high order successive approximations by operator method are considered in the Chap. 2, and here we present the analytical expression for the second order:

Table 5.1 The energy of ground state E_{00} for CQAO $\lambda = 0; \Omega = 1; A = \mu a_{11}; B = \mu; C = 2\mu a_{12}$

a_{11}	μ	E	$a_{12} = 1.0$	$a_{12} = 0.0$	$a_{12} = -0.6$
	0.1	$E^{(0)}$	1.1432	1.1103	1.0889
		$E^{(2)}$	1.1410	1.1083	1.0862
		$E^{(A)}$	1.1409	1.1082	1.0861
0.8	1.0	$E^{(0)}$	1.7066	1.5852	1.5003
		$E^{(2)}$	1.6923	1.5712	1.4776
		$E^{(A)}$	1.6913	1.5689	1.4740
	10	$E^{(0)}$	3.2641	2.9620	2.7444
		$E^{(2)}$	3.2240	2.9208	2.6739
		$E^{(A)}$	3.2200	2.9118	2.6572
	0.1	$E^{(0)}$	1.1226	1.0874	1.0643
		$E^{(2)}$	1.1206	1.0859	1.0621
		$E^{(A)}$	1.1206	1.0859	1.0621
0.4	1.0	$E^{(0)}$	1.6263	1.4859	1.3824
		$E^{(2)}$	1.6130	1.4743	1.3575
		$E^{(A)}$	1.6123	1.4725	1.3545
	10	$E^{(0)}$	3.0597	2.6979	2.4175
		$E^{(2)}$	3.0207	2.6616	2.3266
		$E^{(A)}$	3.0175	2.6538	2.3057

$$E^{(2)}_{MN} = X_1 \left[\frac{1}{4} \left(\frac{1}{\omega_1} - \omega_1 \right) + 2a(2N-1) + c(2M+1) \right]^2 +$$
$$X_2 \left[\frac{1}{4} \left(\frac{1}{\omega_1} - \omega_1 \right) + 2a(2N+3) + c(2M+1) \right]^2 +$$
$$X_3 \left[\frac{1}{4} (\frac{\Omega^2}{\omega_2} - \omega_2) + 2b(2M-1) + c(2N+1) \right]^2 +$$
$$+X_4 \left[\frac{1}{4} \left(\frac{\Omega^2}{\omega_2} - \omega_2 \right) + 2b(2M+3) + c(2N+1) \right]^2 +$$
$$X_5 a^2 + X_6 b^2 + X_7 c^2, \tag{5.38}$$

where

$$a = \frac{A}{4\omega_1^2}; \quad b = \frac{B}{4\omega_2^2}; \quad c = \frac{C}{4\omega_1\omega_2};$$
$$X_1 = \frac{N(N-1)}{Z_{N-2,M}}; X_2 = \frac{(N+1)(N+2)}{Z_{N+2,M}};$$
$$X_3 = \frac{M(M-1)}{Z_{N,M-2}}; X_4 = \frac{(M+1)(M+2)}{Z_{N,M+2}};$$
$$X_5 = \frac{N(N-1)(N-2)(N-3)}{Z_{N-4,M}} + \frac{(N+1)(N+2)(N+3)(N+4)}{Z_{N+4,M}};$$
$$X_6 = \frac{M(M-1)(M-2)(M-3)}{Z_{N,M-4}} + \frac{(M+1)(M+2)(M+3)(M+4)}{Z_{N,M+4}};$$
$$X_7 = \frac{N(N-1)M(M-1)}{Z_{N-2,M-2}} + \frac{(N+1)(N+2)(M+1)(M+2)}{Z_{N+2,M+2}};$$
$$Z_{KL} = -E^{(OM)}_{KL} + E^{(OM)}_{NM}. \tag{5.39}$$

Table 5.1 demonstrates that OM zeroth approximation calculates the energy of the ground state with high accuracy at arbitrary value of the parameters of oscillators. Tables 5.2 and 5.3 prove the effectiveness of OM for calculation of the energy of excited collective states and splitting of levels with different symmetry caused by the interaction.

Table 5.2 The energy of the collective states $E^{\pm}_{02}$ for CQAO $\lambda = 0; \Omega = 1; A = 1; B = 1$

$C/2$	E	E^{+}_{02}	E^{-}_{02}
1.0	$E^{(0)}$	6.527	6.249
	$E^{(2)}$	6.526	6.248
	$E^{(A)}$	6.549	6.214
0.8	$E^{(0)}$	6.434	6.199
	$E^{(2)}$	6.435	6.201
	$E^{(A)}$	6.454	6.173
0.4	$E^{(0)}$	6.227	6.096
	$E^{(2)}$	6.233	6.102
	$E^{(A)}$	6.242	6.083
0.0	$E^{(0)}$	5.985	5.985
	$E^{(2)}$	5.993	5.993
	$E^{(A)}$	5.983	5.983
−0.4	$E^{(0)}$	5.687	5.868
	$E^{(2)}$	5.688	5.869
	$E^{(A)}$	5.657	5.879
−0.8	$E^{(0)}$	5.297	5.752
	$E^{(2)}$	5.260	5.715
	$E^{(A)}$	5.174	5.759
−1.0	$E^{(0)}$	5.046	5.698
	$E^{(2)}$	4.958	5.610
	$E^{(A)}$	4.775	5.690

To illustrate the convergence of OM for the system with several degrees of freedom, the calculation of the energy spectrum of CQAO is presented below. A concise description of OM algorithm has been done in the Chap. 2, and here we make a short outline of it. The following Schrödinger equation has to be resolved:

$$\hat{H}|\Psi_{NM}\rangle = E_{NM}|\Psi_{NM}\rangle \tag{5.40}$$

with Hamiltonian (5.18), and eigenvectors in the form:

$$|\Psi_{NM}\rangle = |N, M\rangle + \sum_{K,L\neq N,M} C^{KL}_{NM}|K, L\rangle. \tag{5.41}$$

Here $|N, M\rangle$ are the eigenvectors obtained from the Eq. (5.20). The operators $\hat{n}_{a,b}$ and their eigenfunctions depend on the transformation parameters and on quantum numbers N, M when the transition to the secondary quantization is performed, as follows from the Eqs. (5.21)–(5.24). As a result, the state vectors $|\Psi^{(0)}_{NM}\rangle \simeq |N, M\rangle$ in OM zeroth approximation are non-orthogonal. However, this non-orthogonality is not related to the expansion (5.41), which uses the fully orthogonal basis unique for various quantum numbers N, M.

Table 5.3 The energy of the collective states $E_{13}^{\pm}$ for CQAO ($\hat{H} = \hat{p}_x^2 + \hat{p}_y^2 + x^2 + y^2 + Cx^2y^2$)

C	E	E_{13}^{+}	E_{13}^{-}
0.1	$E^{(0)}$	10.3495	10.5943
	$E^{(2)}$	10.3440	10.5888
	$E^{(A)}$	10.3439	10.5883
1.0	$E^{(0)}$	12.4137	13.5932
	$E^{(2)}$	12.3339	13.5134
	$E^{(A)}$	12.3323	13.4505
5.0	$E^{(0)}$	16.8013	19.3336
	$E^{(2)}$	16.5758	19.1080
	$E^{(A)}$	16.5965	18.7387
20	$E^{(0)}$	24.1062	28.4908
	$E^{(2)}$	23.6777	28.0622
	$E^{(A)}$	23.7604	27.1386
100	$E^{(0)}$	39.1704	46.9931
	$E^{(2)}$	38.3734	46.1960
	$E^{(A)}$	38.582	44.249
1,000	$E^{(0)}$	82.5140	99.6842
	$E^{(2)}$	80.7340	97.9042
	$E^{(A)}$	81.29	93.2
5,000	$E^{(0)}$	140.5087	169.9703
	$E^{(2)}$	137.4448	166.9065
	$E^{(A)}$	138.39	158.8

The expansion (5.41), being substituted into (5.40), results in the infinite algebraic system of non-linear equations for the energy E_{NM} and coefficients C_{NM}^{KL}, and with the normalization condition corresponding to Brillouin-Wigner perturbation theory (Sect. 1.3):

$$\langle N, M | \Psi_{NM} \rangle = 1.$$

Assuming the Hamiltonian (5.19) of OM zeroth approximation as a principal contributor into exact eigenvalue, the equations above can be solved by using simple iteration procedure, where the operator $\hat{V} = \hat{H} - \hat{H}_0$ is considered as a small value. This procedure is expressed through the following recurrent equations:

$$E_{NM}^{(s)} = E_{NM}^{(0)} + \sum_{K,L \neq N,M} C_{NM}^{KL}(s-1)\langle N, M|\hat{V}|K, L\rangle;$$

$$C_{NM}^{KL}(s) = [E_{NM}^{(s)} - \langle K, L|\hat{H}_0|K, L\rangle]^{-1}$$

$$\{\langle N, M|\hat{V}|K, L\rangle + \sum_{P,Q \neq N,M} \langle N, M|\hat{V}|P, Q\rangle C_{KL}^{PQ}(s-1)\}, \tag{5.42}$$

Table 5.4 The convergence of the iterations $E_{NM}^{(s)}$ for eigenvalues of CQAO $\lambda = 0; \Omega = 1; A = B; C = \mu A$

A	0.1	0.1	1.0	1.0	10	10
μ	1.0	−1.0	1.0	−1.0	1.0	−1.0
$E_{00}^{(0)}$	1.152760	1.085516	1.765877	1.492155	3.108030	2.770058
$E_{00}^{(8)}$	1.150188	1.081282	1.724184	1.443776	3.301210	2.557740
$E_{00}^{(A)}$	1.1502	1.0813	1.7242	1.4438	3.3012	2.5577
$E_{10}^{(0)}$	2.419662	2.231765	3.835767	3.105372	7.610856	5.924771
$E_{10}^{(8)}$	2.414341	2.221196	3.830324	3.066649	7.527043	5.487183
$E_{10}^{(A)}$	2.4143	2.2112	3.8304	3.0666	7.5271	5.4881

where $s = 0, 1, 2 \ldots$ and the initial recurrences are:

$$C_{NM}^{KL}(0) = 0; \quad E_{NM}^{(0)} = E_{NM}^{(OM)}.$$

In a standard perturbation theory, the eigenvalues are calculated as a sum of the successive corrections obtained as a series over the power of selected parameter. This series is usually divergent, also for anharmonic oscillator. In opposite, the OM procedure calculates the energies and wave functions as the limits of the recurrent sequences:

$$\begin{aligned} C_{NM}^{KL} &= \lim_{s \to \infty} C_{NM}^{KL}(s); \\ E_{NM} &= \lim_{s \to \infty} E_{NM}^{(s)}, \end{aligned} \tag{5.43}$$

which converge fast. The numerical data in the Table 5.4 shows that eight iterations of the operator method find the eigenvalues with higher accuracy than ones found in [8] by using more complicated algorithm.

In the conclusion of this section, we would like to discuss one more important application of the operator method. As demonstrated for one-dimensional systems in the Chap. 2, the optimal choice of the artificial parameters for the transition to the secondary quantization is important for the uniform suitability of zeroth approximation. However, in the iteration procedure these parameters influence the speed and not the convergence of the successive approximations. This result valid also for CQAO. The recurrent sequence (5.43) converges in complex space of the parameters $\omega_{1,2}$, which enables the combination of the operator method with the method of complex rotation of coordinates used for the calculation of the energy and the levels width of quasi-stationary states [11]. This problem occurs for the negative values of the parameters A, B in Hamiltonian (5.18).

Table 5.5 shows the real parts $E_{NM}^{(s)}$ of the energy of quasi-stationary levels and imaginary parts $-\Gamma_{NM}^{(s)}/2$ calculated using the Eq. (5.42) after s iterations. As follows from the table, twelve OM iterations were sufficient to obtain six exact digits, which is indeed a good accuracy at effective convergence speed.

Table 5.5 Complex energy values $ReE_{NM}^{(s)} - iImE_{NM}^{(s)}$ for quasi-stationary states of CQAO $\lambda = 0; \Omega = 1; A = B; C = \mu A$

A	−0.1	−0.1	−0.2	−0.2	−0.4	−0.4
μ	1.0	−1.0	1.0	−1.0	1.0	−1.0
$ReE_{00}^{(10)}$	0.753411	0.864702	0.719257	0.798733	0.752215	0.768634
$ImE_{00}^{(10)}$	0.159099	0.046916	0.379077	0.163381	0.642765	0.340915
$ReE_{10}^{(12)}$	1.439368	1.663981	1.458860	1.627066	1.596000	1.651425
$ImE_{10}^{(12)}$	0.571692	0.252742	1.070673	0.540111	1.646405	0.884010

5.3 Regular Perturbation Theory for Two-Electron Atoms

In spite of the long history of the two-electron atom in non-relativistic quantum mechanics, the problem of the development of the effective and universal method for the solution of the Schrödinger equation for such systems is still vital (for example, [12]). This problem is of interest because it could be generalized as well for multi-electron atoms due to the two-particle character of the Coulomb potential.

At present the best accuracy for calculating the energy levels of two-electron atoms is achieved by means of the variational approach based on the special form of a many parameter trial wave function (for example, [13, 14]). However, this method is difficult to apply for numerical analysis of multi-electron atoms. Therefore, a less accurate variational approach with a trial wave function in the form of a product of the one-particle wave functions, the Hartree–Fock (HF) method, is commonly used in this case (cf., [15]). For example, the energy for the He atom (the 1^1S ground state) calculated numerically via the HF method is $E_0^{HF} = -2.86168$ [15], which differs from the more accurate variational result of $E_0^R = -2.90372$ [13, 14] on the value of the correlation energy $\Delta E_c = E_0^R - E_0^{HF} = -0.04204$ (the Coulomb system of units [2] is used in this section). At the same time a very simple analytical approximation of the Coulomb potential with an equal effective charge for each electron leads to $E_0^{eff} = -2,84766$ [2]. This means that the correlation energy is more important than the effect of the self-consistent field $\Delta E_{scf} = E_0^{HF} - E_0^{eff} = -0.01402$ taken into account in the HF approximation.

One can suppose that regular perturbation theory (RPT) on the basis of the independent Coulomb wave functions for each electron will prove to be very effective for taking into account the electron correlation. This basis corresponds to the choice of the state vector by OM zeroth approximation for the considered system (Chap. 3). However, in this case the calculation of the OM further approximations is connected with the calculation of the slowly convergent sums of the two-particle potential matrix elements over the whole spectrum of the intermediate states. Therefore at present only the variational perturbation theory is used for the estimation of these sums by means of a trial function [16]. Unfortunately it is also difficult to generalize this method for the excited states of the two-electron atom and for multi-electron atoms.

The present section demonstrates that the sum over the intermediate states in the second order RPT presented with the OM zeroth approximation basis can be calculated in closed form through the two-particle Coulomb Green's function represented below analytically. This method is applied to both the ground and excited states of the two-electron atom [17]. It is shown that the larger part of the correlation energy proves to be taken into account just in the second order approximation of RPT. Actually our purpose is not only the two-electron atom itself but also development of the approach that could be generalized for calculation of the correlation energy in multi-electron atoms and their interaction with external fields. This approach is expected to be effective because the OM zeroth order approximation allowed us to calculate characteristics of the multi-electron atoms with rather good accuracy [18].

Let us consider the application of RPT for calculation of the ground state energy of the two-electron atom with the following Hamiltonian:

$$\hat{H} = \sum_{i=1,2} \left[-\frac{\Delta_i}{2} - \frac{Z}{r_i} \right] + \frac{1}{|\boldsymbol{r}_1 - \boldsymbol{r}_2|}. \tag{5.44}$$

The potentials with the effective charge Z_e can be used in order to separate the OM zeroth-order Hamiltonian with the known spectrum:

$$\begin{aligned} \hat{H}_0 &= \frac{\hat{p}_1^2}{2} - \frac{Z_e}{r_1} + \frac{\hat{p}_2^2}{2} - \frac{Z_e}{r_2}; \\ E_{mn}^{(0)} &= -\frac{Z_e^2}{2} \left[\frac{1}{(n+1)^2} + \frac{1}{(m+1)^2} \right]; \; n, m = 0, 1, 2, \ldots \end{aligned} \tag{5.45}$$

and the perturbation operator:

$$\hat{V} = -\frac{Z - Z_e}{r_1} - \frac{Z - Z_e}{r_2} + \frac{1}{|\boldsymbol{r}_1 - \boldsymbol{r}_2|}. \tag{5.46}$$

If the ground state is considered, the results for the energy in the zeroth and first-order approximations of the RSPT series (Sect. 1.2) are calculated in a simple way [2]:

$$E_0^{(0)} = -Z_e^2; \; E_0^{(1)} = -2Z_e (Z - Z_e) + \frac{5}{8} Z_e. \tag{5.47}$$

Let us consider the RSPT second-order correction for the same state:

$$E_0^{(2)} = \sum_{\lambda} \sum_{\mu}{}' \frac{\left| \left\langle \lambda\mu \middle| \hat{V} \middle| 00 \right\rangle \right|^2}{-Z_e^2 - E_\lambda^{(0)} - E_\mu^{(0)}}. \tag{5.48}$$

Here λ, μ are the set of quantum numbers for electrons in the intermediate states defined by the vectors $|\lambda\mu\rangle$ that are represented as the product of the Coulomb wave functions including the continuous spectrum that gives important contribution to the whole sum. The term corresponding to the ground state $|00\rangle = \psi_0(\mathbf{r}_1)\psi_0(\mathbf{r}_2)$ should be omitted in the sum. The expression (5.48) can be divided into two terms:

$$E_0^{(2)} = 2\sum_{\lambda\neq 0} \frac{\left|\left\langle \lambda 0 \left| \hat{V} \right| 00 \right\rangle\right|^2}{-\frac{Z_e^2}{2} - E_\lambda} + \sum_{\lambda\neq 0}\sum_{\mu\neq 0} \frac{\left|\left\langle \lambda\mu \left| \hat{V} \right| 00 \right\rangle\right|^2}{-Z_e^2 - E_\lambda^{(0)} - E_\mu^{(0)}} \equiv \Sigma_1 + \Sigma_2. \quad (5.49)$$

The first term can be calculated on the basis of the approach of [19] that was used in the problem of the interaction of hydrogen-like atoms with external fields. In this method, the sum over intermediate states is expressed analytically through one-particle Coulomb Green's function (CGF) and the ground state wave functions:

$$G_E(\mathbf{r}, \mathbf{r}') = \sum_\lambda \frac{\psi_\lambda^*(\mathbf{r})\,\psi_\lambda(\mathbf{r}')}{E - E_\lambda^{(0)}}. \quad (5.50)$$

In this case it leads to the following result:

$$\Sigma_1 = -2\int \psi_0^*(\mathbf{r}_1)\,\psi_0^*(\mathbf{r}_2) \sum_{\lambda\neq 0} \frac{\psi_\lambda^*(\mathbf{r}_1')\,\psi_\lambda(\mathbf{r}_1)}{E - E_\lambda^{(0)}} \times$$

$$\hat{V}(\mathbf{r}_1, \mathbf{r}_2)\psi_0(\mathbf{r}_2)\psi_0^*(\mathbf{r}_2')\hat{V}(\mathbf{r}_1{}', \mathbf{r}_2{}')\psi_0(\mathbf{r}_1{}')\psi_0(\mathbf{r}_2{}')d\mathbf{r}_1 d\mathbf{r}_2 d\mathbf{r}_1{}' d\mathbf{r}_2{}' =$$

$$-2\int \psi_0^*(\mathbf{r}_1)\,\psi_0^*(\mathbf{r}_2)\,\hat{V}(\mathbf{r}_1, \mathbf{r}_2)\psi_0^*(\mathbf{r}_2)\tilde{G}_{-Z_e^2/2}(\mathbf{r}_1, \mathbf{r}_1{}')\psi_0^*(\mathbf{r}_2{}') \times$$

$$\hat{V}(\mathbf{r}_1{}', \mathbf{r}_2{}')\psi_0(\mathbf{r}_1{}')\psi_0(\mathbf{r}_2{}')d\mathbf{r}_1 d\mathbf{r}_2 d\mathbf{r}_1{}' d\mathbf{r}_2{}'. \quad (5.51)$$

Here $\tilde{G}_E(\mathbf{r}_1, \mathbf{r}_1')$ is the reduced CGF:

$$\tilde{G}_E(\mathbf{r}_1, \mathbf{r}_1{}') = G_E(\mathbf{r}_1, \mathbf{r}_1{}') - \frac{\psi_0^*(\mathbf{r}_1)\,\psi_0(\mathbf{r}_1')}{E - E_0}, \quad (5.52)$$

which was calculated analytically in [20]. The total CGF is defined as the sum of the series over the spherical harmonics $Y_{l,m_l}(\Omega)$ (for example, [21]):

$$G_E(\mathbf{r}, \mathbf{r}') = \sum_{l,m_l} \frac{1}{rr'} G_{E,l}(r, r')\, Y_{l,m_l}^*(\Omega)\, Y_{l,m_l}(\Omega');$$

$$G_{E,l}(r, r') = \frac{\nu}{Z}\frac{\Gamma(l+1-\nu)}{\Gamma(2l+2)} M_{\nu,l+1/2}\left(\frac{2Z}{\nu}r_<\right) W_{\nu,l+1/2}\left(\frac{2Z}{\nu}r_>\right), \quad (5.53)$$

with $\nu = Z/\sqrt{-2E}$; M;W are the Whittaker functions [22], $r_<$ and $r_>$ are the minimal and maximal values from r and r' accordingly.

Taking into account that $\psi_0(\boldsymbol{r}) = \psi_0^*(\boldsymbol{r}) = R_{1,0}(r)Y_{0,0}(\theta,\phi)$ (here $R_{1,0}(r) = 2Z^{3/2}e^{-rZ}$ is the radial part, and $Y_{0,0}(\theta,\phi) = \frac{1}{\sqrt{4\pi}}$ is the angular part), one finds as the result of integration over the angular variables:

$$\Sigma_1 = -2\int R_{1,0}(r_1)\,R_{1,0}^2(r_2)\,\hat{V}_0(r_1,r_2)\tilde{G}_{-Z_e^2/2,0}(r_1,r_1{}')R_{1,0}(r_1{}')\times$$
$$R_{1,0}^2(r_2{}')\,\hat{V}_0(r_1{}',r_2{}')r_1r_1'r_2^2r_2'^2dr_1dr_2dr_1{}'dr_2{}', \tag{5.54}$$

with $\hat{V}_0(r_1,r_2) = -(Z-Z_e)\left(\frac{1}{r_1}+\frac{1}{r_2}\right)+\frac{1}{r_>}$; and $r_>$ is the maximal value from r_1, r_2.

Let us use the analogous approach in order to calculate the second term in Eq. (5.51), considering the identity:

$$\int_{-\infty}^{\infty}\frac{d\chi}{(a+i\chi)(b-i\chi)} = -\frac{2\pi}{a+b};\ a<0, b<0. \tag{5.55}$$

The summations over the quantum numbers of different electrons in Σ_2 can be separated:

$$-\int_{-\infty}^{\infty}\frac{d\chi}{2\pi}\sum_{\lambda,\mu\neq 0}\frac{\left\langle\lambda\mu\left|\hat{V}\right|00\right\rangle\left\langle 00\left|\hat{V}\right|\lambda\mu\right\rangle}{\left(-\frac{Z_e^2}{2}-E_\lambda^{(0)}+i\chi\right)\left(-\frac{Z_e^2}{2}-E_\mu^{(0)}-i\chi\right)}. \tag{5.56}$$

and each of the summations can then be calculated by means of CGF's after their analytical continuations to the complex values $E = -Z_e^2/2 \pm i\chi$:

$$\Sigma_2 = -\int_{-\infty}^{\infty}\frac{d\chi}{2\pi}\int\psi_0(\boldsymbol{r}_1)\,\psi_0(\boldsymbol{r}_2)\,\psi_0(\boldsymbol{r}_1{}')\,\psi_0(\boldsymbol{r}_2{}')\,\hat{V}(\boldsymbol{r}_1,\boldsymbol{r}_2)\times$$
$$\tilde{G}_{-Z_e^2/2+i\chi}(\boldsymbol{r}_1,\boldsymbol{r}_1{}')\tilde{G}_{-Z_e^2/2-i\chi}(\boldsymbol{r}_2,\boldsymbol{r}_2{}')\hat{V}(\boldsymbol{r}_1{}',\boldsymbol{r}_2{}')d\boldsymbol{r}_1d\boldsymbol{r}_2d\boldsymbol{r}_1{}'d\boldsymbol{r}_2{}' \tag{5.57}$$

The function

$$\sum_{\lambda,\mu\neq 0}\frac{|\lambda\mu\rangle\,\langle\lambda\mu|}{E-E_\lambda-E_\mu} = 2\tilde{G}_{E/2}(\boldsymbol{r}_1,\boldsymbol{r}_1{}')\psi_0(\boldsymbol{r}_2)\,\psi_0(\boldsymbol{r}_2{}') -$$
$$-\frac{1}{2\pi}\int_{-\infty}^{\infty}\tilde{G}_{E/2+i\chi}(\boldsymbol{r}_1,\boldsymbol{r}_1{}')\tilde{G}_{E/2-i\chi}(\boldsymbol{r}_2,\boldsymbol{r}_2{}')d\chi \tag{5.58}$$

is the two-particle reduced CGF and

$$\sum_{\lambda,\mu} \frac{|\lambda\mu\rangle\,\langle\lambda\mu|}{E - E_\lambda - E_\mu} = -\frac{1}{2\pi} \int\limits_{-\infty}^{\infty} G_{E/2+i\chi}(\boldsymbol{r}_1, \boldsymbol{r}_1{}') G_{E/2-i\chi}(\boldsymbol{r}_2, \boldsymbol{r}_2{}') d\chi \qquad (5.59)$$

is the total two-particle CGF that is the result of convolution of two one-particle CGF's [17, 23].

The one-particle part $-(Z - Z_e)/r_1 - (Z - Z_e)/r_2$ of the perturbation operator doesn't contribute to (5.57) because of the orthogonality of the wave functions. Then Eq. (5.57) has the following form after integration over the angular variables:

$$\Sigma_2 = -\sum_{l=0}^{\infty} \frac{1}{2l+1} \int_{-\infty}^{\infty} \frac{d\chi}{2\pi} \int_0^{\infty} dr_1 dr_2 dr_1' dr_2' \times$$
$$R_{1,0}(r_1)\, R_{1,0}(r_2)\, R_{1,0}(r_1')\, R_{1,0}(r_2') \frac{r_<^l}{r_>^{l+1}} \frac{r'^l_<}{r'^{l+1}_>} \times$$
$$\tilde{G}'_{-Z_e^2/2+i\chi,l}(r_1, r_1{}') \tilde{G}'_{-Z_e^2/2-i\chi,l}(r_2, r_2{}') r_1^2 r_2^2 r'^2_1 r'^2_2. \qquad (5.60)$$

Here the radial part of the reduced CGF is calculated as follows:

$$\tilde{G}'_{-Z_e^2/2\pm i\chi,l}(r, r') = \begin{cases} G_{-Z_e^2/2\pm i\chi,0}(r, r') \pm r r' \frac{\psi_0(r)\psi_0(r')}{i\chi}, & l = 0 \\ G_{-Z_e^2/2\pm i\chi,l}(r, r'), & l > 0 \end{cases}. \qquad (5.61)$$

The second order correction depends analytically on the effective charge. The denominator in Eq. (5.49) is represented in the form $E_0^{(0)} - E_n^{(0)} = Z_e^2(\epsilon_0 - \epsilon_n)$, where ϵ_0 and ϵ_n are the values corresponding to $Z_e = 1$. The perturbation operator is written in the form:

$$\hat{V}(\boldsymbol{r}_1, \boldsymbol{r}_2) = -(Z - Z_e)\, V_1(r_1, r_2) + V_2(\boldsymbol{r}_1, \boldsymbol{r}_2),$$
$$V_1 = (1/r_1 + 1/r_2);\; V_2 = 1/|\boldsymbol{r}_1 - \boldsymbol{r}_2|. \qquad (5.62)$$

If the scale variables $\boldsymbol{x}_{1,2} = Z_e \boldsymbol{r}_{1,2}$ are used for integration in matrix elements, we find that

$$E_0^{(2)} = -a(Z - Z_e)^2 + b(Z - Z_e) - c. \qquad (5.63)$$

Thus, only the parameters a, b, c have to be calculated numerically, which don't depend on Z and Z_e. When calculating a and b, the integration over r_2 and r_2' in

Eq. (5.58) can be performed analytically and numerical integration is only necessary over two variables:

$$a = -8 \int e^{-x_1 - x_1'} \tilde{G}_{-\frac{1}{2},0}\left(x_1, x_1'\right) dx_1 dx_1'; \tag{5.64}$$

$$b = 16 \int e^{-x_1 - 3x_1'} \left(e^{2x_1'} - 1 - x_1'\right) \tilde{G}_{-\frac{1}{2},0}\left(x_1, x_1'\right) dx_1 dx_1'; \tag{5.65}$$

$$c = -8 \int \left(e^{2x_1} - 1 - x_1\right) \left(e^{2x_1'} - 1 - x_1'\right) e^{-3x_1 - 3x_1'}$$

$$\times \tilde{G}_{-\frac{1}{2},0}\left(x_1, x_1'\right) dx_1 dx_1' - \frac{1}{2\pi} \sum_{l=0}^{\infty} \frac{1}{2l+1} \int_{-\infty}^{\infty} d\chi \int 16 e^{-x_1 - x_2 - x_1' - x_2'}$$

$$\times \tilde{G}'_{-1/2+i\chi,l}\left(x_1, x_1'\right) \tilde{G}'_{-1/2-i\chi,l}\left(x_2, x_2'\right) \frac{x_<^l}{x_>^{l+1}} \frac{x_<'^l}{x_>'^{l+1}} x_1 x_2 x_1' x_2' \{d^4 x\};$$

$$\{d^4 x\} = dx_1 dx_2 dx_1' dx_2'. \tag{5.66}$$

The values $a = 1$ and $b = 5/8$ are found with an accuracy of 10^{-8}. The most complicated task is the calculation of the expression in Eq. (5.60) that delivers the coefficient c in Eq. (5.64). The sum over l is found to converge better than $\sum_{l=0}^{\infty} 1/l^3$ and c was calculated as 0.15759 with an accuracy $\sim 3 \cdot 10^{-5}$. Thus, the first three terms of RPT lead to the following result for the ground state energy:

$$E_0^{(0)} + E_0^{(1)} + E_0^{(2)} = -Z^2 + \frac{5}{8} Z - 0.15759. \tag{5.67}$$

This result does not depend on the effective charge and it fits well the result of the variational PT [24], for example, for helium:

$$E_0^{RPT} \approx -2.90759; \ \Delta E = E_0^R - E_0^{RPT} \approx 0.00387, \tag{5.68}$$

which means that RPT converges rather quickly and about 90 % of the correlation energy is taken into account in the second order approximation.

5.4 Energies of the Excited States

Let us now apply RPT to the low excited states (2^1S and 2^3S) of He-like system with configuration $(1s)^1(2s)^1$. If the Hamiltonian $\hat{H}_0$ is chosen in the form (5.45) with the same effective charge for both electrons, the degeneracy of the states with permutation of electrons should be taken into account in the zeroth order approximation:

$$\left|\psi_{\pm}^{(0)}\right\rangle = \frac{1}{\sqrt{2}}\left[\psi_0\left(\boldsymbol{r}_1\right)\psi_1\left(\boldsymbol{r}_2\right) \pm \psi_1\left(\boldsymbol{r}_1\right)\psi_0\left(\boldsymbol{r}_2\right)\right] =$$

$$= \frac{1}{\sqrt{2}}\left[|01\rangle \pm |10\rangle\right], \tag{5.69}$$

where $|\psi_{\pm}\rangle$ are the para- and orthosymmetric states, respectively. Zeroth order approximations for both states are defined by the following formula that replicates the result in [2]:

$$E_{\pm}^{(0)} = -\frac{5}{8}Z_{\pm}^2; \quad Z_{\pm} = Z - \frac{4}{5}(K \pm I); \; Z_+ = 1.8145, Z_- = 1.8497,$$

$$E_{+}^{(0)} = -2.0578, E_{-}^{(0)} = -2.1383. \tag{5.70}$$

Here, the effective charges were calculated from the condition that the RPT first order correction $E_{\pm}^{(1)} = \left\langle \psi_{\pm}^{(0)} \left| \hat{V} \right| \psi_{\pm}^{(0)} \right\rangle$ should be equal to zero [25] and the Coulomb integrals are calculated in the usual way:

$$K = \int \frac{\psi_0^*\left(\boldsymbol{x}_1\right)\psi_1^*\left(\boldsymbol{x}_2\right)\psi_0\left(\boldsymbol{x}_1\right)\psi_1\left(\boldsymbol{x}_2\right)d\boldsymbol{x}_1 d\boldsymbol{x}_2}{\left|\boldsymbol{x}_2 - \boldsymbol{x}_1\right|} = \frac{17}{81};$$

$$I = \int \frac{\psi_0^*\left(\boldsymbol{x}_1\right)\psi_1^*\left(\boldsymbol{x}_2\right)\psi_0\left(\boldsymbol{x}_2\right)\psi_1\left(\boldsymbol{x}_1\right)d\boldsymbol{x}_1 d\boldsymbol{x}_2}{\left|\boldsymbol{x}_2 - \boldsymbol{x}_1\right|} = \frac{16}{729}. \tag{5.71}$$

Following the same ideas as for the ground state, one can write the RSPT second order correction in the following form:

$$E_{\pm}^{(2)} = \left(Z - Z_{\pm}\right)^2 A + \left(Z - Z_{\pm}\right)\left(B \pm \Delta B\right) + \left(C \pm \Delta C\right), \tag{5.72}$$

and after the integration over the angular variables:

$$A = -\int \Big[4e^{-x_1 - x_1'}\tilde{G}_{-\frac{1}{2},0}\left(x_1, x_1'\right) +$$

$$\frac{1}{8}e^{-\frac{1}{2}x_1 - \frac{1}{2}x_1'}(x_1 - 2)(x_1' - 2)\tilde{G}_{-\frac{1}{8},0}\left(x_1, x_1'\right)\Big]dx_1 dx_1'; \tag{5.73}$$

$$B = \int e^{-x_1 - 2x_1'}\left(-8 - 8e^{x_1'} - 6x_1' - 2x_1'^2 - x_1'^3\right) \times$$

$$\tilde{G}_{-1/2,0}\left(x_1, x_1'\right)dx_1 dx_1' + \frac{1}{4}\int e^{-\frac{1}{2}x_2 - \frac{5}{2}x_2'}\left(x_2 - 2\right)\left(x_2' - 2\right) \times$$

$$(e^{2x_2'} - x_2' - 1)\tilde{G}_{-1/8,0}\left(x_2, x_2'\right)dx_2 dx_2';$$

$$\Delta B = -\frac{4}{27}\int e^{-x_1-2x_1'}\left(3x_1'^3 - 4x_1^2 - 4x_1'\right)\tilde{G}_{-1/2,0}\left(x_1, x_1'\right)dx_1dx_1'$$
$$-\frac{4}{27}\int e^{-\frac{1}{2}x_2-\frac{5}{2}x_2'}\left(x_2-2\right)\left(2x_2' + 3x_2'^2\right)\tilde{G}_{-1/8,0}\left(x_2, x_2'\right)dx_2dx_2'; \tag{5.74}$$

$$C = -C_{|00\rangle} - \int e^{-\frac{5}{2}x_1-\frac{5}{2}x_1'}\left[\frac{1}{8}\left(x_1-2\right)\left(x_1'-2\right)\left(e^{2x_1}-1-x_1\right)\times\right.$$
$$\left.\left(e^{2x_1'}-1-x_1'\right) - \frac{32}{729}x_1x_1'(2+3x_1)(2+3x_1')\right]\tilde{G}_{-\frac{1}{8},0}\left(x_1, x_1'\right)dx_1dx_1' -$$
$$\frac{1}{2\pi}\sum_{l=0}^{\infty}\frac{1}{2l+1}\int\limits_{-\infty}^{\infty}d\chi\int\frac{1}{2}e^{-x_1-x_1'-\frac{x_2}{2}-\frac{x_2'}{2}}(x_2-2)(x_2'-2)\frac{x_<^l}{x_>^{l+1}}\frac{x_<'^l}{x_>'^{l+1}}$$
$$\times\tilde{G}'_{-\frac{5}{16}+i\chi,l}\left(x_1, x_1'\right)\tilde{G}'_{-\frac{5}{16}-i\chi,l}\left(x_2, x_2'\right)x_1x_2x_1'x_2'\{d^4x\};$$

$$\Delta C = -C_{|00\rangle} - \frac{4}{27}\int e^{-\frac{5}{2}x_1-\frac{5}{2}x_1'}\tilde{G}_{-\frac{1}{8},0}\left(x_1, x_1'\right)\left(2x_1' + 3x_1'^2\right)\times$$
$$\left(e^{2x_1}-x_1-1\right)\left(x_1-2\right)dx_1dx_1' -$$
$$\frac{1}{2\pi}\sum_{l=0}^{\infty}\frac{1}{2l+1}\int\limits_{-\infty}^{\infty}d\chi\int\frac{1}{2}e^{-x_1-\frac{x_1'}{2}-\frac{x_2}{2}-x_2'}(x_2-2)(x_1'-2)\frac{x_<^l}{x_>^{l+1}}\frac{x_<'^l}{x_>'^{l+1}}\times$$
$$\tilde{G}'_{-\frac{5}{16}+i\chi,l}\left(x_1, x_1'\right)\tilde{G}'_{-\frac{5}{16}-i\chi,l}\left(x_2, x_2'\right)x_1x_2x_1'x_2'\{d^4x\};$$

$$\tilde{G}'_{-5/16\pm i\chi,l}\left(x, x'\right) = \begin{cases} G_{-5/16\pm i\chi,0}\left(x, x'\right) \pm xx'\frac{\psi_0(x)\psi_0(x')}{-\frac{3}{16}\mp i\chi}, & l = 0 \\ G_{-5/16\pm i\chi,l}\left(x, x'\right), & l > 0 \end{cases};$$

$$C_{|00\rangle} = \frac{8}{3}\left|\left\langle 00\left|\frac{1}{|\boldsymbol{x}_2 - \boldsymbol{x}_1|}\right|01\right\rangle\right|^2. \tag{5.75}$$

After performing the numerical integration, we obtain $A \approx -\frac{5}{8}$, $B \approx \frac{17}{81}$, $\Delta B \approx \frac{16}{729}$, $C = -0.08095$, and $\Delta C = -0.03354$, and the second order RPT energy for these states can be written as:

$$E_{2^1S} = -\frac{5}{8}Z^2 + \frac{169}{729} - 0.11449, \tag{5.76}$$

$$E_{2^3S} = -\frac{5}{8}Z^2 + \frac{137}{729} - 0.04741. \tag{5.77}$$

The dependence on the effective charge $Z_\pm$ vanishes in this case, just as for the ground state. For helium we get -2.1716 for the 2^3S state and -2.1508 for 2^1S, which can be compared with the variational results [26] -2.175229 and -2.145973.

It means that the correlation energies for the excited states are taken into account with the same accuracy as for the ground state.

Equations (5.76) and (5.77) can also be compared with the known [27] Z^{-1} perturbation theory expansions:

$$E_{2^1S} = -\frac{5}{8}Z^2 + \frac{169}{729} - 0.114510 + \ldots,$$

$$E_{2^3S} = -\frac{5}{8}Z^2 + \frac{137}{729} - 0.047409 - \ldots.$$

The two-particle Coulomb Green's function found in this chapter allows to fulfil in the closed form the summation of the transition matrix elements from the electron-electron interaction operator over all intermediate states and to develop RPT on this basis. The efficiency of this approach is analyzed when calculating the second order approximation for the energies of two-electron atoms. In this particular case, the low level energies were calculated by other methods with essentially higher accuracy. However the further applications of RPT for higher order approximations, multi-electron systems and analysis of the interaction between atom and external fields are also possible. Let us consider briefly several examples in order to show qualitatively how to extend the method for these cases. The detailed analysis of such system is out of the framework of this book devoted mainly to the basics of OM.

Firstly, let us show how to calculate the further terms of the RPT series. If standard formulas of the perturbation theory [2] are used with the operator (5.46) the third order correction to the ground state energy can be represented in the form analogous to (5.63) with the same scaling transformation:

$$E_0^{(3)} = a_1 Z_e + (b_1 - 2a_1 Z) + \frac{a_1 Z^2 - b_1 Z - c_1}{Z_e}. \tag{5.78}$$

Here the coefficients a_1, b_1, c_1 are expressed in a closed form by means of the two-particle Coulomb Green's functions with the effective charge $Z_e = 1$ and should be calculated numerically. The value $E_0^{(3)}$ can be minimized relatively to Z_e that leads to the following expression:

$$E_0^{(3)} = 2\sqrt{a_1^2 Z^2 - a_1 b_1 Z - a_1 c_1} + b_1 - 2a_1 Z, \tag{5.79}$$

which shows that the RPT successive terms are not reduced to the series over Z^{-1} and it can be useful for analysis of the singularity of the function $E_0(Z)$ [16].

The method described above can be also generalized for many-electron atoms. As for example, Hamiltonian of a three-electron atom can be divided into zeroth order Hamiltonian $\hat{H}_0$ and the perturbation operator $\hat{V}$ in the following way:

$$\hat{H}_0 = \sum_{i=1}^{3} \frac{\hat{p}_i^2}{2} - Z_e^{(1)} \left(\frac{1}{r_1} + \frac{1}{r_2} \right) - \frac{Z_e^{(2)}}{r_3};$$

$$\hat{V} = -(Z - Z_e^{(1)}) \left(\frac{1}{r_1} + \frac{1}{r_2} \right) - \frac{Z - Z_e^{(2)}}{r_3} + \sum_{i=1, j \neq i}^{3} \frac{1}{|\boldsymbol{r}_i - \boldsymbol{r}_j|}, \quad (5.80)$$

with different effective charges $Z_e^{(1,2)}$ for internal and external shells. In the zeroth order approximation, the electrons are considered as distinguishable particles without taking into account the exchange corrections. As shown recently [18], this approach allows one to calculate physical characteristics of multi-electron atoms and ions with rather good accuracy. In order to take into account the permutation symmetry of the wave function, the projection operators (Sect. 2.8) can be used:

$$\hat{T}_{ij}^{(s)} = \frac{1}{\sqrt{2}}(1 + s\hat{P}_{ij}), \ s = (-1)^{S_{ij}}, \quad (5.81)$$

where $\hat{P}_{ij}$ is the permutation operator and S_{ij} is the spin of the electron pair.

The following equation for the wave function $\Psi_0(\boldsymbol{r}_1, \boldsymbol{r}_2, \boldsymbol{r}_3)$ of the three electron atom ground state can be considered instead of the Schrödinger equation from Sect. 2.8:

$$\{\hat{H} - E_0\}(1 - P_{13})(1 + P_{23})\psi_0(\boldsymbol{r}_1, \boldsymbol{r}_2, \boldsymbol{r}_3) = 0;$$

$$\Psi_0(\boldsymbol{r}_1, \boldsymbol{r}_2, \boldsymbol{r}_3) = \frac{1}{2}(1 - P_{13})(1 + P_{23})\psi_0(\boldsymbol{r}_1, \boldsymbol{r}_2, \boldsymbol{r}_3). \quad (5.82)$$

The wave function $\psi_0(\boldsymbol{r}_1, \boldsymbol{r}_2, \boldsymbol{r}_3)$ is calculated by means of RPT with the transformed perturbation operator but without additional symmetrization:

$$\hat{V}_R(E_0) = \hat{V}(1 + \hat{P}) + (\hat{H}_0 - E_0)\hat{P}; \quad \hat{P} = P_{23} - P_{13} - P_{13}P_{23}. \quad (5.83)$$

In this representation, the exchange and correlation corrections are taken into account in the RPT series simultaneously and all sums over the intermediate states are reduced to the two-particle Coulomb Green's functions.

The RPT can be also useful for the analysis of the Rydberg states of helium in order to take into account the polarization effects [28]. In this case Hamiltonian of the system in the scaled variables has the following form [28]:

$$\hat{H}_0(r) = -\left(\nabla_r^2 + \frac{4}{r}\right); \; \hat{H}_0(x) = -\left(\nabla_x^2 + \frac{2}{x}\right);$$

$$\hat{V}(\boldsymbol{x}, \boldsymbol{r}) = \frac{2}{|\boldsymbol{x} - \boldsymbol{r}|} - \frac{2}{x} - \frac{2m}{M_{He} + m} \boldsymbol{\nabla}_r \boldsymbol{\nabla}_x. \tag{5.84}$$

In the RPT second order, the energy shift Δ of the Rydberg level $E(1sNL)$ is represented in a closed form in terms of the reduced Coulomb Green's functions:

$$\Delta = -\int_{-\infty}^{\infty} \frac{d\chi}{2\pi} \int \psi_0(\boldsymbol{r}) \, \Psi_{NLM}^*(\boldsymbol{x}) \, \hat{V}(\boldsymbol{r}, \boldsymbol{x}) \tilde{G}_{-4+i\chi}(\boldsymbol{r}, \boldsymbol{r}') \times$$

$$\tilde{G}_{-1/N^2-i\chi}(\boldsymbol{x}, \boldsymbol{x}') \hat{V}(\boldsymbol{r}', \boldsymbol{x}') \psi_0(\boldsymbol{r}') \, \Psi_{NLM}(\boldsymbol{r}_2') \, d\boldsymbol{r} d\boldsymbol{x} d\boldsymbol{r}' d\boldsymbol{x}'. \tag{5.85}$$

This quantity can be calculated numerically without the expansion of the integrand in a formal power series over $\hat{H}_0(x)/\hat{H}_0(r)$ as in the paper [28].

References

1. M. Born, J.R. Oppenheimer, Ann. Phys. **84**, 457 (1927)
2. L.D. Landau, E.M. Lifshitz, *Quantum Mechanics* (Fizmatgiz, Moscow, 2004)
3. M.R.M. Witwit, J. Phys. A Math. Gen. **24**, 4535 (1991)
4. H. Taseli, R. Eid, J. Phys. A Math. Gen. **29**, 6967 (1996)
5. H.A. Bethe, E.E. Salpeter, *Quantum Mechanics of One- and Two-Electron Atoms* (Plenum, New York, 1977)
6. F.M. Fernandez, Phys. Rev. A **50**, 2953 (1994)
7. A.S. Davydov, *Quantum Mechanics (in Russian)* (BHV-Peterburg, Saint-Peterburg, 2011)
8. F.T. Hioe, D. MacMillen, E.W. Montroll, Phys. Rep. **335**, 307 (1978)
9. I.D. Feranchuk, V.S. Kuz'min, A.P. Ulyanenkov, Chem. Phys. **157**, 61 (1991)
10. A. Messiah, *Quantum Mechanics* (North-Holland, Amsterdam, 1976)
11. D. Farrelly, W.P. Reinhardt, J. Phys. B Atom. Mol. **16**, 2103 (1983)
12. P.E. Grabowski, Phys. Rev. A **81**, 032508 (2010)
13. K. Frankowski, C. Pekeris, Phys. Rev. **146**, 145 (1966)
14. D. Freund, B. Huxtable, J. Morgan III, Phys. Rev. A **29**, 980 (1984)
15. C.F. Fischer, *The Hartree-Fock Method for Atoms. Numerical Approach* (Wiley, New York, 1977)
16. J. Zamastil, Z. Cizek, L. Scala, M. Simanek, Phys. Rev. A **81**, 032118 (2010)
17. I.D. Feranchuk, V.V. Triguk, Phys. Lett. A **375**, 2550 (2011)
18. I.D. Feranchuk, V.V. Triguk, J. Appl. Spectr. **77**, 749 (2011)
19. A. Dalgarno, Proc. R. Soc. **233A**, 70 (1955)
20. L.C. Hostler, Phys. Rev. **178**, 126 (1969)

21. L.C. Hostler, J. Math. Phys. **5**, 591 (1964)
22. I.S. Gradshtein, I.M. Ryzhik, *Tables of Integrals, Sums, Series and Products* (FIZMATGIZ, Moscow, 1963)
23. R. Shakeshaft, Phys. Rev. A **70**, 042704 (2004)
24. C.W. Sherr, R.E. Knight, Rev. Mod. Phys. **35**, 436 (1963)
25. I.D. Feranchuk, L.I. Komarov, I.V. Nechipor, A.P. Ulyanenkov, Ann. Phys. **238**, 370 (1995)
26. C.L. Pekeris, Phys. Rev. **126**, 1470 (1962)
27. V. Yerokhin, K. Pachucki, Phys. Rev. A **81**, 022507 (2010)
28. R.J. Drachman, Phys. Rev. A **26**, 1228 (1982)

Chapter 6
Two-Dimensional Exciton in Magnetic Field with Arbitrary Strength

A two-dimensional exciton in a magnetic field has been of great interest to both theoretical and experimental researchers for many years [1–3] and continues to be after several new and interesting physical effects discovered in recent years [4–7]. The energy spectrum and wave function of exciton in magnetic field, therefore, need to be calculated with increasing precision. Since the 1990s, the perturbation method, the variational method and some other numerical methods have been employed to calculate the energy of this system in weak and strong magnetic fields [1, 3]. The solution of the problem for a medium magnetic field was calculated using extrapolation (see [3] and references therein).

In the last decade, however, solving the problem with higher precision was preferred. In the work [8], the problem was solved using the mixed-basis variational method in combination with the shifted 1/N method, while in the work [9], the asymptotic iteration method was employed. Both of these methods provided solutions with a precision of up to seven decimal places for the ground state $1s$ and the excited states $2p^-$, $3d^-$ only. Numerical results for higher excited states have not been obtained up till now. The increase of the precision and application of these methods to higher excited states are not easy and are inefficient in terms of computing resources. Therefore, the development of a new method to calculate energy and wave functions for not only the ground state but also other high excited states with any given precision is of certain interest for theoretical physics.

In this chapter, we will use the operator method described in Chap. 2 to obtain exact numerical solutions of the Schrödinger equation for a two-dimensional exciton in a constant magnetic field of arbitrary strength. The operator method is applicable to non-perturbation problems that allows considering the problem with arbitrary external magnetic field intensity. Furthermore, the schemes of calculating the high-order terms of corrections allow a finding the exact numerical solutions with a given precision with a high convergence rate. Unfortunately, the operator method cannot be applied directly to atomic systems because the expression of Coulomb interaction, which contains coordinates in the denominator, cannot be calculated

I. Feranchuk et al., *Non-perturbative Description of Quantum Systems*,
Lecture Notes in Physics 894, DOI 10.1007/978-3-319-13006-4_6

using algebraic transformations of the creation and annihilation operators. We will overcome this difficulty by using the Levi-Civita transformation [10]. According to the work [11], a two-dimensional exciton problem through the Levi-Civita transformation is equivalent to a harmonic oscillator problem. This connection makes it possible to obtain the wave functions of a two-dimensional exciton in an algebraic form which is very convenient for calculation. This leads us to the use of the Levi-Civita transformation to transform the problem of a two-dimensional exciton in magnetic field into that of an anharmonic oscillator in order to apply the operator method. The obtained results are also interesting in the meaning that the operator method can be applied to other specific problems of interest in recent years for two-dimensional atomic systems, such as those mentioned in [12, 13].

The special point in the application of the operator method for solving the Schrödinger equation in this chapter is that it transforms the problem into a very simple problem of an anharmonic oscillator. Hence we think about using it to find approximate analytical solutions with high precision uniformly stable over the whole range of magnetic field strength. As mentioned in the previous chapters, we can use the zeroth-order approximation of the operator method as analytical solutions. These solutions are pretty precise in weak and medium magnetic fields. However, their precision decreases dramatically in strong magnetic field. In this chapter, we will improve the calculations by adding to the Schrödinger equation the component of the wave functions representing the asymptotic behavior in a region of strong magnetic field. In this asymptotic component, we also introduce a parameter which is used to significantly improve the precision of our analytical solutions.

The results presented in this chapter have been reported in the papers [14–16].

6.1 The Schrödinger Equation through the Levi-Civita Transformation

In this section, the Levi-Civita transformation [10, 11] is used to transform the Schrödinger equation of a two-dimensional exciton in a magnetic field into that of a two-dimensional anharmonic oscillator. For a two-dimensional exciton in a magnetic field we consider the following Schrödinger equation written in the atomic units:

$$\hat{H}\ \Psi(\mathbf{r}) = E\ \Psi(\mathbf{r}), \tag{6.1}$$

$$\hat{H} = -\frac{1}{2}\left(\frac{\partial^2}{\partial x^2} + \frac{\partial^2}{\partial y^2}\right) - \frac{1}{2} i\,\gamma\left(x\frac{\partial}{\partial y} - y\frac{\partial}{\partial x}\right) + \frac{1}{8}\gamma^2\left(x^2 + y^2\right) - \frac{Z}{r}. \tag{6.2}$$

Here, the effective Rydberg constant: $R^* = m^* e^4/16\pi^2 {\varepsilon_0}^2 \hbar^2$ is an energy unit; the coordinates are measured in unit of the effective Bohr radius: $a^* = 4\pi\varepsilon_0 \hbar^2/m^* e^2$ and the dimensionless parameter γ is defined by the formula $\gamma = \hbar\omega_c/2R^*$ with the cyclotron frequency $\omega_c = eB/2\pi m^*$ and the magnetic field intensity B; m^*, ε are the electron effective mass and the static dielectric constant respectively; Z is the charge of the hole, which equals 1 in this case to compare the obtained results with those in [8,9]. A wide range of γ covering both weak and strong magnetic field regions is considered.

We will now consider the Eqs. (6.1) and (6.2) in another space which is more convenient for calculation through the Levi-Civita transformation:

$$\begin{cases} x = u^2 - v^2, \\ y = 2\,u\,v. \end{cases} \tag{6.3}$$

The transformation (6.3) connects the two real two-dimensional spaces (x, y) and (u, v). We can easily prove the following equalities:

$$dx\,dy = 4(u^2 + v^2)\,du\,dv, r = \sqrt{x^2 + y^2} = u^2 + v^2, \tag{6.4}$$

which will be used in the following calculation.

From Eq. (6.4) we see that the Jacobian of the transformation (6.3) is not a constant but is instead $4(u^2 + v^2)$, so it will appear as a weight in the equation for calculating the scalar product of two state vectors when transforming from the (x, y) space to the (u, v) space. This means that if a certain operator $\hat{K}$ is Hermitian in the (x, y) space then the operator $\tilde{K} = 4(u^2 + v^2)\,\hat{K}$ is also Hermitian in the (u, v) space. Hence, in order to ensure that the Hamiltonian is Hermitian through the transformation (6.3), we need to rewrite Eq. (6.1) as follows:

$$r(\hat{H} - E)\,\Psi(r) = 0.$$

In the (u, v) space, this equation reads:

$$\tilde{H}\,\Psi(u, v) = Z\Psi(u, v)\,, \tag{6.5}$$

with the Hamiltonian:

$$\tilde{H} = -\frac{1}{8}\left(\frac{\partial^2}{\partial u^2} + \frac{\partial^2}{\partial v^2}\right) - \left(E - \frac{\gamma}{2}\,\hat{L}_z\right)(u^2 + v^2) + \frac{\gamma^2}{8}(u^2 + v^2)^3. \tag{6.6}$$

We see in Eqs. (6.5) and (6.6) an interchange of Z and E in the role of the eigenvalue. The energy E is no longer the eigenvalue but is instead a parameter, while Z becomes the eigenvalue of Eq. (6.5). However, we can still solve the Schrödinger equation to find E when keeping $Z = 1$. Furthermore, in the Hamiltonian (6.6) there is also the operator $\hat{L}_z$ which is the orbital angular momentum operator. Because the system under investigation is two-dimensional, this operator is also its projection operator on the direction perpendicular to the plane of motion. In the (u, v) space, this operator reads

$$\hat{L}_z = -\frac{i}{2}\left(u\frac{\partial}{\partial v} - v\frac{\partial}{\partial u}\right). \tag{6.7}$$

We can also prove that $\hat{L}_z$ commutes with the Hamiltonian of Eq. (6.5). This means that angular momentum is conserved in the problem under consideration. We will use this conservation by constructing a basis set for solving Eq. (6.5) which contains the eigenfunctions of the orbital angular momentum operator. Then we replace $\hat{L}_z$ by its eigenvalue in the Hamiltonian (6.6). Now we can see that Eqs. (6.5) and (6.6) represent a two-dimensional anharmonic oscillator. In other words, we have transformed a two-dimensional exciton in magnetic field into an anharmonic oscillator via the Levi-Civita transformation. This result allows the application of the operator method to find the exact numerical solutions of Eqs. (6.5) and (6.6).

6.2 Solving the Schrödinger Equation by the Operator Method

Here, we will present four basic steps of the operator method of solving the Schrödinger equation (6.5).

Step 1 is writing the Hamiltonian in algebraic form:

$$\tilde{H}\left(\frac{\partial}{\partial u}, \frac{\partial}{\partial v}, u, v, \gamma\right) \rightarrow \tilde{H}\left(\hat{a}, \hat{b}, \hat{a}^{+}, \hat{b}^{+}, \gamma\right), \tag{6.8}$$

using the creation and annihilation operators defined as follows:

$$\begin{aligned} \hat{a}(\omega) &= \sqrt{\frac{\omega}{2}}\left(\xi + \frac{1}{\omega}\frac{\partial}{\partial \xi^*}\right), \quad \hat{a}^{+}(\omega) = \sqrt{\frac{\omega}{2}}\left(\xi^* - \frac{1}{\omega}\frac{\partial}{\partial \xi}\right), \\ \hat{b}(\omega) &= \sqrt{\frac{\omega}{2}}\left(\xi^* + \frac{1}{\omega}\frac{\partial}{\partial \xi}\right), \quad \hat{b}^{+}(\omega) = \sqrt{\frac{\omega}{2}}\left(\xi - \frac{1}{\omega}\frac{\partial}{\partial \xi^*}\right), \end{aligned} \tag{6.9}$$

in which the complex coordinates are defined as: $\xi = u + i\,v, \quad \xi^* = u - i\,v$. We can easily check that the operators (6.9) satisfy the well-known commutation relations:

$$\left[\hat{a}\,,\ \hat{a}^{+}\right] = 1, \qquad \left[\hat{b}\,,\ \hat{b}^{+}\right] = 1. \tag{6.10}$$

In definition (6.9), we use complex coordinates for convenience in writing only. The positive real number ω given in Eq. (6.9) is considered a free parameter whose role in the method will be discussed in the next steps.

Plugging (6.9) into (6.6) we obtain the algebraic form of the Hamiltonian as follows:

$$\begin{aligned}\tilde{H}(\hat{a}^+,\hat{b}^+,\hat{a},\hat{b},\gamma) = &\frac{\omega^2-2E+m\gamma}{4\omega}(\hat{a}^+\hat{a}+\hat{b}^+\hat{b}+1)\\ &-\frac{\omega^2+2E-m\gamma}{4\omega}(\hat{a}^+\hat{b}^+ + \hat{a}\,\hat{b})\\ &+\frac{\gamma^2}{64\omega^3}(\hat{a}^+\hat{b}^+ + \hat{a}\,\hat{b}+\hat{a}^+\hat{a}+\hat{b}^+\hat{b}+1)^3.\end{aligned} \tag{6.11}$$

We also have an algebraic form of the orbital angular momentum operator:

$$\hat{L}_z = -\frac{1}{2}(\hat{a}^+\hat{a}-\hat{b}^+\hat{b}), \tag{6.12}$$

which is a neutral operator.

Step 2 is splitting the Hamiltonian (6.12) into two components—the major component and the perturbation one:

$$\tilde{H}(\hat{a}^+,\hat{b}^+,\hat{a},\hat{b},\gamma) = \tilde{H}_0(\hat{a}^+\hat{a},\hat{b}^+\hat{b},\gamma,\omega)+\tilde{V}(\hat{a}^+,\hat{b}^+,\hat{a},\hat{b},\gamma,\omega). \tag{6.13}$$

The separation (6.13) is done based on a completely different principle from that of the perturbation method. Here, the major component contains only neutral operators which are products of equal numbers of creation and annihilation operators:

$$\begin{aligned}\tilde{H}_0(\hat{a}^+\hat{a},\hat{b}^+\hat{b},\gamma,\omega) = &\frac{\omega^2-2E+m\gamma}{4\omega}(\hat{a}^+\hat{a}+\hat{b}^+\hat{b}+1)\\ &+\frac{\gamma^2}{64\omega^3}(\hat{a}^+\hat{a}+\hat{b}^+\hat{b}+1)[(\hat{a}^+\hat{a}+\hat{b}^+\hat{b}+1)\\ &\times(\hat{a}^+\hat{a}+\hat{b}^+\hat{b}+4)+6\hat{a}^+\hat{a}\,\hat{b}^+\hat{b}+2].\end{aligned} \tag{6.14}$$

The rest is the perturbation term:

$$\begin{aligned}\tilde{V}(\hat{a}^+,\hat{b}^+,\hat{a},\hat{b},\gamma,\omega) = &-\frac{\omega^2+2E-m\gamma}{4\omega}(\hat{a}^+\hat{b}^+ + \hat{a}\,\hat{b})\\ &+\frac{\gamma^2}{64\omega^3}\Big[(\hat{a}^+\hat{b}^+)^3+(\hat{a}\,\hat{b})^3+3\hat{a}^+\hat{b}^+\left(\hat{a}^+\hat{a}+\hat{b}^+\hat{b}+1\right)^2\\ &+3(\hat{a}^+\hat{b}^+)^2\left(\hat{a}^+\hat{a}+\hat{b}^+\hat{b}+1\right)+3\left(\hat{a}^+\hat{a}+\hat{b}^+\hat{b}+1\right)(\hat{a}\,\hat{b})^2\\ &+3(\hat{a}^+\hat{b}^+)^2(\hat{a}\,\hat{b})+3\hat{a}^+\hat{b}^+(\hat{a}\,\hat{b})^2+3\left(\hat{a}^+\hat{a}+\hat{b}^+\hat{b}+1\right)^2\hat{a}\,\hat{b}\end{aligned}$$

$$+9\hat{a}^+\hat{b}^+\left(\hat{a}^+\hat{a}+\hat{b}^+\hat{b}+1\right)+9\left(\hat{a}^+\hat{a}+\hat{b}^+\hat{b}+1\right)\hat{a}\,\hat{b}$$
$$+6(\hat{a}^+\hat{b}^+)^2+6(\hat{a}\,\hat{b})^2+6\hat{a}^+\hat{b}^+ + 6\hat{a}\,\hat{b}\Big]. \tag{6.15}$$

We see that the operator $\tilde{H}_0(\hat{a}^+\hat{a},\hat{b}^+\hat{b},\gamma,\omega)$ commutes with the operators $\hat{a}^+\hat{a}$ and $\hat{b}^+\hat{b}$, hence its exact solutions are the wave functions of the harmonic oscillator described by these operators. Furthermore, notice that although the Hamiltonian of the system does not depend on the free parameter ω, the split components $\tilde{H}_0(\hat{a}^+\hat{a},\hat{b}^+\hat{b},\gamma,\omega)$ and $\tilde{V}(\hat{a}^+,\hat{b}^+,\hat{a},\hat{b},\gamma,\omega)$ do. This means that we can adjust the correlation between the major and the perturbation components by changing the value of ω.

Step 3 is finding the approximate zeroth-order energy and wave function using the approximate Hamiltonian $\tilde{H}_0(\hat{a}^+\hat{a},\hat{b}^+\hat{b},\gamma,\omega)$. This Hamiltonian contains only neutral operators $\hat{a}^+\hat{a}$ and $\hat{b}^+\hat{b}$, so its eigenfunctions have the following form:

$$(\hat{a}^+)^j\,(\hat{b}^+)^k\,|0(\omega)\rangle\,, \tag{6.16}$$

in which j,k are non-negative integers and the vacuum state $|0(\omega)\rangle$ is defined from the equations:

$$\hat{a}(\omega)\,|0(\omega)\rangle=0,\quad \hat{b}(\omega)\,|0(\omega)\rangle=0, \tag{6.17}$$

and the normalization equation:

$$\langle 0(\omega)\mid 0(\omega)\rangle=1. \tag{6.18}$$

As stated in Sect. 6.1, the angular momentum is conserved in the problem under consideration. We will use this conservation by constructing a basis set which contains the eigenfunctions of the orbital angular momentum operator $\hat{L}_z$. This operator in the algebraic form (6.12) is also a neutral operator, hence its eigenfunctions are also in the form (6.16). Therefore, the wave function vectors (6.16) are rewritten in normalization form as follows:

$$|n(m)\rangle=\frac{1}{\sqrt{(n-m)!(n+m)!}}(\hat{a}^+)^{n-m}(\hat{b}^+)^{n+m}\,|0(\omega)\rangle\,, \tag{6.19}$$

in which the principal quantum numbers n are non-negative integers: $n = 0,1,2,\ldots$, and the magnetic quantum numbers m are integers satisfying the condition $-n\le m\le n$. Furthermore, we use the notation $n=n_r+|m|$ in which $n_r=0,\ 1,\ 2,\ \ldots$ are radial quantum numbers.

Thus we have the approximate zeroth-order wave function corresponding to the state of principal quantum number n and the magnetic quantum numbers m:

$$\left|\psi_{n(m)}\right\rangle^{(0)} = |n(m)\rangle\,. \tag{6.20}$$

Now we let the Hamiltonian (6.14) act on the wave function (6.20) and consider Eq. (6.5). As a result, we obtain the approximate zeroth-order energy:

$$E^{(0)} = -\frac{2\omega Z}{2n+1} + \frac{1}{2}\omega^2 + \frac{\gamma^2}{16\omega^2}(5n^2 + 5n - 3m^2 + 3) + \frac{1}{2}m\gamma. \tag{6.21}$$

Here the free parameter is determined from the condition $\partial E^{(0)}/\partial\omega = 0$, which leads to the following equation:

$$-\frac{2Z}{2n+1} + \omega - \frac{\gamma^2}{8\omega^3}(5n^2 + 5n - 3m^2 + 3) = 0. \tag{6.22}$$

A numerical analysis of the analytical solutions (6.21) will be given in the next section as a result.

Step 4 is calculating high-order corrections to obtain exact numerical solutions. In principle, we may use various schemes, e.g. the perturbation theory scheme for calculating high-order corrections in order to obtain the energy and the wave function with higher precision. If that scheme leads to a result that converges to a certain value with any given precision then we have the exact numerical solution. In this chapter, we will propose an iteration scheme to calculate the energy and wave function with a given precision.

For convenience we rewrite Eqs. (6.5) and (6.6) as follows:

$$(\tilde{H}^R - E\,\tilde{R})\,|\psi\rangle = 0, \tag{6.23}$$

in which the operator $\tilde{H}^R$ and $\tilde{R}$ takes the following forms:

$$\begin{aligned}\tilde{H}^R &= \frac{\omega^2 + m\gamma}{4\omega}(\hat{a}^+\hat{a} + \hat{b}^+\hat{b} + 1) - \frac{\omega^2 - m\gamma}{4\omega}(\hat{a}^+\hat{b}^+ + \hat{a}\,\hat{b}) \\ &\quad + \frac{\gamma^2}{64\omega^3}(\hat{a}^+\hat{b}^+ + \hat{a}\,\hat{b} + \hat{a}^+\hat{a} + \hat{b}^+\hat{b} + 1)^3 - 1\,, \\ \tilde{R} &= \frac{1}{2\omega}(\hat{a}^+\hat{b}^+ + \hat{a}\,\hat{b} + \hat{a}^+\hat{a} + \hat{b}^+\hat{b} + 1).\end{aligned} \tag{6.24}$$

We will use the basis set of wave functions (6.19) to construct the wave functions of the problem at hands. The matrix elements of the operators (6.24) corresponding to this basis set will be calculated through purely algebraic transformation. In fact, using the commutations (6.10) and Eqs. (6.17) and (6.18), we easily obtain following formula:

$$\begin{aligned} \hat{a}^+\hat{b}^+ \,|n(m)\rangle &= \sqrt{(n+1)^2-m^2}\,|n+1\,(m)\rangle\,, \\ \hat{a}\hat{b}\,|n(m)\rangle &= \sqrt{n^2-m^2}\,|n-1\,(m)\rangle\,, \\ (\hat{a}^+\hat{a}+\hat{b}^+\hat{b})\,|n(m)\rangle &= 2n\,|n\,(m)\rangle\,. \end{aligned} \tag{6.25}$$

from which we calculate the matrix elements as follows:

$$\begin{aligned} H^R_{nn} &= \langle n(m)|\,\tilde{H}^R\,|n(m)\rangle \\ &= \frac{\omega^2-m\gamma}{4\omega}(2n+1)+\frac{\gamma^2}{32\omega^3}(2n+1)(5n^2+5n+3-3m^2)-Z, \\ H^R_{n,n+1} &= \langle n(m)|\,\tilde{H}^R\,|n+1(m)\rangle = \sqrt{(n+1)^2-m^2} \\ &\quad\times\left(-\frac{\omega^2+m\gamma}{4\omega}+\frac{3\gamma^2}{64\omega^3}(5n^2+10n+6-m^2)\right), \\ H^R_{n,n+2} &= \langle n(m)|\,\tilde{H}^R\,|n+2(m)\rangle \\ &= \frac{3\gamma^2}{64\omega^3}(2n+3)\,\sqrt{(n+1)^2-m^2}\sqrt{(n+2)^2-m^2}, \\ H^R_{n,n+3} &= \langle n(m)|\,\tilde{H}^R\,|n+3(m)\rangle \\ &= \frac{\gamma^2}{64\omega^3}\,\sqrt{(n+1)^2-m^2}\sqrt{(n+2)^2-m^2}\sqrt{(n+3)^2-m^2}, \\ R_{nn} &= \langle n(m)|\,\tilde{R}\,|n(m)\rangle = \frac{2n+1}{2\omega}, \\ R_{n,n+1} &= \langle n(m)|\,\tilde{R}\,|n+1(m)\rangle = -\frac{1}{2\omega}\sqrt{(n+1)^2-m^2}. \end{aligned} \tag{6.26}$$

Besides (6.26), we can calculate other non-zero matrix elements using the symmetry properties $H^R_{nk}=H^R_{kn}$, $R_{nk}=R_{kn}$.

We can write the exact wave function as a linear combination of the basic functions (6.19):

$$\left|\psi_{n(m)}\right\rangle = |n(m)\rangle + \sum_{j=|m|,j\neq n}^{+\infty} C_j\,|j(m)\rangle\,, \tag{6.27}$$

and define the approximate wave function at the sth order approximation (sth iteration loop) as:

$$\left|\psi_{n(m)}\right\rangle^{(s)} = |n(m)\rangle + \sum_{j=|m|,j\neq n}^{n+s} C_j^{(s)}\,|j(m)\rangle\,, \tag{6.28}$$

corresponding to the approximate energy $E^{(s)}$. If $\lim_{s\to\infty} C_k^{(s)} = C_k$ then the approximate wave function (6.28) converges to the exact wave function (6.27). If then we have $E^{(s)} \to E^T$ while $s \to +\infty$, we also have the exact energy. We will use the notations for quantum states as in the work [8,9], e.g.: $1s, 2s, 2p^-, 2p^+, 3s, 3p^-, 3p^+, 3d^-, 3d^+$, etc. The first digit stands for the state level $n+1$, in which n is the principal quantum number defined in (6.19); the letters symbolize the orbital quantum numbers $l = |m|$: s corresponding to $l = 0$, p to $l = 1$, d to $l = 2$, f to $l = 3$, g to $l = 4$, etc.; the $\pm$ sign is of the magnetic quantum number m.

By implementing Eq. (6.28) into the Schrödinger equation (6.23), we obtain the expression for the approximate sth order energy:

$$E^{(s)} = \frac{H_{nn}^R + \sum\limits_{k=|m|,k\neq n}^{n+s} C_k^{(s)} H_{nk}^R}{R_{nn} + \sum\limits_{k=|m|,k\neq n}^{n+s} C_k^{(s)} R_{nk}}. \tag{6.29}$$

Here, the coefficients $C_k^{(s)}$ $(k = |m|, |m|+1, \ldots, n-1, n+1, \ldots, n+s)$ are determined by a system of $n+s-|m|$ linear equations:

$$\sum_{k=|m|,\,k\neq n,\,k\neq j}^{n+s} \left(H_{jk}^R - E^{(s-1)}\right) C_k^{(s)} = R_{nj} E^{(s-1)} - H_{nj}^R, \tag{6.30}$$

in which $j = |m|, |m|+1, \ldots, n-1, n+1, \ldots, n+s$.

So, by substituting the solutions of Eq. (6.30) into Eq. (6.29) we obtain the energy of the system at the sth iteration loop, which is called the sth order approximation energy. Numerical results show that with appropriate choice of ω, for the considered problem we obtain a series of approximate energies

$$E^{(0)},\ E^{(1)},\ E^{(2)}, \ldots,\ E^{(s)}, \ldots \tag{6.31}$$

which rapidly converges to a value E^T. We call them the exact numerical solutions because their values can be obtained with any given precision. In this work, we calculate these solutions with precision of up to 20 decimal places due to the limitation of the default precision for real numbers in FORTRAN and the computing speed of computer. However, this is an important progress. As we know, the precision for real numbers in FORTRAN is limited to 15 decimal places. By appropriate programming, we can increase this precision up to 50 decimal places. In our final results for energies and wave functions, we require a precision of up to 20 decimal places in order to avoid the accumulation of errors in calculating procedure. Besides energy, the coefficients $C_k^{(s)}$ also converge rapidly so we obtain not only exact numerical energy but also exact numerical wave functions.

The parameter ω is chosen using the method described in the previous chapter. In principle, this parameter does not affect the exact numerical results. However, investigation shows that the convergence rate to the exact solutions depends significantly on the choice of ω. In this work, the equation $\partial E^{(0)}/\partial\omega = 0$ provides the first value ω_0 which is not the optimal value. For high excited states n_r, such a choice of ω even does not lead to convergence to exact values. We can scan to find the optimal value of parameter around the first value of ω_0. The results in this chapter reconfirm the conclusions of [17, 18] about the existence of the range of parameter such that the convergence rate of (6.28) is highest.

6.3 Exact Numerical Solutions

Table 6.1 presents energies with precision of up to 20 decimal places for the ground state $1s$ $(n = m = 0)$ and some excited states $2p^-(n = 1, m = -1)$, $3d^-(n = 2, m = -2)$. These states were calculated with the precision of up to seven decimal places in [8, 9]. Here, the results in [8, 9] are not shown because all of these seven places overlap with our results shown in Table 6.1. The energy of the $2p^+(n = 1, m = +1)$, $3d^+(n = 2, m = +2)$ states can be calculated from that of the $2p^-$, $3d^-$ states based on the relation $E_{n(m)} = E_{n(-m)} + m\gamma$.

The energies with precision of up to 20 decimal places presented in Table 6.1 are a new record presented for the first time in the paper [14]. By choosing the optimal value of the free parameter, to obtain the ground state energy with seven decimal places, we need $s = 6$ iterations loops for $\gamma' = 0.05$ (weak magnetic field), $s = 8$ iteration loops for $\gamma' = 0.55$ (medium magnetic field), $s = 26$ iteration loops for $\gamma' = 0.95$ (strong magnetic field); and to obtain the precision of 20 decimal places we need $s = 16$, 35, 124 iterations loops for the three cases above, respectively. For excited states, more iteration loops are required but not more than 300. Although the results presented in Table 6.1 are with 20 decimal places, we did attempt to run the program to get up to 50 decimal places for the ground state and some low excited states. Hence, we conclude that the operator method applied in this chapter provides numerical results with any given precision. In other words, they are exact numerical solutions.

The wave function of the system in the form (6.28) is also obtained by calculating the coefficients $C_k^{(s)}$. The more iteration loops are carried out, the more coefficients are obtained. These coefficients converge to a certain value, and we obtain the exact numerical wave function. For example, for the ground state in weak magnetic field $\gamma' = 0.05$, when the energy is obtained with precision of 20 decimal places, the obtained wave function has 18 coefficients $C_k^{(15)}$, three of which are with precision of 20 decimal places and the others with precision ranging from 1 to 19 decimal places. Similarly, for the case $\gamma' = 0.55$, there are 38 coefficients $C_k^{(35)}$ with precision ranging from 1 to 19 decimal places. Furthermore, we can increase the

Table 6.1 Exact energies for the ground state and some low-order excited states with different values of magnetic field

γ'	$1s$	$2p^-$	$3d^-$
0.05	−1.99987017095990613170	−0.24474134557798881450	−0.11440510079734191774
0.10	−1.99942166507712501693	−0.26197520208978872077	−0.13025445178430606781
0.15	−1.99854256186645709957	−0.27410756889059385484	−0.13674173884597401955
0.20	−1.99707966222489033807	−0.28145662051014027268	−0.13659636693603053469
0.25	−1.99482091674920995739	−0.28409801197305611464	−0.13064401380817189943
0.30	−1.99146906712008245351	−0.28179705884272357348	−0.11888179385839621521
0.35	−1.98660128273873874036	−0.27398063451910408445	−0.10076026619266554000
0.40	−1.97960585091255737141	−0.25967324608666750763	−0.07521698217897997805
0.45	−1.96958003578721667516	−0.23736745684644382973	−0.04057485324928431054
0.50	−1.95515968324675916048	−0.20479038588298687027	0.00569412879010892499
0.55	−1.93422334282651768145	−0.15849200397641926789	0.06740516636374326600
0.60	−1.90335295328070946150	−0.09310153718673498788	0.15045496413924365797
0.65	−1.85679038695312423867	0.00008428907312278363	0.26432778727006523882
0.70	−1.78426176250820398205	0.13597809871739040778	0.42515253567577483784
0.75	−1.66598114826115218564	0.34214588246038558324	0.66251797890286951771
0.80	−1.45958713448989797632	0.67521872604493890722	1.03673987328882310035
0.85	−1.05902943085968226100	1.27112332610485384793	1.69141167638720390304
0.90	−0.12110157606243169047	2.55062439464316783261	3.06707775217066545341
0.95	3.23173503617812205039	6.70030514510251978855	7.42860734182398557718

These results are presented with 20 decimal places and overlap with the results after other authors with seven decimal places obtained from the variational method combined with $1/N$ expansion [8] and by the asymptotic perturbation method [9]. For comparison purpose, the magnetic field strength is represented by $\gamma' = \gamma/(\gamma + 1)$

precision of these coefficients up to 20 decimal places by running some more iteration loops.

We notice that the states presented in Table 6.1 and considered in the works [8,9] are special cases only. Indeed, they are $1s$ $(n = 0,\ m = 0)$, $2p^-$ $(n = 1,\ m = -1)$, $3d^-$ $(n = 2,\ m = -2)$ which correspond to the principal quantum number $n = |m|$. It means the radial quantum number $n_r = 0$ for all the states mentioned above. All of them are the lowest states when the magnetic quantum numbers are fixed, so they can be considered the ground states of a one-dimensional motion remaining after the motion related to angular momentum is taken off. For these states, the variational method works well but for states with $n_r \neq 0$, it does not. We have not found any follow-up works of [8,9] on these states. Therefore, our success in obtaining the exact solutions not only for the ground state but also for any excited states corresponding to $n_r = 0$ as well as $n_r \neq 0$ is a significant progress. The computing programs are tested for excited states with principal quantum number of up to $n = 150$. The exact energies for the states with the principal quantum number of up to $n = 4$: $2s$, $3s$, $3p^-$, $4p^-$, $4d^-$, $4f^-$, $5d^-$, $5f^-$, $5g^-$ are presented in Tables 6.2, 6.3 and 6.4 for illustration. These data which are original and are

Table 6.2 Energies of some excited states corresponding to different values of the magnetic field

γ'	$2s$	$3s$	$3p^-$
0.05	−0.21728279396968785397	−0.05137115812285705179	−0.08036721744503109760
0.10	−0.20160591618074023247	0.01598790970267759145	−0.04916051442900800150
0.15	−0.17457048111136615243	0.10955543404170749556	0.00108637565485044813
0.20	−0.13546551574551668372	0.22721834540118963879	0.06738328574548576958
0.25	−0.08298007150252920818	0.37047446264365413831	0.14986742555393444603
0.30	−0.01502952750059654076	0.54292733023387127727	0.25028633092701018244
0.35	0.07142846360607761967	0.75013268675298689141	0.37170668580947103502
0.40	0.18070180520537274524	1.00000000000000000000	0.51867414992190953559
0.45	0.31887814627104300055	1.30366017149884873723	0.69769322374130095062
0.50	0.49467963883695179668	1.67697071075735877489	0.91810396914076738153
0.55	0.72091579831145684540	2.14309105414624009242	1.19360373208799551825
0.60	1.01705668872877975106	2.73703806534873385927	1.54495188613538814666
0.65	1.41405521488835508770	3.51419634203232951294	2.00503229789835964832
0.70	1.96407032991584636290	4.56741386886003820741	2.62903276273533536486
0.75	2.76203075990704173477	6.06476489650834377662	3.51694576840641184652
0.80	4.00000000000000000000	8.34434942653877459541	4.87008864382744499370
0.85	6.13177508892586718321	12.19997778752228732308	7.16153917283062994829
0.90	10.53825335361872298997	20.02986405533713661963	11.82192792999323245234
0.95	24.24854780554921085193	43.93175572792464946928	26.07692664470341686614

Table 6.3 Energies of some high excited states with $n = 3$ corresponding to different values of the magnetic field

γ'	$4p^-$	$4d^-$	$4f^-$
0.05	0.00090050861670000681	−0.03603819112188519016	−0.07763315014992853255
0.10	0.09390796453557639448	0.00747840526168139693	−0.08803520910560353969
0.15	0.21323373403753960833	0.06842568965737196928	−0.08888339654701696510
0.20	0.35668048473475479870	0.14447370643180047832	−0.08315384587850637174
0.25	0.52606940637774516311	0.23628341974950389743	−0.07160076440801957770
0.30	0.72533207396702501971	0.34594127952399409512	−0.05412267756001832601
0.35	0.96036057230843709310	0.47676457059152532284	−0.03006828179726522415
0.40	1.23944041995229360305	0.63351735049084262159	0.00173399981723935616
0.45	1.57415466587235095938	0.82292516365874272605	0.04308672404388300910
0.50	1.98093584143111358970	1.05458031287213984997	0.09667213220019815981
0.55	2.48370977153057734079	1.34249561828604384524	0.16650643696559374134
0.60	3.11856112568506923453	1.70785470976146371028	0.25876268139530569281
0.65	3.94243430849379882861	2.18414967307751588723	0.38332586831977878052
0.70	5.05058514254628752285	2.82749853910929518464	0.55694141965414866451
0.75	6.61507130462100771829	3.73942662403763270927	0.81021701467250590897
0.80	8.98122289598444673098	5.12403034105099778843	1.20531843995110106672
0.85	12.95785147539592703023	7.46015607265407629634	1.88968201934687586146
0.90	20.98177530017893504943	12.19329771792373821144	3.31375703786321820117
0.95	45.30314463505568435467	26.60790678762850209795	7.78159090037994763851

Table 6.4 Energies of some excited states with $n = 4$ corresponding to different values of the magnetic field

γ'	$5d^-$	$5f^-$	$5g^-$
0.05	0.02905749316697708979	−0.01420058007237175099	−0.06007875519489516057
0.10	0.13341069872580500192	0.03570019722639813037	−0.06604657997486724854
0.15	0.26190026516127302629	0.10204170420617281261	−0.06292240231373530831
0.20	0.41356740099965750564	0.18308786847417280362	−0.05344145079515578704
0.25	0.59077510317782206889	0.27975368917130985634	−0.03821140930631437516
0.30	0.79776115680429092118	0.39427787158829185045	−0.01703067582967511760
0.35	1.04063336998574952266	0.53009357416698937372	0.01083372779566949526
0.40	1.32786142397127676506	0.69207101474623163562	0.04663124135459961939
0.45	1.67121288127923035467	0.88704788903212652389	0.09224854476563663044
0.50	2.08732965304714088906	1.12474823987081387279	0.15046790575342597724
0.55	2.60039856986566405608	1.41935251349090562787	0.22543362638168408645
0.60	3.24685460813888191684	1.79227191808221737000	0.32349322334645939986
0.65	4.08414376941622929138	2.27732643588718193308	0.45478350257101208093
0.70	5.20828847495739630265	2.93113687655407534814	0.63643631212786443537
0.75	6.79260902069767290808	3.85605806748978018262	0.89969714506691749633
0.80	9.18472403676149577875	5.25769124636313341999	1.30788909126295026871
0.85	13.19819451070249609938	7.61800718398620959057	2.01085156632785457493
0.90	21.28207427023502943214	12.39055080521797139007	3.46522653749514333096
0.95	45.73489523603524444325	26.89161351897346612252	7.99955272626774688799

presented for the first time in the paper [14] would be interesting for further reference.

We also found an interesting result while analyzing data in Table 6.2. For the case $\gamma' = 0.8$, corresponding to $\gamma = 4$, the energy of the $2s$ state is exactly equal to 4. Similarly, for the $3s$ state and $\gamma' = 0.4$, corresponding to $\gamma = 2/3$, the energy is exactly equal to 1. That means there are some values of magnetic field at which exact analytical solutions of the problem can be derived. This detection needs to be investigated in more details in another research.

The numerical results in the tables are also illustrated in Figs. 6.1 and 6.2. We see that for magnetic field $\gamma > 0.1$, the levels of energies corresponding to the principal quantum numbers begin to disarrange. For example, $2s$ and $2p^+$ levels are higher than $3d^-$ and $4f^-$ levels; $3s$ and $3p^+$ are higher than $4d^-, 4f^-$ levels. That is because we use the principal quantum numbers of the Coulomb problem which is reasonable only in weak magnetic field. In strong magnetic field, Coulomb interactions are considered as a perturbation which eliminates the degenerate Landau levels. Thus, in this case the principal quantum numbers must

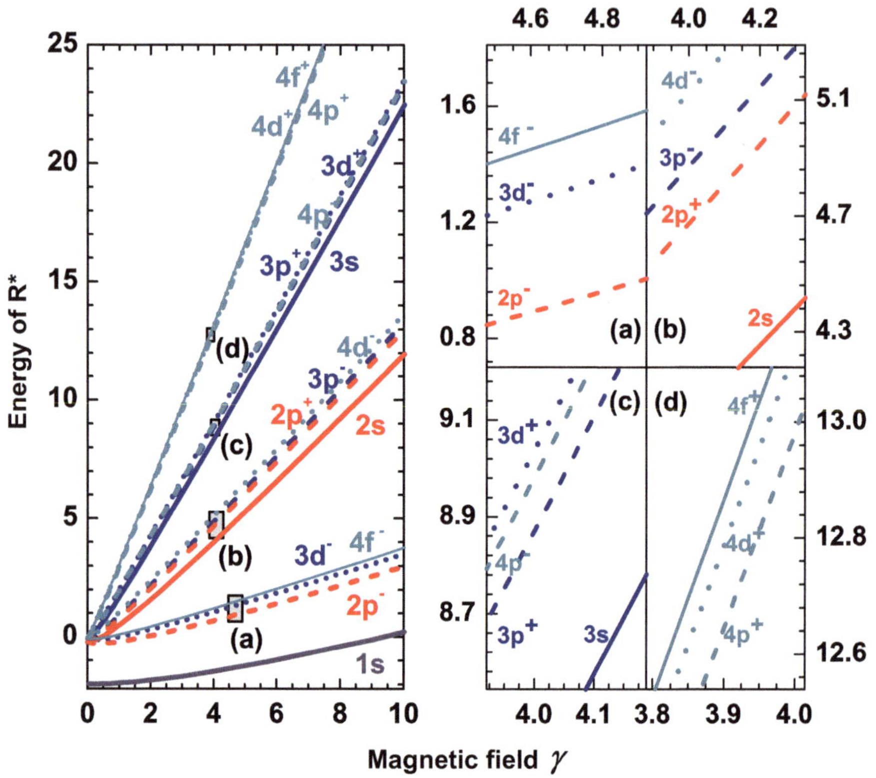

Fig. 6.1 The dependence of energy on magnetic field strength for the states with principal quantum number n=3: $1s$, $2s$, $2p^{\pm}$, $3s$, $3p^{\pm}$, $3d^{\pm}$, $4p^{\pm}$, $4d^{\pm}$, $4f^{\pm}$. We see that in strong magnetic fields, the degenerated Landau levels split out due to Coulomb interaction although they are still close to each other and are magnified in (a), (b), (c), and (d)

follow Landau levels in the problem of motion of electrons in a uniform magnetic field. Our finding of exact solutions for highly excited states with principal quantum number of up to hundreds allows investigation of not only the degenerate separation of Landau levels but also quantum chaos effects. These are some suggestions for further research.

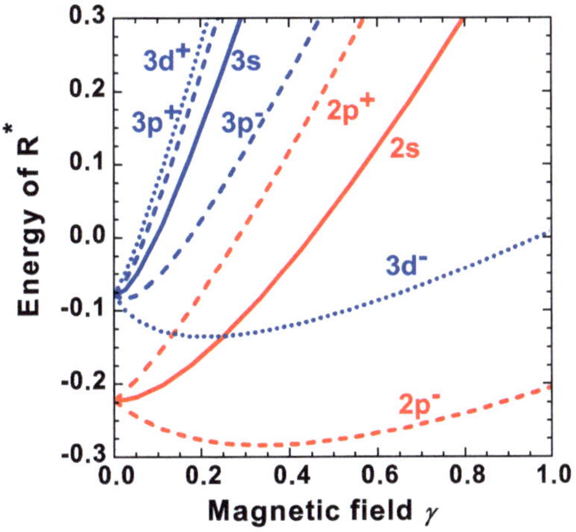

Fig. 6.2 Energy levels in $\gamma \leq 1$ region of magnetic field

6.4 Schrödinger Equation with Asymptotic Components

Figure 6.3 shows the energies of the system corresponding to the ground state and some excited states $1s$, $2s$, $2p^-$, $3d^-$. The dotted line represents the results, which can be considered analytical solutions, obtained with zeroth-order approximation using Eqs. (6.20) and (6.21). The solid line represents the exact numerical results for comparison purposes. These exact results are obtained from the iteration Eqs. (6.26) and (6.27). On this figure, we see that the analytical solutions are pretty precise for the ground and excited states in magnetic field region $\gamma \leq 1$. In stronger magnetic field, this precision decreases while the correlation between the energy levels stays the same.

In order to obtain analytical solutions which are highly precise and are uniformly suitable in the whole range of magnetic field, we need to consider the asymptotic behavior of the wave function in strong magnetic field region. We can easily verify that in strong magnetic field $\gamma >> 1$, the solutions of Eqs. (6.1) and (6.2) take the form:

$$\sim e^{-\frac{1}{4}\gamma(x^2+y^2)}. \tag{6.32}$$

Hence, we can find the solution of Eqs. (6.5) and (6.6) in the form $\Psi(u, v) = e^{-\alpha(u^2+v^2)^2}\psi(u, v)$ in which the parameter α is considered as a second free parameter of the operator method. Here we have used the relation $x^2 + y^2 = (u^2 + v^2)^2$ obtained from the Levi-Civita transformation. $\Psi(u, v)$ can be considered the wave function of the new equation as follows:

$$e^{-\alpha(u^2+v^2)^2}(\tilde{H} \ - Z)\, e^{-\alpha(u^2+v^2)^2}\psi\,(u, v) = 0. \tag{6.33}$$

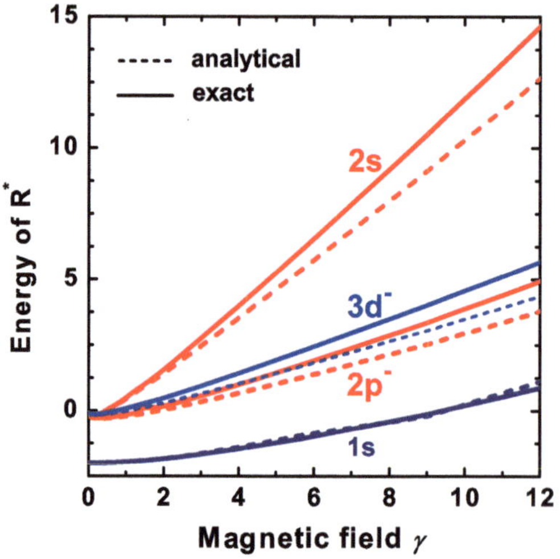

Fig. 6.3 Energies obtained from analytical equations are compared with exact energies of the system for the $1s$, $2s$, $2p^-$, $3d^-$ states. For the ground state, the analytical solution is pretty precise, while its error increases for the excited states in the region of strong magnetic field

By this way, the asymptotic component $e^{-\alpha(u^2+v^2)^2}$ is included in the Hamiltonian of the equation. We have multiplied this asymptotic component to the right side of the operator to make the new Hamiltonian being Hermitian. Equation (6.33) can be rewritten as follows:

$$e^{-2\alpha(u^2+v^2)^2}(\hat{H} - Z)\,\psi\,(u,v) = 0, \tag{6.34}$$

$$\begin{aligned}\hat{H} &= -\frac{1}{8}\left(\frac{\partial^2}{\partial u^2} + \frac{\partial^2}{\partial v^2}\right) - (E - 2\alpha - \frac{m\gamma}{2})(u^2 + v^2) \\ &+ \frac{\gamma^2 - 16\alpha^2}{8}(u^2 + v^2)^3 + \alpha(u^2 + v^2)\left(u\frac{\partial}{\partial u} + v\frac{\partial}{\partial v}\right).\end{aligned} \tag{6.35}$$

Note that Eqs. (6.5) and (6.6) describe a system in which the z-component of the angular momentum is conserved. This conservation still holds when these equations are transformed into (6.34) and (6.35) with the new wave function $\psi(u,v)$, which can easily be checked by calculating the commutation of $\hat{H}$ and $\hat{L}_z$. We can also see that the asymptotic operator $e^{-\alpha(u^2+v^2)^2}$ commutes with $\hat{L}_z$; therefore, if the wave function $\Psi\,(u,v)$ of Eqs. (6.5) and (6.6) is the eigenfunction of $\hat{L}_z$: $\hat{L}_z\,\psi(u,v) = m\,\psi(u,v)$ with the eigenvalues $m = 0, \pm 1, \pm 2, ..$, the same is applicable to the wave function of Eqs. (6.34) and (6.35). We replace $\hat{L}_z$ by its eigenvalues in (6.34) and (6.35) for this reason.

The operators in Eqs. (6.34) and (6.35) can be written in terms of the annihilation and creation operators as follows :

$$
\begin{aligned}
u^2+v^2 &= \frac{1}{2\omega}(\hat{a}^+\hat{b}^+ + \hat{a}\,\hat{b} + \hat{a}^+\hat{a} + \hat{b}^+\hat{b}+1) = \frac{1}{2\omega}(\hat{M}^+ + \hat{M} + \hat{N}+1),\\
\frac{\partial^2}{\partial u^2}+\frac{\partial^2}{\partial v^2} &= 2\omega\,(\hat{a}^+\hat{b}^+ + \hat{a}\,\hat{b} - \hat{a}^+\hat{a} - \hat{b}^+\hat{b}-1)\\
&= 2\omega\,(\hat{M}^+ + \hat{M} - \hat{N}-1), \qquad (6.36)\\
u\frac{\partial}{\partial u}+v\frac{\partial}{\partial v} &= \hat{a}\,\hat{b} - \hat{a}^+\hat{b}^+ - 1 = \hat{M} - \hat{M}^+ - 1,\\
\hat{L}_z = -i\left(u\frac{\partial}{\partial v} - v\frac{\partial}{\partial u}\right) &= -\frac{1}{2}(\hat{a}^+\hat{a} - \hat{b}^+\hat{b}).
\end{aligned}
$$

Here, the operators $\hat{M} = \hat{a}\,\hat{b}$, $\hat{M}^+ = \hat{a}^+\hat{b}^+$, $\hat{N}+1 = \hat{a}^+\hat{a} + \hat{b}^+\hat{b}+1$ are defined for convenience in calculation. These operators commute with the orbital angular momentum operator and make a closed algebra:

$$
\left[\hat{M},\hat{M}^+\right] = \hat{N}+1, \quad \left[\hat{M},\hat{N}+1\right] = 2\hat{M}, \quad \left[\hat{N}+1,\hat{M}^+\right] = 2\hat{M}^+. \qquad (6.37)
$$

Equations (6.34) and (6.35) can be rewritten in algebraic form as follows:

$$
\hat{A}(\hat{H}-Z)\;|\psi\rangle = 0, \qquad (6.38)
$$

in which the Hamiltonian $\hat{H}$ reads :

$$
\begin{aligned}
\hat{H} = &-\left(\frac{\omega}{4}+\frac{E}{2\omega}-\frac{\alpha}{\omega}-\frac{m\gamma}{4\omega}\right)(\hat{M}^+ + \hat{M})\\
&+\left(\frac{\omega}{4}-\frac{E}{2\omega}+\frac{\alpha}{\omega}+\frac{m\gamma}{4\omega}\right)(\hat{N}+1)\\
&+\frac{\gamma^2-16\alpha^2}{64\omega^3}(\hat{M}^+ + \hat{M} + \hat{N}+1)^3\\
&+\frac{\alpha}{2\omega}(\hat{M}^+ + \hat{M} + \hat{N}+1)(\hat{M} - \hat{M}^+ - 1). \qquad (6.39)
\end{aligned}
$$

The asymptotic operator in the equation is written via Fourier transformation as follows:

$$
\hat{A} = e^{-2\alpha(u^2+v^2)^2} = \frac{1}{2\sqrt{2\alpha\pi}}\int_{-\infty}^{+\infty} dk\, e^{-\frac{k^2}{8\alpha}}\, e^{-ik(u^2+v^2)}
$$

in order to be represented in the normal form of the creation and annihilation operators which allows algebraic calculation. This operator can be written in terms of the operators $\hat{M}$, $\hat{M}^+$, $\hat{N}+1$ as follows:

$$\hat{A} = \sqrt{\frac{\chi}{\pi}} \int_{-\infty}^{+\infty} dk\, e^{-\chi k^2} \exp\left\{\frac{-ik}{1+ik}\hat{M}^+\right\}$$
$$\times \exp\left\{(\hat{N}+1)\ln\frac{1}{1+ik}\right\} \exp\left\{\frac{-ik}{1+ik}\hat{M}\right\}, \tag{6.40}$$

in which a new parameter is defined $\chi = \omega^2/2\alpha$. For obtaining the normal form (6.40) of $\hat{A}$, we apply the algebraic equations (6.37) in the following procedure.

Let us consider the exponential component

$$\hat{S} = \exp\left(-ik\,(\hat{M}^+ + \hat{M} + \hat{N} + 1)\right)$$

in the asymptotic operator and convert it into the normal form so that the annihilation operators are on the right, the creation operators on the left and the neutral operators in the middle. To do so, we first define $\tau = ik$ which can be formally considered a real number. We need to find the functions $f_1(\tau)$, $f_2(\tau)$, $f_3(\tau)$ so that:

$$\begin{aligned}\hat{S} &= \exp\left(-\tau\,(\hat{M}^+ + \hat{M} + \hat{N} + 1)\right)\\ &= \exp\left(f_1(\tau)\,\hat{M}^+\right)\exp\left(f_2(\tau)\,(\hat{N}+1)\right)\exp\left(f_3(\tau)\,\hat{M}\right)\end{aligned} \tag{6.41}$$

with the boundary conditions:

$$f_1(0) = 0,\;\; f_2(0) = 0,\;\; f_3(0) = 0. \tag{6.42}$$

The functions can be found by following three steps:

Step 1: Differentiating both sides of (6.41) with respect to τ then multiplying with the inverse operator $\hat{S}^{-1}$ we obtain:

$$\begin{aligned}\hat{M}^+ + \hat{M} + \hat{N} + 1 = {} & f'_1(\tau)\,\hat{M}^+\\ & + f_2'(t)\exp\left(f_1(\tau)\,\hat{M}^+\right)(\hat{N}+1)\exp\left(-f_1(\tau)\,\hat{M}^+\right)\\ & + f'_3(\tau)\exp\left(f_1(\tau)\,\hat{M}^+\right)\exp\left(f_2'(\tau)\,(\hat{N}+1)\right)\\ & \times \hat{M}\exp\left(-f_2'(\tau)\,(\hat{N}+1)\right)\exp\left(-f_1(\tau)\,\hat{M}^+\right).\end{aligned} \tag{6.43}$$

Step 2: Applying the well-known equation below:

$$\exp\left(\hat{X}\right)\ \hat{Y}\ \exp\left(-\hat{X}\right) = \hat{Y} + \left[\hat{X}, \hat{Y}\right] + \frac{1}{2!}\left[\hat{X}, \left[\hat{X}, \hat{Y}\right]\right] + \frac{1}{3!}\left[\hat{X}, \left[\hat{X}, \left[\hat{X}, \hat{Y}\right]\right]\right] + \ldots$$

and the commutation relations (6.37) on (6.43), we obtain the equation:

$$-\hat{M}^{+} - \hat{M} - \hat{N} - 1 = f_1{}'(\tau)\,\hat{M}^{+} + f_2{}'(\tau)\left(\hat{N} + 1 - 2f_1(\tau)\hat{M}^{+}\right) + f'{}_3(\tau)\exp\left(-2f_2{}'(\tau)\right)\left(\hat{M} - f_1(\tau)(\hat{N}+1) + f_1^2(\tau)\hat{M}^{+}\right). \tag{6.44}$$

Step 3: Equating the factors in front of the operators $\hat{M}^{+},\ \hat{M},\ \hat{N}+1$ we have a system of differential equations that have exact solutions. By solving these equations with the use of the boundary conditions (6.42), we arrive at the desired functions:

$$f_1(\tau) = -\frac{\tau}{1+\tau}, \quad f_2(\tau) = -\ln|1+\tau|, \quad f_3(\tau) = -\frac{\tau}{1+\tau},$$

which means we have determined the normal form of the operator as in (6.40).

6.5 Highly Accurate Analytical Solutions

For the approximate analytical solutions we need to plug the wave function (6.19) into Eqs. (6.38) and (6.39) to obtain the approximate zeroth-order energy $E_{nm}(\omega, \chi)$ depending on the two parameters χ and ω. These parameters are determined from the notice that the exact solutions of the equation are independent of this parameter. First, we will calculate the matrix elements of the operators $\hat{H}$ and $\hat{A}$ in the wave function set (6.19) algebraically.

Using the commutation relations (6.37) of the operators $\hat{M},\ \hat{M}^{+},\ \hat{N}+1$ we have the following results for matrix elements of the operator $\hat{H}$:

$$H_{n+s,n} = \langle n+s\,(m)|\,\hat{H}\,|n\,(m)\rangle = \sqrt{\frac{(n+s+m)!(n+s-m)!}{(n+m)!(n-m)!}}\,h_{n+s,n},\ s = 0,1,2,3 \tag{6.45}$$

$$H_{n-s,n} = \langle n-s\,(m)|\,\hat{H}\,|n\,(m)\rangle = \sqrt{\frac{(n+m)!(n-m)!}{(n-s+m)!(n-s-m)!}}\,h_{n-s,n},\quad s = 1,2,3$$

in which

$$
\begin{aligned}
h_{n+3,n} &= \frac{\gamma^2}{64\omega^3} - \frac{\omega}{16\chi^2},\\
h_{n+2,n} &= \frac{3\gamma^2}{64\omega^3}(2n+3) - (2n+3)\frac{3\omega}{16\chi^2} - \frac{\omega}{4\chi},\\
h_{n+1,n} &= -\frac{\omega}{4} - \frac{2E-m\gamma}{4\omega} - \frac{3\omega}{16\chi^2}\alpha_n + \frac{3\gamma^2}{64\omega^3}\alpha_n - \frac{\omega}{2\chi}(n+1),\\
h_{n,n} &= (2n+1)\left(\frac{\omega}{4} - \frac{\omega}{8\chi^2}\beta_n - \frac{2E-m\gamma}{4\omega} + \frac{\gamma^2}{32\omega^3}\beta_n\right),\\
h_{n-1,n} &= -\frac{\omega}{4} - \frac{2E-m\gamma}{4\omega} - \frac{3\omega}{16\chi^2}\alpha_{n-1} + \frac{3\gamma^2}{64\omega^3}\alpha_{n-1} + \frac{\omega}{2\chi}n,\\
h_{n-2,n} &= -\frac{3\omega}{16\chi^2}(2n-1) + \frac{3\gamma^2}{64\omega^3}(2n-1) + \frac{\omega}{4\chi},\\
h_{n-3,n} &= \frac{\gamma^2}{64\omega^3} - \frac{\omega}{16\chi^2}.
\end{aligned}
$$

Here we use the following notations:

$$
\alpha_n = 5n^2 + 10n + 6 - m^2, \quad \alpha_{n-1} = 5n^2 + 1 - m^2, \quad \beta_n = 5n^2 + 5n + 3 - 3m^2.
$$

All other matrix elements of $\hat{H}$ besides (45) are zero. For the operator $\hat{A}$ we have the results:

$$
\begin{aligned}
A_{n+s,n} &= \langle n+s(m)|\,\hat{A}\,|n(m)\rangle\\
&= \sqrt{\frac{(n+m)!(n-m)!}{(n+s+m)!(n+s-m)!}}\,a_{n+s,n}, \quad s = 0,1,2,3,\ldots \qquad (6.46)\\
A_{n-s,n} &= \langle n-s(m)|\,\hat{A}\,|n(m)\rangle\\
&= \sqrt{\frac{(n-m)!(n+m)!}{(n-s-m)!(n-s+m)!}}\,a_{n-s,n}, \quad s = 0,1,2,3,\ldots
\end{aligned}
$$

in which

$$
\begin{aligned}
a_{n+s,n} = &\sum_{h=0}^{n-|m|} \frac{(n+s-m)!(n+s+m)!}{(h+s)!h!(n-m-h)!(n+m-h)!}\\
&\times \sum_{l=0}^{n+[s]} \frac{(-1)^{h+l+[s+1]}(2n+s+1)!}{(2l+\{s\})!(2n+2[s]-2l+1)!}\, I_{2n+s+1}^{h+l+[s+1]}(\chi),
\end{aligned}
$$

$$a_{n-s,n} = \sum_{h=0}^{n-s-|m|} \frac{(n-s-m)!(n-s+m)!}{(h-s)!h!(n-m-h)!(n+m-h)!}$$
$$\times \sum_{l=0}^{n-[s+1]} \frac{(-1)^{h+l-[s]}(2n+s+1)!}{(2l+\{s\})!(2n-2[s+1]-2l+1)!} I_{2n-s+1}^{h+l-[s]}(\chi).$$

Here we use the notations $\{s\} = 0$, $[s] = s/2$ if s is an even number, and $\{s\} = 1$, $[s] = (s-1)/2$ if s is an odd number. Besides, the integral $I_p^q(\chi)$ is defined as follows:

$$I_p^q(\chi) = \frac{2\chi^{p-q}}{\sqrt{\pi}} \int_0^{+\infty} dk \, \frac{e^{-k^2} k^{2q}}{(\chi + k^2)^p}. \tag{6.47}$$

We can use partial integration to lower the power of positive integers p, q and then convert it into the error integral $e^{\chi}\mathrm{erfc}\sqrt{\chi}$ [19].

To calculate the approximate zeroth-order energy, we plug the wave functions (6.19) into Eqs. (6.38) and (6.39) and use the explicit formulae of the matrix elements (6.45) and (6.46). We obtain a formula of energy in the following form, general for every state:

$$E_{nm}(\omega, \chi) = -\frac{1}{2}m\gamma + \omega^2 f_{nm}(\chi) - 2\omega Z \, g_{nm}(\chi) + \frac{\gamma^2}{32\omega^2} h_{nm}(\chi). \tag{6.48}$$

In (6.48) and from here on, we use the additional to ω parameter χ. We note that the functions $f(\chi)$, $g(\chi)$, $h(\chi)$ depend on χ only. These functions are different for each state; hence the indices of these functions in (6.48) are the characteristic quantum numbers of particular states. In the following equations, however, we neglect these indices for convenience. These functions have the following explicit form:

$$f(\chi) = \left[2(2n+1)a_{n,n} - 2a_{n,n+1} - 2a_{n,n-1} - \frac{d_n}{8\chi^2}\right.$$
$$\left. - \frac{1}{2\chi}(a_{n,n+2} + 2(n+1)a_{n,n+1} - 2na_{n,n-1} - a_{n,n-2})\right]$$
$$\times \left[a_{n,n+1} + a_{n,n-1} + (2n+1)a_{n,n}\right]^{-1}, \tag{6.49}$$
$$g(\chi) = \frac{a_{n,n}}{a_{n,n-1} + a_{n,n+1} + (2n+1)a_{n,n}},$$
$$h(\chi) = \frac{d_n}{a_{n,n+1} + a_{n,n-1} + (2n+1)a_{n,n}},$$

with

$$d_n(\chi) = a_{n,n-3} + 3(2n-1)a_{n,n-2} + 3(5n^2+1-m^2)a_{n,n-1}$$
$$+2(2n+1)\beta_n a_{n,n} + 3\alpha_n a_{n,n+1} + 3(2n+3)a_{n,n+2} + a_{n,n+3}.$$

The parameters $\omega,\ \chi$ in (6.48) are determined from the following equations:

$$\partial E(\omega,\chi)/\partial\omega = 0, \tag{6.50}$$

$$\partial E(\omega,\chi)/\partial\chi = 0. \tag{6.51}$$

Equation (6.50) is obtained from the fact that the exact solutions of the equation are independent of the free parameter. Although χ looks like a variational parameter, Eq. (6.51) is obtained by the same principle for Eq. (6.50). Indeed, both the parameters χ and ω define certain coordinate representation of the wave function whose choice definitely does not affect on the exact energy of the system. That leads to Eqs. (6.50) and (6.51) in the zeroth approximation.

After substituting (6.48) into (6.50) and (6.51) and rearranging, we obtain the following equations for ω and χ:

$$\omega = \frac{g(\chi)h'(\chi) - 2h(\chi)g'(\chi)}{2\,(h(\chi)f(\chi))'}, \tag{6.52}$$

$$\frac{\gamma^2}{\omega^3} = \frac{g(\chi)f'(\chi) - 2f(\chi)g'(\chi)}{2\,(h(\chi)f(\chi))'}. \tag{6.53}$$

We see that the parameter ω can be eliminated from the equations that leads to the analytical expressions of $\gamma(\chi)$, $E_{nm}(\chi)$, representing the dependence of energy on the magnetic field strength via the parameter χ. The explicit expressions of the functions $f(\chi)$, $g(\chi)$, $h(\chi)$ given in (6.49) allow us to obtain explicit expressions of energy for any excited states. Numerical analysis shows that the obtained analytical expressions have high precision which is uniformly stable in the whole range of the magnetic field γ.

Results for the Ground State

The energy of the ground state is:

$$E\,(\chi,\omega,\gamma) = \frac{\gamma^2}{16\omega^2}\frac{-\chi(1+\chi)+(3+2\chi)\chi I(\chi)}{(-1+I(\chi))}$$
$$+\omega^2\frac{-2+I(\chi)}{4\chi\,(-1+I(\chi))} + \omega Z\frac{I(\chi)}{\chi\,(-1+I(\chi))}, \tag{6.54}$$

in which ω can be written in terms of χ:

$$\begin{aligned}\omega(\chi) = & [8\chi(1+\chi) - 2\chi(13+8\chi)I(\chi) \\ & + \left(-3+18\chi+8\chi^2\right)I^2(\chi) + 6I^3(\chi)\big] \\ & \times [2(-2+\chi)\chi + (7-4\chi)\chi I(\chi) \\ & + \left(1-5\chi+2\chi^2\right)I^2(\chi) + 2\chi I^3(\chi)\big]^{-1}.\end{aligned} \tag{6.55}$$

We can also write the magnetic field strength in terms of χ as follows:

$$\begin{aligned}\frac{1}{4\omega^3}\gamma^2 = & \left[8\chi - 14\chi I(\chi) + (-3+6\chi)I^2(\chi) + 2I^3(\chi)\right]\chi^{-2} \\ & \times [2(-2+\chi)\chi + (7-4\chi)\chi I(\chi) \\ & + \left(1-5\chi+2\chi^2\right)I^2(\chi) + 2\chi I^3(\chi)\big]^{-1}.\end{aligned} \tag{6.56}$$

Here, we use the function $I(\chi)$ defined as follows:

$$I(\chi) = \frac{2\chi}{\sqrt{\pi}}\int_0^{+\infty} d\tau \frac{e^{-\tau^2}}{\chi+\tau^2} = \sqrt{\pi\chi}e^{\chi}\mathrm{erfc}(\sqrt{\chi}),$$

in which the function erfc(x) is an error integral. So we obtain the explicit expression for the energy of the ground state depending on the magnetic field strength in terms of χ. This energy can be rewritten as follows:

$$\begin{aligned}E(\chi) = & \Big[128\chi^2(1+\chi)^2 - 16\chi^2(42+73\chi+31\chi^2)I(\chi) \\ & +4\chi(-24+274\chi+473\chi^2+180\chi^3)I^2(\chi) \\ & -4\chi(-99+108\chi+314\chi^2+116\chi^3)I^3(\chi) \\ & +(18-483\chi-360\chi^2+228\chi^3+112\chi^4)I^4(\chi) \\ & +6(-9+24\chi+40\chi^2+8\chi^3)I^5(\chi) + 36(1+\chi)I^6(\chi)\Big](2\chi)^{-1} \\ & \times\left(2(-2+\chi)\chi + (7-4\chi)\chi I(\chi) + (1-5\chi+2\chi^2)I^2(\chi) + 2\chi I^3(\chi)\right)^{-2}.\end{aligned} \tag{6.57}$$

Figure 6.4a representing Eq. (6.55) shows the dependence of the parameter ω on χ. We see that ω takes values from 1.6 to 2.0 when χ varies from 0.0 to 150.0 (which can be considered infinitely large). Figure 6.4b represents the dependence of the magnetic field strength on the parameter χ. This dependence is one-fold, i.e. for each value of χ there is only one value of γ and from (6.57) we have one corresponding value of energy. In other words, we have obtained the dependence of the ground state energy on the magnetic field strength $E(\gamma)$, which is shown in

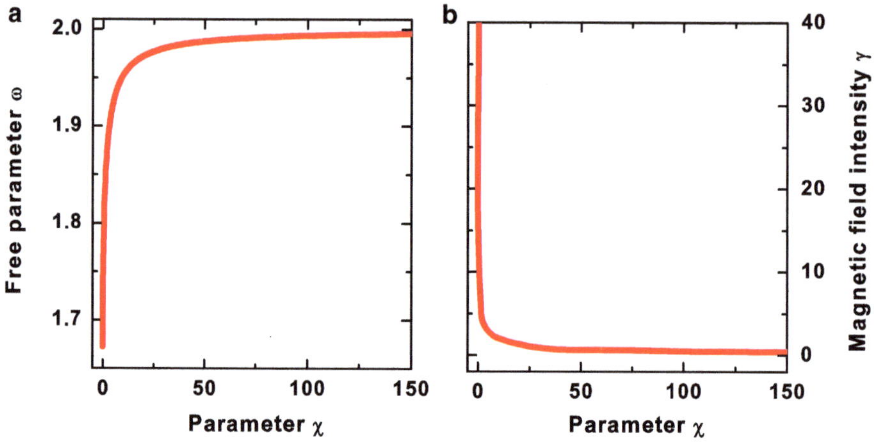

Fig. 6.4 The dependence of ω and magnetic field strength γ on χ

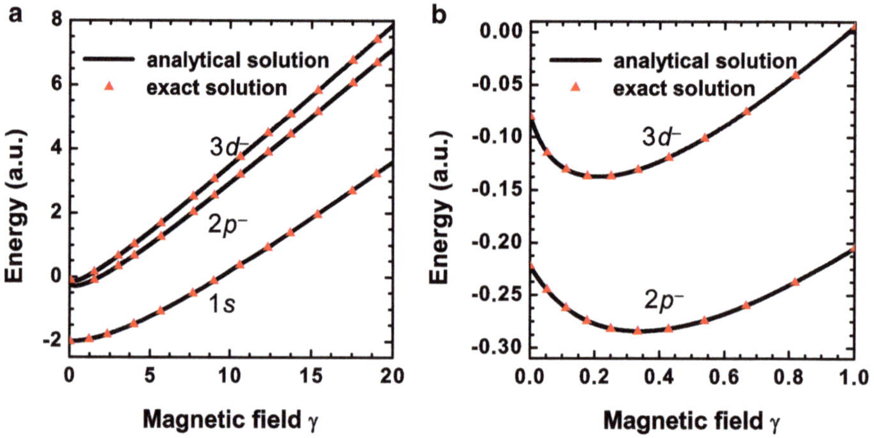

Fig. 6.5 Energies for some states as functions of magnetic field: (**a**) for a large range of magnetic field; (**b**) for weak magnetic field

Fig. 6.5a. The exact values of energy are also shown in this figure for comparison. In Table 6.5, specific numerical data is given and is compared with the exact values of energy to show the precision of the analytical expression (6.57). We see that this precision is very high and uniform over the whole range of the magnetic field for up to three decimal places. Note that we need to use the asymptotic expression of the error integral:

$$\mathrm{erfc}(x) = \frac{e^{-x^2}}{x\sqrt{\pi}}\left(1 - \frac{1}{2x^2} + \frac{3}{2^2x^4} - \frac{5\times 3}{2^3x^6} + \frac{7\times 5\times 3}{2^4x^8} - \dots\right),$$

Table 6.5 The analytical solution for the ground state $1s(n = 0, m = 0)$ in comparison with the exact numerical solution, assuming $\gamma' = \gamma/(\gamma + 1)$

γ'	Analytical solution	Exact numerical solution
0.05	−1.999870	−1.99987017095990613170
0.10	−1.999421	−1.99942166507712501693
0.15	−1.998542	−1.99854256186645709957
0.20	−1.997079	−1.99707966222489033807
0.25	−1.994819	−1.99482091674920995739
0.30	−1.991466	−1.99146906712008245351
0.35	−1.986595	−1.98660128273873874036
0.40	−1.979593	−1.97960585091255737141
0.45	−1.969555	−1.96958003578721667516
0.50	−1.955114	−1.95515968324675916048
0.55	−1.934144	−1.93422334282651768145
0.60	−1.903220	−1.90335295328070946150
0.65	−1.856577	−1.85679038695312423867
0.70	−1.783933	−1.78426176250820398205
0.75	−1.665490	−1.66598114826115218564
0.80	−1.458876	−1.45958713448989797632
0.85	−1.058024	−1.05902943085968226100
0.90	−0.119709	−0.12110157606243169047
0.95	3.233664	3.23173503617812205039

when calculating numerically the expression (6.57) with χ larger than 150.0 in the case of weak magnetic field to avoid calculating the exponential function with very large value.

We see in Fig. 6.4b that when $\gamma \to 0$ then $\chi \to +\infty$ and vice versa. Expanding (6.54) in terms of χ, we obtain the analytical expression for the asymptotic cases as follows:

$$E = 0.5\,\gamma - 1.253\,\sqrt{\gamma} - 0.688 - 0.463\frac{1}{\sqrt{\gamma}} + O\left[\gamma^{-3/2}\right], \quad \gamma \gg 1 \tag{6.58}$$

$$E = -2 + \frac{3}{64}\gamma^2 - \frac{153}{65536}\gamma^4 + O[\gamma^6], \quad \gamma \ll 1. \tag{6.59}$$

In order to demonstrate the accuracy of our analytical solutions for the whole range of the field intensity, we also calculate several terms of the exact asymptotic series of solution of Eqs. (6.1) and (6.2) by the conventional asymptotic method [20] in the cases $\gamma \ll 1$ (perturbation theory) and $\gamma \gg 1$ (strong coupling series). For the both cases, the first two terms of the series (6.58) and (6.59) coincide with that obtained by the conventional asymptotic method. The following terms are different between the two methods but the numerical analysis shows that this difference is very small. So we can say that our analytical solutions describe exactly the system in the asymptotic cases while the magnetic field is either very strong or very weak.

Results for Excited States

We can find the explicit expressions for the energy of excited states similar to (6.54), (6.55) and (6.56) for the ground state. More exactly, we have obtained the explicit expressions for the energy of the $2p^-$ and $3d^-$ states similar to (6.57) although they are longer and are not reported here. We consider these states as examples for excited states. Table 6.6 presents the energy in the whole range of the magnetic field strength in comparison with the exact values to demonstrate the high precision of the analytical solutions for these excited states. Figure 6.5a, b also show the values in Table 6.6. We see that for all three considered states, the analytical energies obtained have uniform precision of up to four decimal places in the whole range of the magnetic field strength.

Table 6.6 Analytical solution for the excited states $2p^-(n = 1, m = -1)$ and $3d^-(n = 2, m = -2)$ in comparison with the exact numerical solution, assuming $\gamma' = \gamma/(\gamma + 1)$

	$2p^-$		$3d^-$	
γ'	Analytical solution	Exact numerical solution	Analytical solution	Exact numerical solution
0.05	−0.244740	−0.24474134557798881450	−0.114399	−0.11440510079734191774
0.10	−0.261966	−0.26197520208978872077	−0.130243	−0.13025445178430606781
0.15	−0.274086	−0.27410756889059385484	−0.136728	−0.13674173884597401955
0.20	−0.281422	−0.28145662051014027268	−0.136581	−0.13659636693603053469
0.25	−0.284053	−0.28409801197305611464	−0.130628	−0.13064401380817189943
0.30	−0.281742	−0.28179705884272357348	−0.118865	−0.11888179385839621521
0.35	−0.273918	−0.27398063451910408445	−0.100743	−0.10076026619266554000
0.40	−0.259604	−0.25967324608666750763	−0.075200	−0.07521698217897997805
0.45	−0.237293	−0.23736745684644382973	−0.040557	−0.04057485324928431054
0.50	−0.204711	−0.20479038588298687027	0.005711	0.00569412879010892499
0.55	−0.158409	−0.15849200397641926789	0.067422	0.06740516636374326600
0.60	−0.093015	−0.09310153718673498788	0.150472	0.15045496413924365797
0.65	0.000173	0.00008428907312278363	0.264345	0.26432778727006523882
0.70	0.136069	0.13597809871739040778	0.425170	0.42515253567577483784
0.75	0.342239	0.34214588246038558324	0.662535	0.66251797890286951771
0.80	0.675315	0.67521872604493890722	1.036757	1.03673987328882310035
0.85	1.271222	1.27112332610485384793	1.691429	1.69141167638720390304
0.90	2.550724	2.55062439464316783261	3.067092	3.06707775217066545341
0.95	6.700411	6.70030514510251978855	7.428620	7.42860734182398557718

The Wave Function in Algebraic Form

The approximate zeroth-order wave function with asymptotic component can also be obtained in algebraic form as follows:

$$|\Psi_{nm}\rangle = \sqrt{\frac{\chi}{\pi}} \int_{-\infty}^{+\infty} dk\, e^{-\chi k^2} \exp\left\{\frac{-ik}{1+ik}\hat{M}^+\right\}$$
$$\times \exp\left\{(\hat{N}+1)\ln\frac{1}{1+ik}\right\} \exp\left\{\frac{-ik}{1+ik}\hat{M}\right\} |n(m)\rangle\,.$$

Expanding the operators in exponential form, applying them on the state function, using commutation relations (6.37) and definition of the vacuum state (6.17), we obtain the wave function in the following form:

$$|\Psi_{nm}\rangle = \sum_{l=0}^{+\infty}\sum_{j=0}^{n-|m|} \frac{\sqrt{(n-m)!(n+m)!(n-j+l+m)!(n-j+l-m)!}}{l!j!(n-j+m)!(n-j-m)!}$$
$$\times \sqrt{\frac{2\chi}{\pi}} \int_{-\infty}^{+\infty} dk\, e^{-2\chi k^2} \frac{(-ik)^{j+l}}{(1+ik)^{2n-j+l+1}} |n-j+l(m)\rangle\,. \tag{6.60}$$

Note that when calculating the matrix elements using the wave function (6.60), the integrals in the final results are similar to (6.47) and can be converted to the error integral.

6.6 OM Application to Complex Two-Dimensional Atomic Systems

In the previous sections, we have shown that the operator method can be successfully applied to the problem of two-dimensional exciton in a magnetic field with arbitrary strength. We obtain not only exact numerical solutions but also approximate analytical solutions with a very high precision in the whole range of the magnetic field strength. In this section, we will show the application can be performed for other more complex two-dimensional atomic systems: (1) two-dimensional screened donor state in a magnetic field; (2) two-dimensional negatively charged exciton.

Two-Dimensional Screened Donor State in a Magnetic Field

For the problem of two-dimensional screened donor state, the Hamiltonian is different from the one of an exciton in a magnetic field only in the term of interaction. Instead of the Coulomb interaction, the term $-Ze^{-\lambda r}/r$ is used with the screening parameter λ [8, 9]. While transforming the Schrödinger equation into the space with variables u, v by the Levi-Civita transformation this term becomes:

$$\hat{S} = \exp\left[-\lambda(\mathrm{u}^2 + \mathrm{v}^2)\right]. \tag{6.61}$$

The system with the screened potential remains the conservation of the orbital angular momentum. Therefore, the way of solving the problem is the same as in Sects. 6.1, 6.2 and 6.3 for the exciton in a magnetic field. The only difference is additional calculation of matrix elements of the operator (6.61).

The operator $\hat{S}$ in the form of the annihilation and creation operators is easy to transform into the normal form by the procedure similar to (6.41)–(6.44):

$$\begin{aligned}\hat{S} = \exp\left\{-\frac{\lambda}{\lambda + 2\omega}\hat{M}^{+}\right\} \exp\left\{\left(\ln\frac{2\omega}{\lambda + 2\omega}\right)(\hat{N} + 1)\right\} \\ \times \exp\left\{-\frac{\lambda}{\lambda + 2\omega}\hat{M}\right\}.\end{aligned} \tag{6.62}$$

For simplicity, we use the notation $\mu = \lambda/2\omega$. Matrix elements of the operator (6.62) can be calculated algebraically; as a result we have:

$$\begin{aligned}S_{n,n+s} &= \langle n(m)|\,\hat{S}\,|n + s(m)\rangle \\ &= \frac{1}{s!}\sqrt{\frac{(n+s-m)!(n+s+m)!}{(n-m)!(n+m)!}}\,\frac{(-\mu)^s}{(1+\mu)^{2n+s+1}}F_s(n,m,\mu^2),\end{aligned} \tag{6.63}$$

with $F_s(n,m,x)$ is the Gaussian hypergeometric function, defined as follows:

$$\begin{aligned}F_s(n,m,x) &= {}_2F_1(m-n,-m-n;s+1;x) \\ &= \sum_{k=0}^{n-|m|}\frac{s!(n-m)!(n+m)!}{k!(k+s)!(n-m-k)!(n+m-k)!}x^k.\end{aligned}$$

A computing program was constructed by FORTRAN that allows to calculate the wave function and energy for any state with a precision up to 20 decimal places. The program is checked for value of the screening parameter up to $\lambda = 5.0$, and for the excited states with the principal quantum numbers up to $n = 150$, $n_r = 3$.

Two-Dimensional Negatively Charged Exciton

Negatively charged exciton is a very important subject for both theoretical and experimental investigations [21–23]. The solutions of the Schrödinger equation for three-body systems in two-dimensional space are only reported for helium [24, 25]. However, in those works there is no discussion about the negatively charged exciton. Formally, a negatively charged exciton is similar to the ion H^- or the helium atom but its bound energy is much smaller. Besides, the effective masses of electron and of hole are in the same order of size in principle, therefore we cannot use the Born-Oppenheimer approximation for movement of the hole. Thus, the negatively charged exciton is not similar to the helium at all.

The Schrödinger equation for a negatively charged exciton in two-dimensional space has the form :

$$\hat{H}\Psi(\mathrm{r}_1,\mathrm{r}_2) = E\,\Psi(\mathrm{r}_1,\mathrm{r}_2), \tag{6.64}$$

$$\hat{H} = -\frac{1}{2}\Delta_1 - \frac{1}{2}\Delta_2 - \alpha_h\left(\frac{\partial^2}{\partial x_1\partial x_2} + \frac{\partial^2}{\partial y_1\partial y_2}\right) - \frac{Z}{r_1} - \frac{Z}{r_2} + \frac{1}{|\mathrm{r}_1 - \mathrm{r}_2|}. \tag{6.65}$$

This is the equation describing the relative movement of two electrons with respect to the hole. The movement of center of mass was separated and not considered here. The quantities $r_1 = \sqrt{x_1{}^2 + y_1{}^2}$, $r_2 = \sqrt{x_2{}^2 + y_2{}^2}$ are correspondingly the relative distances from electron to the hole ; $r_{12} = |\mathrm{r}_1 - \mathrm{r}_2| = \sqrt{(x_1 - x_2)^2 + (y_1 - y_2)^2}$ is the distance between two electrons. We use the notations:

$$\Delta_1 = \frac{\partial^2}{\partial x_1{}^2} + \frac{\partial^2}{\partial y_1{}^2},\quad \Delta_2 = \frac{\partial^2}{\partial x_2{}^2} + \frac{\partial^2}{\partial y_2{}^2},\quad \alpha_h = \frac{1}{1 + m_h^*/m_e^*},$$

in which m_e^*, m_h^* are the effective masses of electron and hole in superconductors.

Here we use the atomic unit system, in which the distance and energy units are the effective Bohr radius $a_0{}^* = 4\pi\varepsilon_0\hbar^2/\mu e^2$ and the effective Rydberg constant $R_y{}^* = \mu e^4/16\pi^2\varepsilon_0{}^2\hbar^2$ correspondingly. Because the relative movement between electron and hole is considered, the reduced mass $\mu = m_e^* m_h^*/(m_e^* + m_h^*)$ is used. For example, for the semiconductor GaAs/AlGaAs, from the data $m_e^* \simeq 0.067\ m_e$ and $m_h^* \simeq 0.45\ m_e$, we have the reduced mass $\mu \simeq 0.058\ m_e^*$. Therefore, the effective Bohr radius increases by about 17 times against the Bohr radius; and the effective Rydberg constant is about 17 times smaller than the Rydberg constant. The other difference between the negatively charged exciton and helium atom is the additional term in the Hamiltonian (6.65):

$$\hat{J} = \alpha_h\left(\frac{\partial^2}{\partial x_1\partial x_2} + \frac{\partial^2}{\partial y_1\partial y_2}\right), \tag{6.66}$$

which is neglected in the case of helium atom because of the smallness of the coefficient $\alpha_h \simeq 0.00014$. For the negatively charged exciton in the semiconductor GaAs/AlGaAs, $\alpha_h \simeq 0.13$ so the term (6.66) of the Hamiltonian could not be neglected.

Fortunately, including the term (6.66) in the Hamiltonian, the orbital angular momentum is still conserved. Therefore, the Schrödinger equation (6.64) should be solved with the following equations:

$$\hat{L}_z \Psi(\mathrm{r}_1, \mathrm{r}_2) = m\, \Psi(\mathrm{r}_1, \mathrm{r}_2), \tag{6.67}$$

$$\hat{L}_z = i\left(x_1 \frac{\partial}{\partial y_1} - y_1 \frac{\partial}{\partial x_1} + x_2 \frac{\partial}{\partial y_2} - y_2 \frac{\partial}{\partial x_2}\right), \tag{6.68}$$

with $m = 0, \pm 1, \pm 2, \ldots$ is a magnetic quantum number.

The algebraic method will be used for calculation via the annihilation and creation operators, defined as follows:

$$\begin{aligned} \hat{a}_s^+ &= \sqrt{\frac{\omega}{2}}\left(x_s - \frac{1}{\omega}\frac{\partial}{\partial x_s}\right), \quad \hat{a}_s = \sqrt{\frac{\omega}{2}}\left(x_s + \frac{1}{\omega}\frac{\partial}{\partial x_s}\right), \\ \hat{b}_s^+ &= \sqrt{\frac{\omega}{2}}\left(y_s - \frac{1}{\omega}\frac{\partial}{\partial y_s}\right), \quad \hat{b}_s = \sqrt{\frac{\omega}{2}}\left(y_s + \frac{1}{\omega}\frac{\partial}{\partial y_s}\right). \end{aligned} \tag{6.69}$$

Here, index $s = 1, 2$; the free parameter ω is a positive real number.

The operators defined in (6.69) and having the same nature (creation operator if having the sign $(+)$; annihilation operator if not having the sign $(+)$), commute each other. Otherwise, they satisfy the following commutation relations:

$$\left[\hat{a}_s, \hat{a}_t^+\right] = \delta_{st}, \left[\hat{b}_s, \hat{b}_t^+\right] = \delta_{st}. \tag{6.70}$$

For diagonalization of the operator $\hat{L}_z$, we rewrite the creation and annihilation operators by the canonical transformations:

$$\begin{aligned} \hat{u}_s^+ &= \frac{1}{\sqrt{2}}\left(\hat{a}_s^+ + i\hat{b}_s^+\right), \hat{u}_s = \frac{1}{\sqrt{2}}(\hat{a}_s - i\hat{b}_s), \\ \hat{v}_s^+ &= \frac{1}{\sqrt{2}}\left(\hat{a}_s^+ - i\hat{b}_s^+\right), \hat{v}_s = \frac{1}{\sqrt{2}}\left(\hat{a}_s + i\hat{b}_s\right). \end{aligned} \tag{6.71}$$

The new operators satisfy the same commutation relations (6.70). Now the operators in Hamiltonian (6.65) are rewritten in term of annihilation and creation operators (6.71):

$$\Delta_1 = -\frac{\omega}{2}(\hat{M}_1^+ - \hat{N}_1 + \hat{M}_1), \qquad \Delta_2 = -\frac{\omega}{2}(\hat{M}_2^+ - \hat{N}_2 + \hat{M}_2),$$

$$\begin{aligned} &{x_1}^2 + {y_1}^2 = \frac{1}{2\omega}(\hat{M}_1^+ + \hat{N}_1 + \hat{M}_1),\ {x_2}^2 + {y_2}^2 = \frac{1}{2\omega}(\hat{M}_2^+ + \hat{N}_2 + \hat{M}_2), \\ &(x_1 - x_2)^2 + (y_1 - y_2)^2 = \frac{1}{2\omega}(\hat{M} + \hat{M}^+ + \hat{N} - 2\hat{m}^+ - 2\hat{m} - 2\hat{n}), \end{aligned} \tag{6.72}$$

$$\frac{\partial^2}{\partial x_1 \partial x_2} + \frac{\partial^2}{\partial y_1 \partial y_2} = \frac{1}{2\omega}(\hat{m}^+ + \hat{m} - \hat{n}).$$

In (6.72) the notations $\hat{M}^+ = \hat{M}_1^+ + \hat{M}_2^+$, $\hat{M} = \hat{M}_1 + \hat{M}_2$, $\hat{N} = \hat{N}_1 + \hat{N}_2$, $\hat{n} = \hat{n}_1 + \hat{n}_2$ are used for the operators defined as follows :

$$\begin{aligned} &\hat{M}_1^+ = 2\hat{u}_1^+\hat{v}_1^+,\ \hat{M}_1 = 2\hat{u}_1\hat{v}_1,\ \hat{M}_2^+ = 2\hat{u}_2^+\hat{v}_2^+,\ \hat{M}_2 = 2\hat{u}_2\hat{v}_2, \\ &\hat{N}_1 = 2\hat{u}_1^+\hat{u}_1 + 2\hat{v}_1^+\hat{v}_1 + 2, \quad \hat{N}_2 = 2\hat{u}_2^+\hat{u}_2 + 2\hat{v}_2^+\hat{v}_2 + 2, \\ &\hat{m}^+ = \hat{v}_1^+\hat{u}_2^+ + \hat{u}_1^+\hat{v}_2^+, \qquad \hat{m} = \hat{v}_1\hat{u}_2 + \hat{u}_1\hat{v}_2, \\ &\hat{n}_1 = \hat{u}_2^+\hat{u}_1 + \hat{v}_2^+\hat{v}_1, \qquad \hat{n}_2 = \hat{u}_1^+\hat{u}_2 + \hat{v}_1^+\hat{v}_2. \end{aligned} \tag{6.73}$$

Also the operator of orbital angular momentum has the form:

$$\hat{L}_z = \hat{u}_1^+\hat{u}_1 - \hat{v}_1^+\hat{v}_1 + \hat{u}_2^+\hat{u}_2 - \hat{v}_2^+\hat{v}_2 \tag{6.74}$$

that is a neutral operator, commuting with all operators in (6.73). It is important for algebraic calculations that the operators (6.73) build a closed algebra with the following commutation relations:

$$\begin{aligned} &\left[\hat{M}_1,\ \hat{N}_1\right] = 4\hat{M}_1,\ \left[\hat{M}_1, \hat{M}_1^+\right] = 2\hat{N}_1,\ \left[\hat{N}_1,\ \hat{M}_1^+\right] = 4\hat{M}_1^+, \\ &\left[\hat{M}_2,\ \hat{N}_2\right] = 4\hat{M}_2,\ \left[\hat{M}_2,\ \hat{M}_2^+\right] = 2\hat{N}_2,\ \left[\hat{N}_2,\ \hat{M}_2^+\right] = 4\hat{M}_2^+, \\ &\left[\hat{m},\ \hat{m}^+\right] = \tfrac{1}{2}(\hat{N}_1 + \hat{N}_2), [\hat{n}_1,\ \hat{n}_2] = -\tfrac{1}{2}(\hat{N}_1 - \hat{N}_2), \\ &\left[\hat{N}_1,\ \hat{m}^+\right] = 2\hat{m}^+,\ \left[\hat{N}_1,\ \hat{m}^+\right] = 2\hat{m}^+,\ \left[\hat{m}, \hat{N}_1\right] = 2\hat{m},\ \left[\hat{m}, \hat{N}_2\right] = 2\hat{m}, \\ &[\hat{m},\ \hat{n}_1] = \hat{M}_1, [\hat{m},\ \hat{n}_2] = \hat{M}_2,\ \left[\hat{n}_1,\ \hat{m}^+\right] = \hat{M}_2^+, \left[\hat{n}_2,\ \hat{m}^+\right] = \hat{M}_1^+, \\ &\left[\hat{M}_1,\ \hat{n}_1\right] = 0, \left[\hat{M}_2,\ \hat{n}_1\right] = 2\hat{m}, \left[\hat{M}_1,\ \hat{n}_2\right] = 2\hat{m}, \left[\hat{M}_2,\ \hat{n}_2\right] = 0, \\ &\left[\hat{n}_2, \hat{M}_1^+\right] = 0, \left[\hat{n}_2, \hat{M}_2^+\right] = 2\hat{m}^+, \left[\hat{n}_1, \hat{M}_1^+\right] = 2\hat{m}^+, \left[\hat{n}_1, \hat{M}_2^+\right] = 0, \\ &\left[\hat{m},\ \hat{M}_1^+\right] = 2\hat{n}_2, \left[\hat{m},\ \hat{M}_2^+\right] = 2\hat{n}_1, \left[\hat{M}_1, \hat{m}^+\right] = 2\hat{n}_1, \left[\hat{M}_2, \hat{m}^+\right] = 2\hat{n}_2, \\ &\left[\hat{N}_1,\ \hat{n}_1\right] = -2\hat{n}_1, \left[\hat{N}_2,\ \hat{n}_1\right] = 2\hat{n}_1, \left[\hat{N}_1,\ \hat{n}_2\right] = -2\hat{n}_2, \left[\hat{N}_2,\ \hat{n}_2\right] = 2\hat{n}_2. \end{aligned} \tag{6.75}$$

Apart from the relations (6.75), other commutation relations are equal zero. Indeed, beside the operators $\hat{n}_1,\ \hat{n}_2$ and $\hat{N}_1,\ \hat{N}_2$, other operators with the sign $(+)$ are creation operators, without the sign $(+)$ are annihilation operators. All creation operators (or annihilation operators) commute each to other. Besides, the operators with the same index (1 or 2) commute each to other, except for operators $\hat{n}_1,\ \hat{n}_2$.

In the Hamiltonian (6.65) there are several terms related to the Coulomb interaction that contain coordinates in denominator. We will use the Laplace transformation to lead these terms into the form suitable for algebraic calculations. The basic transformation is as follows:

$$\hat{U} = \frac{1}{r} = \frac{1}{\sqrt{\pi}} \int\limits_0^{+\infty} \frac{e^{-tr^2}}{\sqrt{t}}\, dt. \tag{6.76}$$

By using Eqs. (6.76) and (6.72) we rewrite the Coulomb interaction terms:

$$\begin{aligned}
\hat{V}_1 &= -\frac{Z}{r_1} = -\frac{Z}{\sqrt{\pi}} \int\limits_0^{+\infty} \frac{dt}{\sqrt{t}} \exp\left[\frac{-t}{2\omega}\left(\hat{M}_1^+ + \hat{N}_1 + \hat{M}_1\right)\right], \\
\hat{V}_2 &= -\frac{Z}{r_2} = -\frac{Z}{\sqrt{\pi}} \int\limits_0^{+\infty} \frac{dt}{\sqrt{t}} \exp\left[\frac{-t}{2\omega}\left(\hat{M}_2^+ + \hat{N}_2 + \hat{M}_2\right)\right], \\
\hat{V}_{12} &= \frac{1}{|\mathrm{r}_1 - \mathrm{r}_2|} = \frac{Z}{\sqrt{\pi}} \int\limits_0^{+\infty} \frac{dt}{\sqrt{t}} \\
&\quad \times \exp\left[-\frac{t}{2\omega}(\hat{M} + \hat{M}^+ + \hat{N} - 2\hat{m}^+ - 2\hat{m} - 2\hat{n})\right].
\end{aligned} \tag{6.77}$$

The closed algebra (6.75) allows us to use the procedure similar to (6.42)–(6.44) to have the normal form of the operators:

$$\begin{aligned}
\hat{V}_1 &= -\frac{\sqrt{2\omega} Z}{\sqrt{\pi}} \int\limits_0^{+\infty} dt \frac{1}{\sqrt{t}} \exp\left(-\frac{t}{1+2t}\hat{M}_1^+\right) \\
&\quad \times \exp\left(-\hat{N}_1 \ln\sqrt{1+2t}\right) \exp\left(-\frac{t}{1+2t}\hat{M}_1\right), \\
\hat{V}_2 &= -\frac{\sqrt{2\omega} Z}{\sqrt{\pi}} \int\limits_0^{+\infty} dt \frac{1}{\sqrt{t}} \exp\left(-\frac{t}{1+2t}\hat{M}_2^+\right) \\
&\quad \times \exp\left(-\hat{N}_2 \ln\sqrt{1+2t}\right) \exp\left(-\frac{t}{1+2t}\hat{M}_2\right),
\end{aligned} \tag{6.78}$$

$$\hat{V}_{12} = \frac{\sqrt{2\omega}}{\sqrt{\pi}} \int_0^{+\infty} \frac{dt}{\sqrt{t}} \exp\left(-\frac{t}{1+4t}\hat{M}^+\right) \exp\left(\frac{2t}{1+4t}\hat{m}^+\right)$$
$$\times \exp\left(-\frac{1}{2}\ln\sqrt{1+4t}\;\hat{N}\right) \exp\left(\ln\sqrt{1+4t}\;\hat{n}\right)$$
$$\times \exp\left(-\frac{t}{1+4t}\hat{M}\right) \exp\left(\frac{2t}{1+4t}\hat{m}\right).$$

Now we will construct the basis set of wave functions in the algebraic representation. From the eigenfunctions of the neutral operators $\hat{u}_1^+\hat{u}_1,\ \hat{u}_2^+\hat{u}_2,\ \hat{v}_1^+\hat{v}_1,\ \hat{v}_2^+\hat{v}_2$, we can easily build the basis set as follows :

$$|j_1 j_2 j_3 j_4\rangle = \frac{1}{\sqrt{j_1!j_2!j_3!j_4!}} (\hat{u}_1^+)^{j_1} (\hat{v}_1^+)^{j_2} \hat{u}_2^{+j_3} (\hat{v}_2^+)^{j_4} |0\,(\omega)\rangle\,, \qquad (6.79)$$

with $j_1,\ j_2,\ j_3,\ j_4$ are the non-negative integer. The vacuum state is a solution of the equations:

$$\hat{u}_1\,|0(\omega)\rangle = 0,\quad \hat{u}_2\,|0(\omega)\rangle = 0,\quad \hat{v}_1\,|0(\omega)\rangle = 0,\quad \hat{v}_2\,|0(\omega)\rangle = 0, \qquad (6.80)$$

with the normalization condition: $\langle 0(\omega)\ |0(\omega)\rangle = 1$.

As mentioned above, due to the conservation of the orbital angular momentum, we should construct basis set from the eigenfunctions of the operator $\hat{L}_z$. The operator $\hat{L}_z$ is diagonal so the eigenvalue is $m = j_1 + j_3 - j_2 - j_4$. We determine the principal quantum number $N = j_1 + j_2 + j_3 + j_4$ and see that $N = 2j_1 + 2j_3 - m = 2j_2 + 2j_4 + m$. Because N is a non-negative integer, it should be $N = 2n + |m|$, with $n = 0,\ 1,\ 2, \ldots$ related to the numbers $j_1,\ j_2,\ j_3,\ j_4$ as follows: (a) $n = j_1 + j_3$ if $m < 0$, (b) $n = j_2 + j_4$ if $m \geq 0$. For convenience, we use four quantum numbers $n,\ j_1,\ j_2,\ m$ for characterization of the states in the basis set and rewrite (6.79) as follows:

$$|n, j_1, j_2, m\rangle = \frac{1}{\sqrt{j_1!j_2!(n+|m|-j_1)!(n-j_2)!}}$$
$$\times (\hat{u}_1^+)^{j_1} (\hat{v}_1^+)^{j_2} (\hat{u}_2^+)^{n+|m|-j_1} (\hat{v}_2^+)^{n-j_2} |0\,(\omega)\rangle\,, \qquad (6.81)$$

for the case of $m \geq 0$, and

$$|n, j_1, j_2, m\rangle = \frac{1}{\sqrt{j_1!j_2!(n-j_1)!(n+|m|-j_2)!}}$$
$$\times (\hat{u}_1^+)^{j_1} (\hat{v}_1^+)^{j_2} (\hat{u}_2^+)^{n-j_1} (\hat{v}_2^+)^{n+|m|-j_2} |0\,(\omega)\rangle\,, \qquad (6.82)$$

for the case of $m < 0$. Here, n is a non-negative integer; m is an integer; j_1, j_2 are two non-negative integers satisfying the conditions: $n + |m| \geq j_1 \geq 0$ and $n \geq j_2 \geq 0$ in the case of $m \geq 0$; or $n \geq j_1 \geq 0$, $n + |m| \geq j_2 \geq 0$ in the case of $m < 0$.

We consider here a specific case when $m = 0$. Note that the Hamiltonian (6.65) contains only the operators given in (6.73), we construct the new basis set as a combination of (6.81) and (6.82) in order to have the form:

$$|n, j_1, j_2\rangle = (\hat{M}_1^+)^{j_1} (\hat{M}_2^+)^{j_2} (m^+)^{n-j_1-j_2} |0(\omega)\rangle . \tag{6.83}$$

The use of new basis set (6.83) will save much more computing resource than using the basis set (6.81) and (6.82). The matrix elements of Hamiltonian can be calculated by the algebraic method based on the commutation relation (6.70) for the basic wave functions (6.81) and (6.82); or based on the commutation relation (6.75) for the basic wave functions (6.83). Here we give some formulae for calculation of matrix elements with the basis set (6.83) as follows:

$$\begin{aligned}
\hat{N}_1 |n, j_1, j_2\rangle &= 2(n + j_1 - j_2 + 1) |n, j_1, j_2\rangle , \\
\hat{N}_2 |n, j_1, j_2\rangle &= 2(n - j_1 + j_2 + 1) |n, j_1, j_2\rangle , \\
\hat{n}_1 |n, j_1, j_2\rangle &= (n - j_1 - j_2) |n, j_1, j_2 + 1\rangle + 2j_1 |n, j_1 - 1, j_2\rangle , \\
\hat{n}_2 |n, j_1, j_2\rangle &= (n - j_1 - j_2) |n, j_1 + 1, j_2\rangle + 2j_2 |n, j_1, j_2 - 1\rangle , \\
\hat{M}_1 |n, j_1, j_2\rangle &= 4j_1(n - j_2) |n - 1, j_1 - 1, j_2\rangle \\
&\quad + (n - j_1 - j_2)(n - j_1 - j_2 - 1) |n - 1, j_1, j_2 + 1\rangle , \qquad (6.84) \\
\hat{M}_2 |n, j_1, j_2\rangle &= 4j_2(n - j_1) |n - 1, j_1, j_2 - 1\rangle \\
&\quad + (n - j_1 - j_2)(n - j_1 - j_2 - 1) |n - 1, j_1 + 1, j_2\rangle , \\
\hat{m} |n, j_1, j_2\rangle &= (n + j_1 + j_2 + 1)(n - j_1 - j_2) |n - 1, j_1, j_2\rangle \\
&\quad + 4j_1 j_2 |n - 1, j_1 - 1, j_2 - 1\rangle .
\end{aligned}$$

Consequently, we have constructed a closed algebra including the quadratic operators of the annihilation and creation operators and then have successfully expressed the Hamiltonian of the negatively charged exciton via these operators. This result allows us to use purely algebraic calculations while solving the Schrödinger equation for the considered system by the operator method. The basis set of wave functions was also constructed via the annihilation and creation operators and the useful formulae are established for calculation of matrix elements. All calculations in this section are enough for applying the operator method for the negatively charged exciton.

References

1. W. Edelstein, Phys. Rev. B **39**, 7697 (1989)
2. A.H. MacDonald, D.S. Ritchie, Phys. Rev. B **33**, 8336 (1986)
3. J.-L. Zhu, Y. Cheng, J.-J. Xiong, Phys. Rev. B **41**, 10792 (1990)
4. G.V. Astakhov, D.R. Yakovlev, V.V. Rudenkov, P.C.H. Christianen, T. Barrick, S.A. Gooker, A.B. Dzyubenko, W. Ossau, J.C. Maan, G. Karczewshi, T. Wojtowicz, Phys. Rev. B **71**, 201312 (2005)
5. A. Bruno-Alfonso, L. Candido, G.Q. Haiz, J. Phys. Condens. Matter **22**, 125801 (2010)
6. A. Poszwa, Phys. Scr. **84**, 055002 (2011)
7. D. Nandi, A.D.K. Finck, J.P. Eisenstein, L.N. Pfeiffer, K.W. West, Nature **488**, 481 (2012)
8. V.M. Villalba, R. Pino, Physica B **315**, 289 (2002)
9. A. Soylu, I. Boztosun, Physica E **40**, 443 (2008)
10. T. Levi-Civita, Acta Math. **30**, 305 (1906)
11. V.-H. Le, T.-G. Nguyen, J. Phys. A Math. Gen. **26**, 1409 (1993)
12. J. Zhu, S.L. Ban, S.H. Ha, Phys. Stat. Sol. B **248**, 384 (2011)
13. F. Milota, J. Sperling, A. Nemeth, T. Mancal, H.F. Kauffmann, Acc. Chem. Res. **42**, 1364 (2009)
14. N.-T. Hoang-Do, D.-L. Pham, V.-H. Le, Physica B **423**, 31 (2013)
15. N.-T. Hoang-Do, V.-H. Hoang, V.-H. Le, J. Math. Phys. **54**, 052105 (2013)
16. N.-T. Hoang-Do, Q.-G. Le, T.-M. Nguyen, V.-H. Le, J. Sci. HCMC UP Nat. Sci. Tech. **43**, 23 (2013)
17. C.Z. An, I.D. Feranchuk, L.I. Komarov, L.I. Nakhamchik, J. Phys. A Math. Gen. **19**, 1583 (1986)
18. Q.-K. Hoang, V.-H. Le, L.I. Komarov, Proc. Natl. Acad. Sci. Belarus Ser Phys. Math. Sci. **3**, 71 (1997)
19. M. Abramowitz, I.A. Stegun, *Handbook of Mathematical Functions with Formulas, Graphs, and Mathematical Tables*, 9th edn. (Dover, New York, 1972)
20. F.T. Hioe, D. MacMillen, E.W. Montroll, Phys. Rep. **43**, 305 (1978)
21. L.V. Keldysh, A.N. Kozlov, Zh. Eksp. Teor. Fiz. **54**, 978 (1968)
22. B.P. Zakharchenya, Usp. Fiz. Nauk **164**, 345 (1994)
23. S. Takeyama, Y. Natori, Y. Hirayama, E. Kojima, Y. Arishima, H. Mino, G. Karczewski, T. Wojtowicz, J. Kossut, J. Phys. Soc. Jpn. **77**, 044702 (2008)
24. L. Hilico, B. Gremaud, T. Jonckheere, N. Billy, D. Delande, Phys. Rev. A **66**, 022101 (2002)
25. S.H. Patil, Eur. J. Phys. **29**, 517 (2008)

Chapter 7
Atoms in the External Electromagnetic Fields

In this chapter, the operator method is applied to analyze the following models intensively used for the description of the interaction of atomic systems with external fields: (1) hydrogen-like atom in electric and magnetic fields, and (2) two-level atom in resonant electromagnetic field.

Hydrogen-like atom in electric field or/and in magnetic field is an old problem involving many investigations resulted in discovery of extremely important quantum effects such as the Zeeman effect as well as the Stark effect [1–4]. For a long time, the problem of hydrogen-like atom in a constant magnetic field has been considered as a model for testing various methods of solving the dynamical equations of microscopic world [5–9]. Recently, this problem is of great interest because of the new observations in astrophysics related to the measurement of spectrum of atoms in a very strong magnetic field [10, 11]. From other side, the problem of hydrogen-like atom in electric field remains interesting because of the new trend of research about interaction of ultra-short intense laser pulses with atoms and molecules. Particularly, the dependence of photo-ionization rate in the parameters of electric field is of interest recently [12, 13].

The application of OM to the problem of hydrogen-like atom in a strong magnetic field was given in the works [14–16]. It was shown that the zero approximation of OM gives a high accuracy and the precision is uniformly suitable for the whole range of the magnetic intensity. In the work [16] the possibility of calculating higher corrections by the iteration scheme was also given. The advantage of OM is that the calculation process is general enough for applying to electromagnetic fields of various configurations. In the work [17], OM was used for calculation of highly excited Rydberg states of hydrogen atom in an electric field. The obtained results fitted well to the experimental data [18].

To apply the algebraic method with using annihilation and creation operators for the atomic problems, the Kustaanheimo-Stiefel transformation [19] was used in [20–23]. The problem of hydrogen-like atom thus can be transformed to the harmonic oscillator in four-dimensional space. In this chapter we will recall

I. Feranchuk et al., *Non-perturbative Description of Quantum Systems*,
Lecture Notes in Physics 894, DOI 10.1007/978-3-319-13006-4_7

the main points of this connection, especially the algebraic calculus, and show the application of the method for hydrogen atom in magnetic and in electric fields.

The model of a two-level system in a one-mode quantum field (TLS) is one of the simplest but widespread and effective model used for consideration of the qualitative characteristics of the interaction between a quantum system and a resonant external field (see, for example, [24] and papers cited therein). It is of great interest both as a mathematical problem and as a good physical model for consideration of the non-adiabatic transitions [25], the squeezed states [26], the quantum chaotic system [27] and other effects.

Accurate eigenvalues and eigenfunctions of TLS interacting with a single-mode quantum field are calculated analytically and numerically by means of special iteration procedure based on OM. This procedure permits one to consider the solution within the wide range of the Hamiltonian parameters and to find the uniformly approximating analytical formula for the eigenvalues [28]. The characteristic features of the model such as the level intersections, the population of the field states and the TLS evolution on the basis of OM were considered in the recent publications [29–31].

7.1 Hydrogen-Like Atom and Harmonic Oscillator

We will consider the dimensionless Schrödinger equation for four-dimensional harmonic oscillator. For convenience in notation and calculation, the two-dimensional complex coordinates are used instead of the four-dimensional real coordinates. We will call it the ξ-space and the equation in the ξ-space is as follows:

$$\hat{H}\,\psi(\xi) = Z\,\psi(\xi), \tag{7.1}$$

$$\hat{H} = -\frac{1}{2}\frac{\partial^2}{\partial\xi_s\partial\xi_s^*} + \frac{1}{2}\omega^2\xi_s\xi_s^*. \tag{7.2}$$

In the Hamiltonian (7.1) and further on the repeating of indices s, t mean a summation over all values of these indices $s, t = 1, 2$. Here we also use the transformation of coordinates:

$$\begin{cases} x_\lambda = \xi_s^*(\sigma_\lambda)_{st}\,\xi_t, \\ \phi = \arg(\xi_1), \end{cases} \tag{7.3}$$

with $\lambda = 1, 2, 3$ corresponding to three coordinates x_1, x_2, x_3. The extra variable ϕ could contain certain physics interpretation [32], however, it does not have any special meaning in our further calculations and the physical wave functions will be required to be independent on this variable. In Eq. (7.3) we use the Pauli matrices in a following explicit form:

$$\sigma_1 = \begin{pmatrix} 0 & 1 \\ 1 & 0 \end{pmatrix}, \quad \sigma_2 = \begin{pmatrix} 0 & -i \\ i & 0 \end{pmatrix}, \quad \sigma_3 = \begin{pmatrix} 1 & 0 \\ 0 & -1 \end{pmatrix}.$$

Via the transformation (7.3), Eqs. (7.1) and (7.2) written in ξ-space becomes an equation in the real $x,\ y,\ z,\ \phi$ coordinates:

$$\left\{ -\frac{1}{2} r \left(\frac{\partial}{\partial x_\lambda} + A_\lambda(\mathbf{r}) \frac{\partial}{\partial \phi} \right) \left(\frac{\partial}{\partial x_\lambda} + A_\lambda(\mathbf{r}) \frac{\partial}{\partial \phi} \right) \right.$$
$$\left. - \frac{1}{8r} \frac{\partial^2}{\partial \phi^2} + \frac{1}{2} \omega^2 r \right\} \psi(\mathbf{r}, \phi) = Z \psi(\mathbf{r}, \phi). \tag{7.4}$$

We are interested only on the physically meaningful wave function $\psi(\mathbf{r})$, independent on the variable ϕ, then Eq. (7.4) becomes the one for a hydrogen-like atom:

$$\left(-\frac{1}{2} \frac{\partial^2}{\partial x_\lambda \partial x_\lambda} - \frac{Z}{r} \right) \psi(\mathbf{r}) = E \psi(\mathbf{r}), \tag{7.5}$$

with the energy in the discrete region $E = -\omega^2/2$. Here and further on the repeating of Greek indices means summation over them in a whole region: $\lambda = 1, 2, 3$.

We note that to obtain Eq. (7.5) from Eq. (7.4) the last should be divided by r. To explain this circumstance we look at the identity:

$$\frac{1}{8r} dx_1 dx_2 dx_3 d\phi = d\xi_1' d\xi_1'' d\xi_2' d\xi_2'',$$

for transforming the ξ-space into the space of coordinates $\mathbf{r}, \phi$. Here we use notations: $\xi_s' \equiv \mathrm{Re}\xi_s$, $\xi_s'' \equiv \mathrm{Im}\xi_s$. Because we use the definition of the scalar product of two wave functions for each space as follows:

$$\langle \psi_a(\mathbf{r}) | \, \psi_b(\mathbf{r}) \rangle = \int_{-\infty}^{+\infty} dx_1 \int_{-\infty}^{+\infty} dx_2 \int_{-\infty}^{+\infty} dx_3 \, \psi_a^*(\mathbf{r}) \, \psi_b(\mathbf{r}), \tag{7.6}$$

$$\langle \psi_a(\xi) | \, \psi_b(\xi) \rangle = \int_{-\infty}^{+\infty} d\xi_1' \int_{-\infty}^{+\infty} d\xi_1'' \int_{-\infty}^{+\infty} d\xi_2' \int_{-\infty}^{+\infty} d\xi_2'' \psi_a^*(\xi) \, \psi_b(\xi), \tag{7.7}$$

we can see that if operator $\hat{A}$ is hermitic in the ξ then the operator $\hat{A}/r$ will be hermitic in the $\mathbf{r}$-space. Instead of solving the Schrödinger (7.5) for hydrogen-like atom we can deal with the Eqs. (7.1) and (7.2) for harmonic oscillator in two-dimensional complex space. The wave function has to be independent on the extra variable ϕ, which means the wave functions in ξ-space should satisfy the following equation:

$$\frac{d}{d\phi}\psi(\mathbf{r}) = 0 \Leftrightarrow \left(\xi_s^* \frac{\partial}{\partial \xi_s^*} - \xi_s \frac{\partial}{\partial \xi_s}\right)\psi(\xi) = 0. \tag{7.8}$$

For the problem of hydrogen-like atom in electromagnetic field $\mathbf{A}(\mathbf{r})$, $\varphi(\mathbf{r})$, we consider the following equation in ξ -space:

$$\left\{-\frac{1}{2}\frac{\partial^2}{\partial \xi_s \partial \xi_s^*} + \frac{1}{2}\omega^2 \xi_s \xi_s^* - \frac{1}{2}B_s^* \frac{\partial}{\partial \xi_s} - \frac{1}{2}B_s \frac{\partial}{\partial \xi_s^*} \right.$$
$$\left. - \frac{1}{4}\frac{\partial B_s^*}{\partial \xi_s} - \frac{1}{4}\frac{\partial B_s}{\partial \xi_s^*} - \frac{1}{2}B_s^* B_s \right\} \psi(\xi) = Z\psi(\xi). \tag{7.9}$$

Here we have the connection between $(\mathbf{A}, \varphi) \Leftrightarrow \left(B_s, B_s^*\right)$:

$$\begin{cases} A_\lambda(\mathbf{r}) = \dfrac{i}{2r}(\sigma_\lambda)_{st}(\xi_s^* B_t^* - \xi_s B_t), \\ \varphi(\mathbf{r}) = \dfrac{1}{8r^2}(\sigma_\lambda)_{st}(\xi_s^* B_s^* + \xi_s B_s)^2. \end{cases} \tag{7.10}$$

From (7.10) we can get equations for inversion of the fields:

$$\begin{cases} B_s(\xi) = i(\sigma_\lambda)_{ts}\xi_t^* A_\lambda + \xi_s^* \sqrt{2\varphi}, \\ B_s^*(\xi) = -i(\sigma_\lambda)_{st}\xi_t A_\lambda + \xi_s \sqrt{2\varphi}. \end{cases} \tag{7.11}$$

For the problem of hydrogen atom in electromagnetic fields we can consider the Eq. (7.9) in ξ-space with the potentials defined by Eq. (7.11) and with the wave function satisfying also the Eq. (7.8).

As has been shown in the works [20, 22, 23], the algebraic method of solving Schrödinger equation based on the connection between the Coulomb problem and the problem of harmonic oscillator is possible and this fact will be used hereinafter. The readers interested in a relativistic are referred to [21].

Based on the fundamentals described above, the wave functions of a hydrogen-like atom can be extracted from the wave functions of the harmonic oscillator in ξ-space by enforcing them to satisfy the condition (7.8). For the hydrogen-like atom with nuclear charge Z and energy $\mathrm{e} = -\omega^2/2$ we consider the Schrödinger equation (7.1) and (7.2) for the harmonic oscillator in ξ-space in the form:

$$-\frac{1}{2}\omega\,(\hat{a}_s^+ \hat{a}_s + \hat{b}_s^+ \hat{b}_s + 2)\,|\psi\rangle = Z\,|\psi\rangle\,, \tag{7.12}$$

where the annihilation and creation operators are defined as follows:

$$\begin{cases} \hat{a}_s(\omega) = \sqrt{\frac{\omega}{2}}\left(\xi_s + \frac{1}{\omega}\frac{\partial}{\partial \xi_s^*}\right), & \hat{a}_s^+(\omega) = \sqrt{\frac{\omega}{2}}\left(\xi_s^* - \frac{1}{\omega}\frac{\partial}{\partial \xi_s}\right), \\ \hat{b}_s(\omega) = \sqrt{\frac{\omega}{2}}\left(\xi_s^* + \frac{1}{\omega}\frac{\partial}{\partial \xi_s}\right), & \hat{b}_s^+(\omega) = \sqrt{\frac{\omega}{2}}\left(\xi_s - \frac{1}{\omega}\frac{\partial}{\partial \xi_s^*}\right), \end{cases} \tag{7.13}$$

and the vacuum state $|0(\omega)\rangle$ is determined by the equations:

$$\hat{a}_s\,|0(\omega)\rangle = 0\,,\quad \hat{b}_s\,|0(\omega)\rangle = 0, \tag{7.14}$$

with the normalization condition: $\langle 0(\omega) \mid 0(\omega)\rangle = 1$. Operators (7.13) satisfy the following standard commutation relations:

$$\left[\hat{a}_s(\omega), \hat{a}_t^+(\omega)\right] = \delta_{st}\,,\quad \left[\hat{b}_s(\omega), \hat{b}_t^+(\omega)\right] = \delta_{st} \tag{7.15}$$

and other commutation relations are equal to zero.

The wave functions of the harmonic oscillator in ξ-space, which are in fact the solutions of the Eq. (7.12), can be represented in the form:

$$|n_1, n_2, n_3, n_4\,, \omega\rangle = \frac{1}{\sqrt{n_1!n_2!n_3!n_4!}}(\hat{a}_1^+)^{n_1}(\hat{a}_2^+)^{n_2}(\hat{b}_1^+)^{n_3}(\hat{b}_2^+)^{n_4}\,|0(\omega)\rangle\,, \tag{7.16}$$

with n_1, n_2, n_3, n_4 are non-negative integers. From (7.16), we will now establish the state vectors for the hydrogen-like atom in the discrete region of energy. For this purpose, the state vectors (7.16) have to satisfy the condition (7.8), which has the following simple form in the algebraic representation of annihilation and creation operators:

$$(\hat{a}_s^+\hat{a}_s - \hat{b}_s^+\hat{b}_s)\,|\psi\rangle = 0. \tag{7.17}$$

Substituting the wave vector (7.16) into Eqs. (7.12) and (7.17), we obtain the following equations for eigenvalues:

$$-\frac{1}{2}\omega\,(n_1 + n_2 + n_3 + n_4 + 2) = Z, \tag{7.18a}$$

$$n_1 + n_2 = n_3 + n_4. \tag{7.18b}$$

Equation (7.18b) is the condition for the state vector (7.16) to be the wave function of the hydrogen-like atom. This condition can be rewritten in the form: $n_a = n_b \equiv n$, which means that in the wave function (7.16) the quantity $n_a = n_1 + n_2$ of the operators $\hat{a}_s^+$ equals to the one of the operators $\hat{b}_s^+$: $n_b = n_3 + n_4$. Equation (7.18b) leads to the energy level of hydrogen-like atom:

$$\omega_n = \frac{Z}{n+1}, \quad E_n = -\frac{Z^2}{2(n+1)^2}. \tag{7.19}$$

Thus the matrix elements of any physical operator $\hat{A}$:

$$\begin{aligned}\langle\psi_j(\mathbf{r})|\,\hat{A}\,|\psi_k(\mathbf{r})\rangle &= \int\limits_{-\infty}^{+\infty} dx_1 \int\limits_{-\infty}^{+\infty} dx_2 \int\limits_{-\infty}^{+\infty} dx_3\, \psi_j^*(\mathbf{r})\hat{A}\,\psi_k(\mathbf{r}) \\ &= \langle j,\omega_j|\, r\hat{A}\,|k,\omega_k\rangle\end{aligned} \tag{7.20}$$

can be calculated algebraically. The possibility of purely algebraic calculation follows from several circumstances. First, there exists a unitary transformation which transforms the wave function from certain frequency ω to another frequency ω':

$$|\psi(\omega')\rangle = U(\omega',\omega)\,|\psi(\omega)\rangle\,\hat{a}_s(\omega') = U(\omega',\omega)\,\hat{a}_s(\omega)\,U^{-1}(\omega',\omega), \tag{7.21}$$

where $U(\omega',\omega) = \exp\left\{\frac{1}{2}\ln\frac{\omega'}{\omega}\left[\hat{a}_s(\omega)\hat{b}_s(\omega) - \hat{a}_s^+(\omega)\hat{b}_s^+(\omega)\right]\right\}$ or in the normal form:

$$\begin{aligned}U(\omega',\omega) = &\frac{4\omega\omega'}{(\omega+\omega')^2}\exp\left[-\frac{\omega'-\omega}{\omega'+\omega}\hat{M}^+(\omega)\right] \\ &\times\exp\left[-\ln\frac{\omega'+\omega}{2\sqrt{\omega'\omega}}\hat{N}(\omega)\right]\exp\left[\frac{\omega'-\omega}{\omega'+\omega}\hat{M}(\omega)\right].\end{aligned} \tag{7.22}$$

For convenience, in Eq. (7.22) and henceforth, we use the notations:

$$\begin{aligned}&\hat{M} = \hat{a}_s\hat{b}_s, \quad \hat{M}^+ = \hat{a}_s^+\hat{b}_s^+, \quad \hat{N}+2 = \hat{a}_s^+\hat{a}_s + \hat{b}_s^+\hat{b}_s + 2\,, \\ &\hat{m}_\lambda = (\sigma_\lambda)_{st}\hat{a}_t\hat{b}_s, \qquad \hat{m}_\lambda^+ = (\sigma_\lambda)_{st}\hat{a}_s^+\hat{b}_t^+, \\ &\hat{n}_\lambda^a = (\sigma_\lambda)_{st}\hat{a}_s^+\hat{a}_t, \qquad \hat{n}_\lambda^b = (\sigma_\lambda)_{st}\hat{b}_t^+\hat{b}_s.\end{aligned} \tag{7.23}$$

Second, all operators corresponding to physical quantities can be expressed via the combinations (7.23) of annihilation and creation operators. We write here, for example, some operators:

$$r = \xi_s^*\xi_s = \frac{1}{2\omega}(\hat{M} + \hat{M}^+ + \hat{N} + 2),$$

$$x_\lambda = \xi_s^*(\sigma_\lambda)_{st}\xi_t = \frac{1}{2\omega}(\hat{m}_\lambda + \hat{m}_\lambda^+ + \hat{n}_\lambda^a + \hat{n}_\lambda^b),$$

$$r\mathbf{p}^2 = -\frac{\partial^2}{\partial\xi_s\partial\xi_s^*} = \frac{\omega}{2}(\hat{N}+2-\hat{M}-\hat{M}^+),\ \ rp_\lambda = -\frac{i}{2}(\hat{m}_\lambda-\hat{m}_\lambda^+), \tag{7.24}$$

$$\hat{l}_\lambda = \frac{1}{2}(\sigma_\lambda)_{st}\left(\xi_t\frac{\partial}{\partial\xi_s} - \xi_s^*\frac{\partial}{\partial\xi_t^*}\right) = \frac{1}{2}(\hat{n}_\lambda^a - \hat{n}_\lambda^b),$$

$$\hat{L}^2 = \hat{l}_\lambda\hat{l}_\lambda = \frac{1}{4}\hat{N}(\hat{N}+2) - \hat{M}^+\hat{M},$$

$$x_\lambda p_\lambda = -\frac{i}{2}\left(\xi_s\frac{\partial}{\partial\xi_s} + \xi_s^*\frac{\partial}{\partial\xi_s^*}\right) = \frac{1}{2}\omega\,(\hat{N}+2-\hat{M}^+-\hat{M}^+).$$

The fifteen operators in (7.23) form a closed algebra, isomorphic to the algebra of the Lie group SO(4,2) of the dynamic symmetry of hydrogen-like atom due to the commutation relations:

$$\begin{aligned}
&\left[\hat{m}_\lambda,\hat{m}_\mu^+\right] = \delta_{\lambda\mu}(\hat{N}+2) - i\varepsilon_{\lambda\mu\nu}(\hat{n}_\nu^a - \hat{n}_\nu^b),\\
&\left[\hat{n}_\lambda^a,\hat{n}_\mu^a\right] = 2i\varepsilon_{\lambda\mu\nu}\hat{n}_\nu^a, && \left[\hat{n}_\lambda^b,\hat{n}_\mu^b\right] = -2i\varepsilon_{\lambda\mu\nu}\hat{n}_\nu^b,\\
&\left[\hat{m}_\lambda,\hat{n}_\mu^a+\hat{n}_\mu^b\right] = 2\delta_{\lambda\mu}\hat{M}, && \left[\hat{m}_\lambda,\hat{n}_\mu^a-\hat{n}_\mu^b\right] = 2i\varepsilon_{\lambda\mu\nu}\hat{m}_\nu,\\
&\left[\hat{n}_\mu^a+\hat{n}_\mu^b,\hat{m}_\lambda^+\right] = 2\delta_{\lambda\mu}\hat{M}^+, && \left[\hat{n}_\mu^a-\hat{n}_\mu^b,\hat{m}_\lambda^+\right] = -2i\varepsilon_{\lambda\mu\nu}\hat{m}_\nu^+,\\
&\left[\hat{M},\hat{M}^+\right] = \hat{N}+2,\quad \left[\hat{M},\hat{N}+2\right] = 2\hat{M},\quad \left[\hat{N}+2,\hat{M}^+\right] = 2\hat{M}^+,\\
&\left[\hat{m}_\lambda,\hat{M}^+\right] = \hat{n}_\lambda^a+\hat{n}_\lambda^b, && \left[\hat{M},\hat{m}_\lambda^+\right] = \hat{n}_\lambda^a+\hat{n}_\lambda^b,\\
&\left[\hat{M},\hat{n}_\lambda^a+\hat{n}_\lambda^b\right] = 2\hat{m}_\lambda, && \left[\hat{n}_\lambda^a+\hat{n}_\lambda^b,\hat{M}^+\right] = 2\hat{m}_\lambda^+,\\
&\left[\hat{M},\hat{n}_\lambda^a-\hat{n}_\lambda^b\right] = 0, && \left[\hat{n}_\lambda^a-\hat{n}_\lambda^b,\hat{M}^+\right] = 0,\\
&\left[\hat{m}_\lambda,\hat{N}+2\right] = 2\hat{m}_\lambda, && \left[\hat{N}+2,\hat{m}_\lambda^+\right] = 2\hat{m}_\lambda^+,\\
&\left[\hat{n}_\lambda^a+\hat{n}_\lambda^b,\hat{N}+2\right] = 0, && \left[\hat{n}_\lambda^a-\hat{n}_\lambda^b,\hat{N}+2\right] = 0.
\end{aligned} \tag{7.25}$$

Here, $\delta_{\lambda\mu}$ is the Kronecker delta symbol; $\varepsilon_{\lambda\mu\nu}$ is the Levi-Civita symbol.

Now we construct the wave functions of hydrogen-like atoms with given quantum numbers. Taking into account the fact that the angular momentum and its projection in the z-axis are the integrals of motion we first build wave functions in the parabolic coordinates. The wave functions belong to eigenvalues of the operators $\hat{l}_3$, i.e. satisfy the following equations:

$$\hat{l}_3\,|\psi\rangle = \frac{1}{2}(\hat{a}_1^+\hat{a}_1 - \hat{a}_2^+\hat{a}_2 - \hat{b}_1^+\hat{b}_1 + \hat{b}_2^+\hat{b}_2)\,|\psi\rangle = m\,|\psi\rangle\,. \tag{7.26}$$

Here the integer m is the magnetic quantum number. Substituting the states vectors (7.16) into Eq. (7.26) we have the equation:

$$n_1 - n_2 - n_3 + n_4 = 2m. \tag{7.27}$$

Taking into account Eqs. (7.18b) and (7.27) we can have the wave functions are derived:

$$|n_1, n_2, m\rangle = \frac{1}{\sqrt{n_1!(n_1+|m|)!n_2!(n_2+|m|)!}} \times (\hat{a}_1^+\hat{b}_1^+)^{n_1}(\hat{a}_2^+\hat{b}_2^+)^{n_2}(\hat{m}_\pm^+)^{|m|}\,|0(\omega)\rangle\,, \tag{7.28}$$

where we use $\hat{m}_+^+ = \hat{m}_1^+ + i\hat{m}_2^+ = \hat{a}_1^+\hat{b}_2^+$ for the case when $m > 0$; and $\hat{m}_-^+ = \hat{m}_1^+ - i\hat{m}_2^+ = \hat{a}_2^+\hat{b}_1^+$ for the case when $m < 0$. In both cases the principal quantum number: $n = n_1 + n_2 + |m|$.

We can also establish the wave functions with two given quantum numbers: the angular momentum quantum number l and the magnetic quantum number m, i. e. the wave function in spherical coordinates. We require the wave function to satisfy the equation:

$$\hat{L}^2\,|\psi\rangle = \left[\frac{1}{4}\hat{N}(\hat{N}+2) - \hat{M}^+\hat{M}\right]\,|\psi\rangle = l(l+1)\,|\psi\rangle\,. \tag{7.29}$$

We will find the wave function with given quantum numbers l, m in the combination of state vectors (7.28):

$$|l, m\rangle = (\hat{m}_\pm^+)^{|m|}\sum_{k=0}^{N} C_k(\hat{m}_3^+)^k(\hat{M}^+)^{N-k}\,|0(\omega)\rangle\,, \tag{7.30}$$

where C_k are coefficients we need to find in order to have the wave function (7.30) being a solution of Eq. (7.29). By substituting (7.30) into (7.29) and using the commutation relations (7.25) as well as Eq. (7.14), we obtain recurrent equations defining the coefficients C_k. The calculations performed by purely algebraic method described above result in:

$$|jlm\rangle = N_{jlm}(\hat{M}^+)^j(\hat{m}_\pm^+)^{|m|} \times \sum_{k=0}^{l-|m|}\frac{(-1)^k(l+|m|+k)!}{k!(k+|m|)!(l-|m|-k)!}(\hat{a}_1^+\hat{b}_1^+)^k(\hat{M}^+)^{l-|m|-k}\,|0(\omega)\rangle\,.$$

7.2 Analytical Estimate for Rydberg States of a Hydrogen Atom in an Electric Field

The increase of the computer facilities does not reduce the significance of analytical calculations permitting to make a qualitative analysis of the physical effects, and to estimate the range of parameters which are of interest in the specific experimental conditions. One of the possible ways of performing such analytical investigations can be based on using the operator method for solution of the Schrödinger equation. In the previous chapters, it has been shown that the OM zeroth approximation gives a good estimation of the Hamiltonian eigenvalues in the entire range of its parameters and define correctly the relative positions of the energy levels depending on the quantum numbers of the system.

In the last years of the 1980s, a number of experimental and theoretical papers were devoted to the investigation of highly excited states of the hydrogen atom in a homogeneous electric field (see, for example, [18, 33–35]). A particular interest was evoked by the quasi-steady Stark states with vanishingly small energy. This section will recall some main results of the paper [17] in which the simplified approximation of the zeroth-order OM energy formula is used to find a sufficiently accurate analytical estimation of the energies and widths of the system levels. Further on, we also investigate these quantities depending on the quantum numbers and on the field amplitude.

The initial Hamiltonian of the system can be written as follows:

$$\hat{H} = \frac{1}{2}\hat{p}^2 - \frac{1}{r} + \beta x_3. \tag{7.31}$$

Here atomic units are used; the x_3-axis is directed along the electric field amplitude **F** and the dimensionless parameter β is:

$$\beta = \frac{F}{F_0}, \qquad F_0 = \frac{{m_e}^2 e^5}{\hbar^4}.$$

In accordance with Sect. 7.1, the Schrödinger equation with Coulomb singularity in the Hamiltonian can be considered on the basis of the two-dimensional complex space where the operator (7.31) reduces to the anharmonic oscillator one. First let us perform the following scale transformation of the coordinates and the energy of the system:

$$x_\lambda \to \varepsilon x_\lambda\,, \qquad E = -\frac{1}{2\varepsilon^2}.$$

The coordinate x_λ is considered as a real value and this transformation for complex x_λ is analogous to the complex coordinate rotation used for the description of the quasi-stationary states [18]. The complex coordinates ξ_s, ξ_s^* $(s = 1, 2)$ of the two-dimensional complex space are connected with the Cartesian coordinates

x_λ ($\lambda = 1, 2, 3$) by the transformation (7.3). Then the Schrödinger equation is modified to:

$$r(\hat{H} - E)\,|\psi\rangle = 0$$

and can be written in the form:

$$\tilde{H}(\xi)\,|\psi\rangle = \varepsilon\,|\psi\rangle\,,$$
$$\tilde{H}(\xi) = -\frac{1}{2}\frac{\partial^2}{\partial\xi_s\partial\xi_s^*} + \frac{1}{2}\xi_s\xi_s^* + \beta\varepsilon^3\left[(\xi_1\xi_1^*)^2 - (\xi_2\xi_2^*)^2\right]. \qquad (7.32)$$

In order to obtain the OM zeroth approximation, the operator has to be transformed to the second quantized form:

$$\begin{cases} \hat{a}_s = \sqrt{\dfrac{\omega_s}{2}}\left(\xi_s + \dfrac{1}{\omega_s}\dfrac{\partial}{\partial\xi_s^*}\right), & \hat{a}_s^+ = \sqrt{\dfrac{\omega_s}{2}}\left(\xi_s^* - \dfrac{1}{\omega_s}\dfrac{\partial}{\partial\xi_s}\right), \\ \hat{b}_s = \sqrt{\dfrac{\omega_s}{2}}\left(\xi_s^* + \dfrac{1}{\omega_s}\dfrac{\partial}{\partial\xi_s}\right), & \hat{b}_s^+ = \sqrt{\dfrac{\omega_s}{2}}\left(\xi_s - \dfrac{1}{\omega_s}\dfrac{\partial}{\partial\xi_s^*}\right), \end{cases} \qquad (7.33)$$
$$\left[\hat{a}_s, \hat{a}_t^+\right] = \delta_{st}\,, \quad \left[\hat{b}_s, \hat{b}_t^+\right] = \delta_{st}\,,$$

with arbitrary parameters ω_1, ω_2. Then one has to select from the operator $\tilde{H}$ the part which commutes with the particle number operators $\hat{n}_1^a = \hat{a}_1^+\hat{a}_1$ $\hat{n}_2^a = \hat{a}_1^+\hat{a}_1$, $\hat{n}_1^b = \hat{b}_1^+\hat{b}_1$ $\hat{n}_2^b = \hat{b}_2^+\hat{b}_2$:

$$\begin{aligned} \tilde{H}_0 = \sum_{s=1,2} \Big\{ &(\omega_s + 1/\omega_s)(\hat{n}_s^a + \hat{n}_s^b + 1) - \\ &-(-1)^s(\beta\varepsilon^3/\omega_s^2)\left[2 + 3(\hat{n}_s^a + \hat{n}_s^b) + (\hat{n}_s^a + \hat{n}_s^b)^2 + 2\hat{n}_s^a\hat{n}_s^b\right]\Big\}. \end{aligned} \qquad (7.34)$$

In this approximation the eigenfunctions of the system are defined by the quantum numbers n_1^a n_2^a, n_1^b n_2^b. There are only three independent quantum numbers because the vector $|\psi\rangle$ for the physically meaningful state does not have to depend on the supplementary variable ϕ of the considered space. This condition is fulfilled if $|\psi\rangle$ is the solution of the Eq. (7.17) that leads to:

$$n_1^a - n_1^b + n_2^a - n_2^b = 0.$$

Another integral of motion is defined by the angular momentum projection $\hat{l}_3$ on the electric field direction that leads to:

$$n_1^a - n_1^b - n_2^a + n_2^b = 2m.$$

Here m is the magnetic quantum number. The other parabolic quantum numbers n_1 and n_2 as shown in (7.28) are:

$$n_1 = \frac{1}{2}(n_1^a + n_1^b) - |m| \,, \quad n_2 = \frac{1}{2}(n_2^a + n_2^b) - |m| \,, \quad |m| \le n_1, n_2 .$$

Then the zeroth-order OM energy formula is:

$$\varepsilon_{n_1 n_2 m} = \frac{1}{4} \sum_{s=1,2} \left[(\omega_s + 1/\omega_s)(2n_s + |m| + 1) - (-1)^s (\beta\varepsilon^3/\omega_s^2) B_s \right], \tag{7.35}$$

where:

$$B_s = 2 + 3(2n_s + |m|) + \frac{3}{2}(2n_s + |m|)^2 - \frac{1}{2}m^2 .$$

As demonstrated in previous chapters, the OM successive approximations for Eq. (7.32) converge to the exact complex eigenvalues E with different quantum numbers. But we consider only the OM zeroth approximation because we are interested in the analytical investigation of the problem. The equations for calculating the parameters ω_1, ω_2 are obtained from the conditions of independence of the eigenvalues on the choice of the wave function representation and in the considered approximation they are:

$$\begin{aligned} \frac{\partial \varepsilon_{n_1 n_2 m}}{\partial \omega_s} &= 0 \,, \quad (s = 1,\, 2) \\ {\omega_s}^3 - \omega_s + (-1)^s 2\beta\varepsilon^3 \frac{B_s}{2n_s + |m| + 1} &= 0. \end{aligned} \tag{7.36}$$

The system of simple algebraic equations (7.35), (7.36) permits one to calculate the energy levels with accuracy 1 % typical for the OM zeroth approximation. Let us consider the analytical solution of these equations in the limit $n_1 \gg n_2$ which corresponds to the quasi-stationary states with vanishingly small energies investigated experimentally [18]. In this limit one can find the following expansions for the parameters:

$$\begin{aligned} &\omega_1 \simeq \rho^{1/3} + \frac{1}{3}\rho^{-1/3} + \ldots, \quad |\rho| \gg 1 \\ &\rho = 2\beta\varepsilon^3 \frac{B_1}{n_0} \simeq 3\beta\varepsilon^3 \left(n_0 + \frac{1 - m^2}{3n_0} \right), \quad n_0 = 2n_1 + |m| + 1, \end{aligned} \tag{7.37}$$

$$\begin{aligned} &\omega_2 \equiv \omega'_2 + i\omega''_2 = \frac{\sqrt{3}}{2}(A_+ + A_-) + \frac{1}{2} i (A_+ - A_-), \\ &A_\pm = \left(\sqrt{\Delta^2/4 + 1/27} \pm \Delta/2 \right)^{1/3}, \\ &|\omega_2| \ll |\omega_1| , \end{aligned} \tag{7.38}$$

where

$$\Delta = 3\beta|\varepsilon_0|^3\left(2n_2 + |m| + 1 + \frac{1 - m^2}{3(2n_1 + |m| + 1)}\right),$$

$$\varepsilon_0{}^2 = -\frac{n_0}{4\nu(3n_0\nu/8 - 1)}, \quad \nu = \left[3\beta\left(n_0 + \frac{1 - m^2}{3n_0}\right)\right]^{1/3}.$$

Other roots of the cubic equations (7.36) give the solutions which do not satisfy the inequalities, which are essential for the existence of the canonical transformation (7.33):

$$0 < \omega_s',\ E'' < 0,\ E \equiv E' + i\,E''.$$

Let us substitute expansions (7.37), (7.38) into formula (7.35) and find the energy of the quasi-stationary state with the corresponding set of quantum numbers:

$$E^{(0)}_{n_1 n_2 m} = -\frac{1}{2\varepsilon^2_{n_1 n_2 m}} = E_1 + i\Gamma/2,$$

$$E_1 = \frac{(3n_0\nu/8 - 1)(n_0/2\nu + 2\rho_1)}{(n_0/2\nu + 2\rho_1)^2 + 4\rho_2{}^2}, \qquad \Gamma = \frac{4\rho_2(3n_0\nu/8 - 1)}{(n_0/2\nu + 2\rho_1)^2 + 4\rho_2{}^2}, \tag{7.39}$$

where

$$\rho_1 = -\frac{2n_2 + |m| + 1}{8}\left(3 - \frac{1}{|\omega_2|^2}\right)\omega''_2\,|\varepsilon_0|\ ,$$

$$\rho_2 = -\frac{2n_2 + |m| + 1}{8}\left(3 + \frac{1}{|\omega_2|^2}\right)\omega'_2\,|\varepsilon_0|\,.$$

We use here the reduced amplitude of the electric field [34]:

$$f = \beta N_0^4\,, \quad N_0 = n_0/2,$$

and take into account the estimation:

$$\frac{\nu\rho_{1,2}}{n_0} \sim \left(\frac{n_2}{n_1}\right)^{4/3} \ll 1.$$

Then the formulae (7.39) simplify to:

$$2N_0^2 E_1 \simeq A\left[1 - (2/N_0^2)\,q\rho_1\right], \qquad N_0^2\Gamma \simeq (4A/N_0^2)\,q\rho_2, \tag{7.40}$$

with

$$q = \left[6(N_0^*/N_0)\,f\right]^{1/3}, \quad A = 2q(3q/4 - 1), \quad N_0^* = \frac{1}{2}\left(n_0 + \frac{1 - m^2}{3n_0}\right).$$

Formulae (7.40) deliver a good estimation for the parameters which define the spectrum of the system near $E_1 = 0$. For example, the limiting curve which corresponds to $n_1 \to \infty$ passes through zero when $f = f_0 = 2^5/3^4 \approx 0.395$. The distance between the levels near $E_1 = 0$ is defined by the formula:

$$\left.\frac{\partial E_1}{\partial N_1}\right|_{E_1=0} = 3(2\beta^3)^{1/4} = 3.568\ \beta^{3/4}.$$

The corresponding values found in [34] by means of the Bohr-Sommerfeld quantization condition are:

$$f_0 \approx 0.384, \qquad \left.\frac{\partial E_1}{\partial N_1}\right|_{E_1=0} = 3.708\ \beta^{3/4}.$$

The analytical expression for the parabolic quantum number n_{1c} of the level with zero energy is:

$$n_{1c} \simeq \frac{2^{5/4}}{3}\beta^{3/4} - \frac{1+|m|}{2},$$

which is in a good accordance with the experimental results from [18] and with the theoretical values from [36]. Let us also note that the field amplitude F and the parabolic quantum number n_1 are included in the values E_1 and γ in such combinations which define the scaling in the Stark effect for Rydberg atoms when $n_1 \gg 1$. This result has been recently described in [37] using another method.

In order to give a notion for the accuracy of our estimations we compare them in Table 7.1 with the numerical results from [33] for some values of the quantum numbers and amplitudes of the field. The analytical formulae (7.40) ensure a relative accuracy of 5 % while calculating the width of the levels and $\sim$ 10 % for their energies. It is essential to note that the OM accuracy of calculating the matrix

Table 7.1 Numerical estimation for our analytical formulae for high excited Rydberg state energies in comparison with other numerical calculations

							$E_1 \times 10^4$		$\Gamma \times 10^4$	
	N_1	N_2	$\|M\|$	n_1	n_2	$\|m\|$	this work	[33]	this work	[33]
$F = 6.5$ kV/cm	48	0	0	24	0	0	1.277	1.433	0.390	0.417
	48	2	0	24	1	0	1.355	1.523	1.332	1.366
	49	1	1	24	0	1	1.741	2.087	0.969	0.964
	50	0	0	25	0	0	2.432	2.687	0.509	0.521
$F = 8.0$ kV/cm	48	0	0	24	0	0	3.118	3.383	0.631	0.638
	48	2	0	24	1	0	3.265	3.433	1.981	1.955
	49	1	1	24	0	1	3.748	4.090	1.391	1.370
	50	0	0	25	0	0	4.524	4.764	0.734	0.719

elements of the physical operators is of the same magnitude as for the energy levels [38, 39]. Therefore the method described in this section can be useful for the calculation of the atomic photo-ionization cross-section [18], too.

7.3 Iterative Calculation of Energy for Quasi-Stationary States

For the hydrogen atom in a constant electric field we have only quasi-stationary states and correspondingly the energy has complex value. The direct numerical method of solving the Schrödinger equation leads to additional difficulties. Therefore, some modifications of calculation method were given; particularly the method of using complex coordinates is used for some integration [40]. The energy levels calculation for Rydberg quasi-stationary states, therefore, become a complicated problem, especially, when the high accuracy of the results is required [41].

In the previous chapters we proved that the iteration scheme of calculating the high orders of OM approximations is effective for numerical calculation of energy and wave functions with any given precision. The calculation is expected to be effective also for quasi-stationary states, and we will apply this scheme of OM for Eq. (7.32).

According to the OM scheme, the Hamiltonian (7.32) is divided into two parts:

$$\tilde{H}(\varepsilon) = \tilde{H}_0 + \sum_{\alpha=1,2} \tilde{V}_\alpha,$$
$$\tilde{V}_\alpha = \frac{1-\omega_\alpha^2}{4\omega_\alpha}(\hat{M}_\alpha + \hat{M}_\alpha^+)$$
$$- \frac{(-1)^\alpha \beta\varepsilon^3}{4{\omega_\alpha}^2}\left[2(2+\hat{N}_\alpha)\hat{M}_\alpha + 2\hat{M}_\alpha^+(2+\hat{N}_\alpha) + \hat{M}_\alpha^2 + \hat{M}_\alpha^{+2}\right], \tag{7.41}$$

where $\hat{N}_\alpha = \hat{n}_\alpha^a + \hat{n}_\alpha^b$, $\hat{M}_\alpha = \hat{a}_\alpha \hat{b}_\alpha$, $\hat{M}_\alpha^+ = \hat{a}_\alpha^+ \hat{b}_\alpha^+$. The diagonal part $\tilde{H}_0$ which commutes with the particle number operators is given in (7.34). For calculation process we need matrix elements of the non-diagonal part $\tilde{V}_\alpha$ of the Hamiltonian:

$$\begin{aligned}\hat{M}_1 \left|n_1 n_2 m\right\rangle &= \frac{1}{2}\sqrt{{n_1}^2 - m^2}\ \left|n_1 - 2, n_2 m\right\rangle, \\ \hat{M}_1^+ \left|n_1 n_2 m\right\rangle &= \frac{1}{2}\sqrt{(n_1+2)^2 - m^2}\ \left|n_1 + 2, n_2 m\right\rangle,\end{aligned} \tag{7.42}$$

where $|n_1 n_2 m\rangle$ for $n_1, n_2 = 0, 1, 2, \ldots$ $|m| \le n_1, n_2$ are the basis state vectors described in (7.28) for ξ-space corresponding to the basis set of the parabolic wave functions. The formulae for $\alpha = 2$ are similar to (7.42) for the case $\alpha = 1$. Note that there is the separation of coordinates in the Hamiltonian $\tilde{H}(\varepsilon)$ but the two oscillators corresponding to $\alpha = 1$ and $\alpha = 1$ are connected to each other via the parameter ε^3. The eigenfunctions of Eq. (7.32) with quantum numbers $n_1,\ n_2,\ m$ will be found in the form:

$$|\psi_{n_1 n_2 m}\rangle = |n_1 n_2 m\rangle + \sum_{k=0}^{+\infty} \sum_{\substack{j=0 \\ jk \neq n_1 n_2}}^{+\infty} C_{jk} \, |jkm\rangle \,, \tag{7.43}$$

with the normalization condition: $\langle n_1 n_2 m \mid \psi_{n_1 n_2 m}\rangle = 1$.

In the expansion (7.43) the magnetic number m is not changed because of the conservation of the angular momentum projection. It should be noted that the wave function expansion (7.43) has local property because each term in this expansion belongs to different basis set. Indeed, each state vector $|n_1 n_2 m\rangle$ dependent on the frequencies $\omega_1,\ \omega_2$ which are the function of the quantum numbers n_1, n_2, m, see Eq. (7.36), and thus $\langle n_1 n_2 m \mid n'_1 n'_2 m\rangle \neq 0$. Nevertheless, we will fix the parameters $\omega_1,\ \omega_2$ in the expansion (7.43), i.e. $|jkm, \omega_{1,2}(n_1 n_2 m)\rangle$. By this way we have a set of orthogonal and normalized state vectors for the expansion of wave functions. We will use the iteration scheme of OM which has the following form for the specified considered problem:

$$\begin{aligned}
\varepsilon^{(s)\alpha}_{n_1 n_2 m} &= \langle n_1 n_2 m|\, \tilde{H}_\alpha(\varepsilon^{(s-1)}_{n_1 n_2 m})\, |n_1 n_2 m\rangle \\
&\quad + \sum_{j,k \neq n_1 n_2} \langle jkm|\, \tilde{H}_\alpha(\varepsilon^{(s-1)}_{n_1 n_2 m})\, |n_1 n_2 m\rangle\, C^{(s-1)}_{jk}, \\
C^{(s)}_{jk} &= \left[\varepsilon^{(s-1)}_{n_1 n_2 m} - \langle jkm|\, \tilde{H}_\alpha(\varepsilon^{(s-1)}_{n_1 n_2 m})\, |jkm\rangle\right]^{-1} \\
&\quad \times \Bigg(\langle jkm|\, \tilde{H}_\alpha(\varepsilon^{(s-1)}_{n_1 n_2 m})\, |jkm\rangle \\
&\quad + \sum_{\substack{j'k' \neq jk \\ j'k' \neq n_1 n_2}} \langle j'k'm|\, \tilde{H}_\alpha(\varepsilon^{(s-1)}_{n_1 n_2 m})\, |jkm\rangle\, C^{(s)}_{j'k'} \Bigg),
\end{aligned} \tag{7.44}$$

$$\varepsilon^{(s)}_{n_1 n_2 m} = \sum_{\alpha=1,2} \varepsilon^{(s)\alpha}_{n_1 n_2 m}, \quad \varepsilon_{n_1 n_2 m} = \lim_{s\to\infty} \varepsilon^{(s)}_{n_1 n_2 m}, \quad C_{jk} = \lim_{s\to\infty} C^{(s)}_{jk}.$$

All matrix elements in (7.44) can be calculated using the formulae (7.42). The iteration scheme (7.44) is quickly convergent for any amplitude of electric field. Table 7.2 compares the results by the operator method [42] with calculations by other methods [41].

Table 7.2 Quasi-stationary energy levels $E = E' + iE''$ of the hydrogen atom in an electric field

$\|n_1, n_2, m>$	β	$-E'$	$-E''$
(0,0,0)	0.1	2.054836	0.014538
(0,0,0)	0.25	1.170062	0.188576
(2,0,0)	0.004	0.114305	0.000001
(2,0,0)	0.01	0.103895	0.001640
(2,0,0)	0.02	0.088984	0.015446
(0,2,0)	0.012	0.175060	0.009628
(0,2,0)	0.024	0.220676	0.041892
(4,0,0)	0.003	0.035205	0.001681
(4,0,0)	0.004	0.029577	0.004197
(0,4,0)	0.008	0.156605	0.002160
(0,4,0)	0.01	0.166094	0.005443
(0,4,0)	0.014	0.183546	0.014372

7.4 Operator Method for Hydrogen Atom in Magnetic Field

Let us now consider the problem of a hydrogen atom in a magnetic field described by the following Schrödinger equation:

$$\left\{-\frac{1}{2}\frac{\partial^2}{\partial x_\lambda \partial x_\lambda} - \frac{Z}{r} - \frac{i}{2}\gamma\left(x_1\frac{\partial}{\partial x_2} - x_2\frac{\partial}{\partial x_1}\right) + \frac{1}{8}\gamma^2(x_1^2 + x_2^2)\right\}\psi(\mathbf{r}) = E\psi(\mathbf{r}). \tag{7.45}$$

The equation is written in the atomic system of units when $\gamma = \hbar^3 B/m_e^2 ce^3$ is a dimensionless magnetic intensity; Z is nuclear charge. Instead of Eq. (7.45) we will use the equation written in the ξ-space with essentially simplified algebraic structure. The equation in ξ-space for the considered problem is as follows:

$$\begin{aligned}(\tilde{H}(\xi) - Z)\,\psi(\xi) \equiv \Bigg\{&-\frac{1}{2}\frac{\partial^2}{\partial\xi_s\partial\xi_s^*} + 2\gamma^2\xi_1^*\xi_1\xi_2^*\xi_2\xi_s^*\xi_s - E\xi_s^*\xi_s - Z \\ &+\gamma\xi_s^*\xi_s\left(\xi_1^*\frac{\partial}{\partial\xi_1^*} + \xi_2\frac{\partial}{\partial\xi_2} - \xi_1\frac{\partial}{\partial\xi_1} - \xi_2^*\frac{\partial}{\partial\xi_2^*}\right)\Bigg\}\psi(\xi)=0.\end{aligned} \tag{7.46}$$

As described in Sect. 7.1, Eq. (7.46) can be written in the presentation of annihilation and creation operators that is useful for applying the OM:

$$\tilde{H} = \frac{\omega}{2}(\hat{N} + 2 - \hat{M} - \hat{M}^+) - \frac{E + m\gamma/2}{2\omega}(\hat{N} + 2 + \hat{M} + \hat{M}^+) +$$

$$+\frac{1}{16\omega^3}\gamma^2(\hat{N}+2+\hat{M}+\hat{M}^+)$$
$$\times(\hat{N}_1+1+\hat{M}_1+\hat{M}_1^+)(\hat{N}_2+1+\hat{M}_2+\hat{M}_2^+). \quad (7.47)$$

In (7.47) we replace the operator

$$\hat{l}_3=\frac{i}{2}\left(\xi_1^*\frac{\partial}{\partial\xi_1^*}+\xi_2\frac{\partial}{\partial\xi_2}-\xi_1\frac{\partial}{\partial\xi_1}-\xi_2^*\frac{\partial}{\partial\xi_2^*}\right)$$

by the magnetic quantum number m because for the considered system the angular momentum projection is conserved. The diagonal part of the Hamiltonian (7.47) that commutes with the particle number operators is obtained:

$$\hat{H}_0=\left(\frac{\omega}{4}-\frac{E+m\gamma/2}{2\omega}\right)(\hat{N}+2)$$
$$+\frac{1}{16\omega^3}\gamma^2\Big[(\hat{N}+2)(\hat{N}_1+1)(\hat{N}_2+1)$$
$$+2\hat{M}_1^+\hat{M}_1(\hat{N}_2+1)+2\hat{M}_2^+\hat{M}_2(\hat{N}_1+1)\Big]. \quad (7.48)$$

We can calculate the energy in zero order approximation by OM with the results shown in the Table 7.3. Several formulae required for calculation of the matrix elements with respect to the parabolic wave functions are presented here:

$$\left(\hat{M}_1\right)^j|n_1,n_2,m\rangle=\sqrt{\frac{n_1!\,(n_1+|m|)!}{(n_1-j)!\,(n_1+|m|-j)!}}\,|n_1-j,n_2,m\rangle\,,$$
$$\left(\hat{M}_1^+\right)^j|n_1,n_2,m\rangle=\sqrt{\frac{(n_1+j)!\,(n_1+|m|+j)!}{n_1!\,(n_1+|m|)!}}\,|n_1+j,n_2,m\rangle\,,$$
$$\left(\hat{N}_1+1\right)|n_1,n_2,m\rangle=(2n_1+|m|+1)\,|n_1,n_2,m\rangle \quad (7.49)$$

Let us use now the OM zeroth approximation with some modifications to obtain an analytical description of a hydrogen atom in a magnetic field of arbitrary strength. Taking into account the asymptotic behavior of the wave function in the region of strong magnetic field we consider the wave function of hydrogen atom in magnetic field in the form:

$$\psi(\mathbf{r})=e^{-\frac{1}{2}\upsilon\,({x_1}^2+{x_2}^2)}\Psi(\mathbf{r}), \quad (7.50)$$

where υ is a parameter to be defined latter. Substituting the wave function (7.50) into the Eq. (7.45) we have the following equation for the new wave function $\Psi(\mathbf{r})$:

Table 7.3 Energy of ground state calculated by OM in zeroth order approximation compared with exact numerical calculations

γ	$E_{1s}^{(0)}$	$E_{1s}^{(0)} + \Delta E_{1s}^{(2)}$	[43]
0.01	−0.499 975 001	−0.499 975 002 760	−0.499 975 002 759
0.02	−0.499 900 020	−0.499 900 044 109	−0.499 900 044 089
0.04	−0.499 600 319	−0.499 600 703 002	−0.499 600 701 769
0.06	−0.499 101 611	−0.499 103 536 095	−0.499 103 522 564
0.1	−0.497 512 316	−0.497 526 738 745	−0.497 526 480 401
0.2	−0.490 188 942	−0.490 392 003 224	−0.490 381 565 035
0.4	−0.460 508 812	−0.464 505 768 398	−0.464 605 379 868
0.6	−0.421 077 143	−0.428 043 598 749	−0.427 462 287 757
1	−0.308 761 166	−0.331 972 429 129	−0.331 168 896 733
2	0.067 442 249	−0.015 366 630 637	−0.022 213 907 665
4	1.0	0.769 419 885 036	0.719 201 983 948
10	4.273 986 768	3.537 250 502 446	3.252 202 836 286
20	10.247 642 981	8.604 763 540 674	7.784 601 484 567
40	22.834 813 500	19.304 961 929 265	17.198 970 175 222
100	62.172 884 749	52.800 667 751 171	46.210 195 763 695

$$e^{-\upsilon\,({x_1}^2+{x_2}^2)}\left\{-\frac{1}{2}\frac{\partial^2}{\partial x_\lambda \partial x_\lambda}-\frac{Z}{r}+\frac{1}{2}i\gamma\left(x_1\frac{\partial}{\partial x_2}-x_2\frac{\partial}{\partial x_1}\right)-E+\upsilon\right.$$
$$\left.+\upsilon\left(x_1\frac{\partial}{\partial x_1}+x_2\frac{\partial}{\partial x_2}+1\right)+\frac{1}{8}(\gamma^2-4\upsilon^2)(x_1^2+x_2^2)\right\}\Psi(\mathbf{r})=0. \qquad (7.51)$$

Here we multiply in the left-hand side of the Schrödinger equation (7.45) by the factor $\exp\left[-\upsilon\ ({x_1}^2+{x_2}^2)/2\right]$ in order to have the Hamiltonian of Eq. (7.51) being hermitic. Equation (7.51) is equivalent to the equation in ξ-space as follows:

$$\hat{A}\left(\hat{H}-Z\right)\Psi(\xi)=0, \qquad (7.52)$$

where

$$\hat{H}(\xi)=-\frac{1}{2}\frac{\partial^2}{\partial\xi_s\partial\xi_s^*}-\left(E-\upsilon+\frac{1}{2}m\gamma\right)\xi_s\xi_s^*+\frac{1}{2}\left(\gamma^2-4\upsilon^2\right)\xi_s\xi_s^*\xi_1\xi_1^*\xi_2\xi_2^*$$
$$+2\upsilon\xi_1\xi_1^*\left(\xi_2\frac{\partial}{\partial\xi_2}+\xi_2^*\frac{\partial}{\partial\xi_2^*}\right)+2\upsilon\xi_2\xi_2^*\left(\xi_1\frac{\partial}{\partial\xi_1}+\xi_1^*\frac{\partial}{\partial\xi_1^*}\right),$$
$$\hat{A}=e^{-\upsilon\,({x_1}^2+{x_2}^2)}=e^{-4\upsilon\xi_1\xi_1^*\xi_2\xi_2^*}.$$

For the use in further calculation we write the normal form of the factor $\hat{A}$:

$$\begin{aligned}
e^{-4\upsilon\,\xi_1\xi_1^*\,\xi_2\xi_2^*} &= \frac{1}{2\pi}\int_0^{+\infty} dt \int_{-\infty}^{+\infty} dk\, e^{-ikt} e^{-4\upsilon t\xi_1\xi_1^*)} e^{ik\xi_2\xi_2^*} \\
&= \frac{1}{2\pi}\int_0^{+\infty} dt \int_{-\infty}^{+\infty} dk\, e^{-ikt} \exp\left(-\frac{t}{t+x}\hat{a}_1^+\hat{b}_1^+ + \frac{ik}{1-ik}\hat{a}_2^+\hat{b}_2^+\right) \\
&\times \exp\left[(\hat{a}_1^+\hat{a}_1+\hat{b}_1^+\hat{b}_1)\ln\frac{x}{t+x}+(\hat{a}_2^+\hat{a}_2+\hat{b}_2^+\hat{b}_2+1)\ln\frac{1}{1-ik}\right] \\
&\times \exp\left(-\frac{t}{t+x}\hat{a}_1\hat{b}_1 + \frac{ik}{1-ik}\hat{a}_2\hat{b}_2\right), \qquad (7.53)
\end{aligned}$$

where we use the new parameter $x = \omega^2/\upsilon$ instead of the parameter υ. We will use the state vectors (7.28) (parabolic wave function) for the wave function of our system in the zeroth order of approximation. Therefore, all matrix elements can be calculated using the formulae (7.49), and here the matrix elements for the operator $\hat{A}$ are:

$$\begin{aligned}
A_{n_1n_2;n'_1n'_2} &= \langle n_1n_2m|\, e^{-4\upsilon\,\xi_1\xi_1^*\,\xi_2\xi_2^*}\, |n'_1n'_2m\rangle \\
&= (-1)^{n_2+n'_2+m} x^{n_1+n'_1+n_2+n'_2+2m+1} \\
&\times \sqrt{n_1!(n_1+m)!n_2!(n_2+m)!n'_1!(n'_1+m)!n'_2!(n'_2+m)!} \\
&\times \sum_{p=0}^{\min(n_1,n'_1)} \sum_{s=0}^{\min(n_2,n'_2)} \frac{(n_2-n'_2-2s)!}{p!(p+m)!s!(s+m)!} \\
&\times \frac{1}{(n_1-p)!(n'_1-p)!(n'_2-s)!} \\
&\times \sum_{q=0}^{n_2-n'_2-2s} \frac{(n_1+n'_1+n_2+n'_2+m-2p-q)!}{q!(n_2+n'_2-2s-q)!(n_2+n'_2+m-q)!} \\
&\times \sum_{l=0}^{n_1+n'_1+n_2+n'_2+m-2p-q} \frac{1}{x^{l+q}} \\
&\times \frac{(-1)^l}{l!(n_1+n'_1+n_2+n'_2+m-2p-q-l)!} \\
&\times \int_0^{+\infty} dt \frac{e^{-t}}{(t+x)^{n_1+n'_1+m+1-l}}. \qquad (7.54)
\end{aligned}$$

All integrals appearing in (7.54) finally lead to the integral:

$$I(x) = \int_0^{+\infty} dt \frac{e^{-t}}{t+x},$$

which is well investigated.

We note that the Eq. (7.26) is symmetric with respect to the transformation $\xi_1 \to \xi_2$, $\xi_2 \to \xi_1$. Therefore we will consider the wave function in the form:

$$\Psi^{\pm}_{n_1 n_2 m} = |n_1 n_2 m\rangle \pm |n_2 n_1 m\rangle\,. \tag{7.55}$$

Substituting the wave function (7.55) into Eq. (7.52) we obtain energy in the zeroth order approximation of OM which is considered as analytical energy:

$$E^{anl}_{n_1 n_2 m}(\omega, x, \gamma) = \frac{\omega^2}{2} g(x) - Z\omega p(x) + \frac{\gamma^2}{\omega^2} h(x), \tag{7.56}$$

where $g(x)$, $p(x)$, $h(x)$ are elementary functions of variables x and $I(x)$. The parameters x and ω can be defined as the OM parameters, and they satisfy the following equations:

$$\frac{\partial E}{\partial x} = 0\,, \quad \frac{\partial E}{\partial \omega} = 0,$$

which lead to the equations:

$$\frac{\omega^2}{2} g'(x) - Z\omega p'(x) + \frac{\gamma^2}{\omega^2} h'(x) = 0, \tag{7.57}$$

$$\frac{\omega^2}{2} g'(x) - Z\omega p'(x) + \frac{\gamma^2}{\omega^2} h'(x) = 0. \tag{7.58}$$

Equations (7.57) and (7.58) can be solved analytically with the solution for the parameter ω:

$$\omega = Z \frac{p'(x)h(x) + p(x)h'(x)}{g'(x)h(x) + g(x)h'(x)}. \tag{7.59}$$

Therefore, we obtain the energy as an analytical function of magnetic amplitude γ via the parameter x: $E(x)$, $\gamma(x)$.

For demonstration of the advantages of the method we present the calculations for the ground state as follow:

$$\begin{aligned} E(x) &= \frac{\omega^2}{2x\,[1 - xI(x)]} - \omega Z \frac{I(x)}{1 - xI(x)} \\ &+ \frac{\gamma^2}{8\omega^2} \frac{x\,[1 + x - x(2+x)I(x)]}{1 - xI(x)}, \end{aligned}$$

Table 7.4 Numerical estimation for highly accurate analytical energy obtained by OM

γ	$E^{(0)}+E^{(2)}$ [15]	E_{exact} [44]	$E_{analytical}$ (7.60)
1	−0.33029	−0.33114	−0.329
2	-1.6146×10^{-2}	-2.2215×10^{-2}	-1.761×10^{-2}
20	8.4222	7.7848	7.8243
200	107.76	95.273	95.424

$$\begin{aligned}\omega &= Z^2(2(1+x)-(1+9x+4x^2)I(x)\\&\quad +x(2+7x+2x^2)I^2(x)+2x^2I^3(x))\\&\quad \times(-1+x-x(1+2x)I(x)+x(2+2x+x^2)I^2(x))^{-1},\\ \frac{\gamma^2}{4\omega^3} &= Z\frac{2-(1+3x)I(x)+x^2I^2(x)}{x^2\left[-1+x-x(1+2x)I(x)+x(2+2x+x^2)I^2(x)\right]}.\end{aligned} \tag{7.60}$$

Essentially, that the analytical formula (7.60) gives the energy of ground state suitable for the whole range of magnetic field. In Table 7.4 some numerical results by the operator method are shown compared with other method [15, 44]. The expansion in the asymptotic region of magnetic field gives the results exactly coinciding with the well-known results:

$$\begin{aligned}E(\gamma) &= \frac{1}{2}\gamma - Z^2\ln^2\gamma + .. \quad \gamma \gg 1,\\ E(\gamma) &= -\frac{1}{2}\gamma - \frac{1}{4Z^2}\gamma^2 + .. \quad \gamma \ll 1.\end{aligned} \tag{7.61}$$

7.5 Two Level System in a Single-Mode Quantum Field

The most popular representation of the Hamiltonian of TLS in a single-mode quantum field is connected with Jaynes-Cummings [45] model. It is analytically solvable due to the so called rotating wave approximation (RWA). But the strict analysis of the RWA validity is not usually considered in the concrete applications and a range of the system parameters where the results are correct remains uncertain. Moreover, the exact isolated solutions for TLS were found by several authors [46] and it was proved that RWA did not describe peculiarities of the accurate energy spectrum in dependence on the atom-field coupling constant.

Therefore, it is of great interest to analyze the accurate numerical solution of the TLS problem within a wide range of the Hamiltonian parameters in comparison with the RWA results. Such kind of studies is useful for the determination of validity limits of RWA as well as for describing of physical systems with a rather big coupling constant (e.g. the interaction of condensed matter with high-power laser radiation or the processes occurring at the long-wave coherent radiation).

There is the continued-fraction algorithm used by a number of authors for numerical study of the spectrum and the time evolution of a system [47]. But this method is rather ineffective for a large coupling constant and quantum numbers because of the effect of the exponentially diverging solution of the Schrödinger equation.

In this section we use OM for the numerical and analytical analysis of the TLS problem. We will see, that the OM algorithm permits one to find the eigenvalues and eigenfunctions of the TLS Hamiltonian with any required accuracy and to analyze various characteristics of the system almost with the same efficiency as on the basis of the analytical solution. It is found that the RWA results coincide qualitatively with the accurate ones only in the range of small coupling constant and when the low energy levels are excited. The OM zeroth approximation for the TLS problem is shown to lead to the analytical formulae which uniformly fit the accurate energy spectrum for any parameters of the system [28]. The analogous results were also obtained later in [48, 49].

The Hamiltonian of the considered model has a simple dimensionless form:

$$\hat{H} = \frac{1}{2}E\sigma_3 + a^+a + f(\sigma_+ + \sigma_-)(a + a^+), \tag{7.62}$$

where a^+ and a are the photon creation and annihilation operators; E is the atomic level separation energy; f is the atom-field coupling constant proportional to the dipole moment of the transition; $\sigma_3, \sigma_\pm = \frac{1}{2}(\sigma_1 \pm i\sigma_2)$ are standard Pauli matrices with the commutation relations:

$$[\sigma_i, \sigma_j] = 2i\varepsilon_{ijk}\sigma_k$$

and the energy units are chosen in such a way that the photon frequency equals 1. The Hamiltonian (7.62) leads to the exact solution only if the counter-rotating terms σ_+a^+, σ_-a are omitted.

The exact integral of motion (combined parity) is assumed to exist in the system; it can be written in the form [the representation corresponding to the Hamiltonian (7.62)]:

$$\hat{P} = \sigma_3\hat{S} = \sigma_3 e^{i\pi a^+a}. \tag{7.63}$$

Thus, the exact state vector of the system depends on the two quantum numbers and it is the joint solution to the following equations:

$$\begin{aligned} \hat{H}|\Psi_{np}\rangle &= \mathcal{E}_{np}|\Psi_{np}\rangle, \\ \hat{P}|\Psi_{np}\rangle &= p|\Psi_{np}\rangle. \end{aligned} \tag{7.64}$$

Here the numbers $p = \pm 1$ define the parity and $n = 0, 1, 2, \ldots$ are the energy quantum numbers for the steady-state eigenvalues $\mathcal{E}_{np}$.

In accordance with the OM prescription, one has to use a complete set of the basic vectors depending on arbitrary parameters which take into account the variation of system states due to the interaction. In the considered case the atom-field coupling leads to a shift of the equilibrium position of field oscillators. This shift can be described by means of the following canonical transformation in the operator form:

$$
\begin{aligned}
&a = -u + b, a^{+} = -u + b^{+}, \\
&b = \hat{R}^{-1} a \hat{R}, \quad \hat{R} = e^{u(a^{+}-a)} = e^{-u^2/2} e^{ua^{+}} e^{-ua},
\end{aligned} \tag{7.65}
$$

where b^{+} and b are the new creation and annihilation operators, parameter u will be defined later. Then the relevant basic set can be chosen by means of eigenfunctions of the operators $\hat{n} = b^{+}b$ and σ_1:

$$
\begin{aligned}
&|\Phi_{ns}\rangle = |n, u\rangle \chi_s, \\
&\hat{n} = n|n, u\rangle, \quad \sigma_1 \chi_s = s\chi_s \quad (s = \pm).
\end{aligned} \tag{7.66}
$$

The state $|n, u\rangle$ corresponds to the n-quantum excitation of the field coherent state and connects with the photon vacuum $|0\rangle$ as follows:

$$
|n, u\rangle = \frac{(a^{+} + u)^n}{\sqrt{n!}} \sum_{k=0}^{\infty} \frac{u^k}{k!} (a^{+})^k |0\rangle e^{-u^2/2}. \tag{7.67}
$$

The amplitude u of the classical component of the field to be determined defines the choice of the presentation for the wave function of the system stationary states. The functions χ_s are the following linear combinations of the atom ground and excited states:

$$
\chi_{\pm} = \frac{1}{\sqrt{2}} (\chi_{\uparrow} \pm \chi_{\downarrow}).
$$

The transformation of Eq. (7.65) leads to presenting the Eq. (7.64) in the following form:

$$
\begin{aligned}
&\left[\frac{1}{2} E\sigma_3 + u^2 - 2uf\sigma_1 + (-u + f\sigma_1)(b + b^{+}) + b^{+}b \right] |\Psi_{np}\rangle = \mathcal{E}_{np} |\Psi_{np}\rangle, \\
&\sigma_3 e^{-2u^2} e^{2ub^{+}} e^{i\pi b^{+}b} e^{2ub} |\Psi_{np}\rangle = p|\Psi_{np}\rangle,
\end{aligned} \tag{7.68}
$$

and the state vector is expanded into the series of the basic set functions:

$$
|\Psi_{np}\rangle = \sum_{k=0}^{\infty} \sum_{s=\pm} C_{ks}^{np} |k, u\rangle \chi_s. \tag{7.69}
$$

Finally, the algebraic recurrence relations for coefficients C_{k+}^{np} and C_{k-}^{np} can be found as:

$$(k+u^2-2uf-\mathcal{E}_{np})C_{k+}^{np}+(f-u)(\sqrt{k}C_{k-1,+}^{np}+\sqrt{k+1}C_{k+1,+}^{np})+\frac{1}{2}EC_{k-}^{np}=0,$$
$$C_{k-}^{np}=p\sum_m S_{km}(u)C_{m+}^{np}, \tag{7.70}$$

where

$$S_{km}(u)=S_{mk}(u),$$
$$S_{km}(u)=(-1)^m e^{-2u^2}\sqrt{\tfrac{m!}{k!}}(2u)^{k-m}L_m^{k-m}(4u^2), \quad k\geq m, \tag{7.71}$$

here $L_n^\alpha(x)$ are the Laguerre polynomials [50].

The equation for C_{k-}^{np} also follows from (7.68) but it is the linear combination of the Eq. (7.70) when $\mathcal{E}_{np}$ coincides with the exact eigenvalue. The further numerical calculations show that the connection between C_{k-}^{np} and C_{k+}^{np} due to the exact integral of motion $\hat{P}$ is more important for the convergence of the OM series than the connection of these coefficients through the Eq. (7.68).

The most effective algorithm for the calculation of the OM successive approximation is based on simple iterations within the system of equations for unknown eigenvalue and the coefficients of the wave function expansion. This scheme provides the diagonal part of the total Hamiltonian is taken into account exactly in every iteration order (Chap. 2). Applying this algorithm to the Eq. (7.70) we arrive at the following recurrence relations for the successive approximations to the accurate values $\mathcal{E}_{np}$ and C_{k+}^{np}:

$$\begin{aligned}\mathcal{E}_{np}(t)&=n+u^2-2uf+\frac{1}{2}pES_{nn}(u)+(f-u)(\sqrt{n}C_{n-1,+}^{np}(t-1)\\&+\sqrt{n+1}C_{n+1,+}^{np}(t-1))\\&+\frac{1}{2}pE\sum_{m(\neq n)}S_{nm}(u)C_{m+}^{np}(t-1),\end{aligned}$$

$$\begin{aligned}C_{k+}^{np}(t)&=\delta_{kn}-(1-\delta_{kn})[k+u^2-2uf+\frac{1}{2}pES_{kk}(u)-\mathcal{E}_{np}(t-1)]^{-1}\\&\{(f-u)[\sqrt{k}C_{k-1,+}^{np}(t-1)+\sqrt{k+1}C_{k+1,+}^{np}(t-1)]+\\&+\frac{1}{2}pE\sum_{m(\neq k)}S_{km}(u)C_{m+}^{np}(t-1)]\}.\end{aligned} \tag{7.72}$$

and the values $\mathcal{E}_{np}$ and C_{k+}^{np} are defined as the limits of the sequences:

$$\mathcal{E}_{np} = \lim_{t\to\infty} \mathcal{E}_{np}(t), \quad C_{k+}^{np} = \lim_{t\to\infty} C_{k+}^{np}(t). \tag{7.73}$$

One can use the OM zeroth-order approximation in order to find the initial elements of the sequences:

$$C_{k+}^{np}(0) = \delta_{nk}, \quad \mathcal{E}_{np}(0) = n + u^2 - 2uf + \frac{1}{2}pES_{nn}(u). \tag{7.74}$$

By the definition, the exact eigenvalues don't depend on the parameter u which is related to the choice of the wave function representation only and the accurate numerical calculations confirm this statement. But the rate of convergence of the sequences (7.73) depends on this value and it proves to be maximal when

$$u = f. \tag{7.75}$$

This choice is optimal for the OM zeroth approximation as well (see below).

The iteration scheme described above permits one to find the solution to the Schrödinger equation for the TLS with any required accuracy within the entire range of the Hamiltonian parameters. In this sense we shall consider this solution as an exact one and it proves to be as effective for the analysis of the system characteristics as an analytical solution.

Let us consider some features of the exact solution which are qualitatively differed from the results obtained in the limits of RWA for this model. Figure 7.1a, b compare the accurate energy spectrum of the system with its asymptotic approximations: RWA in the case of $f \ll 1$ and $\mathcal{E}_{np} \simeq n - f^2$ in the limit of strong coupling. One can see that the intersections of the levels with different parities theoretically described earlier [46] lead to the formation of peculiar "plaits" on the diagram of the levels. The real spectrum structure can be approximated by the spectrum of RWA only in the range of small enough coupling constant until to the first intersection of the levels with different parities. The width of this range decreases for high-excited states of the system proportionally to $\frac{1}{\sqrt{n}}$. Figure 7.1 illustrates this statement and shows that our numerical solution is effective for any quantum numbers.

Figure 7.2 shows the contribution of various harmonics of the field in the formation of the accurate stationary states. We remind that in the RWA limits the nth state is a superposition of two field states only. However, as follows from Fig. 7.2, quite a lot of field quanta contribute to the formation of the first system excited state even for the coupling constant $f \simeq 1$.

We also verified the conjectured analytical solution by [51, 52] for those parameters which were considered in these papers. It is also possible to deduce the approximate but analytical formulae which interpolate the energy and other characteristics of the TLS uniformly within the entire range of the coupling constant

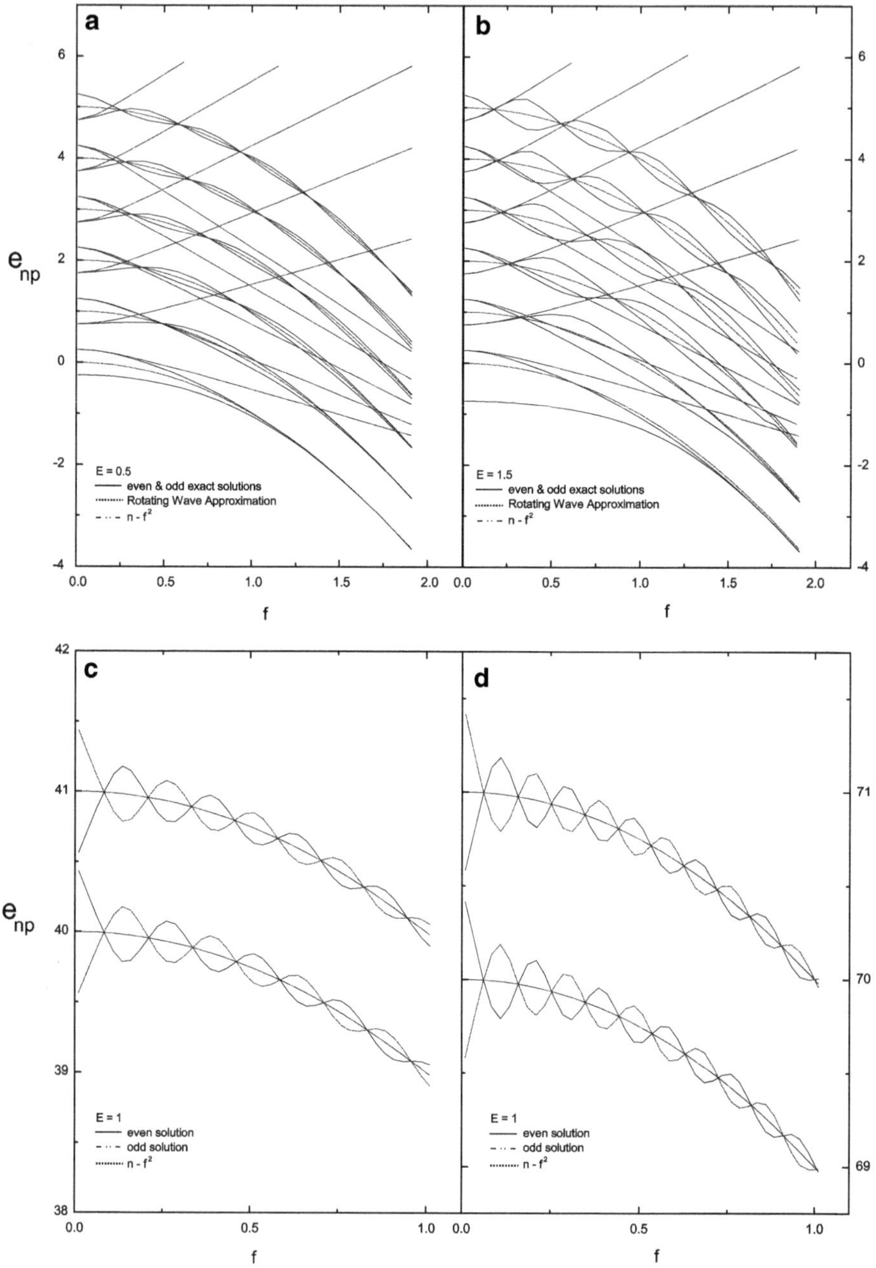

Fig. 7.1 **(a, b)**—exact and approximate eigenvalues of the TLS Hamiltonian as the functions of the coupling constant and the separation energy; **(c, d)**—highly excited states of the TLS

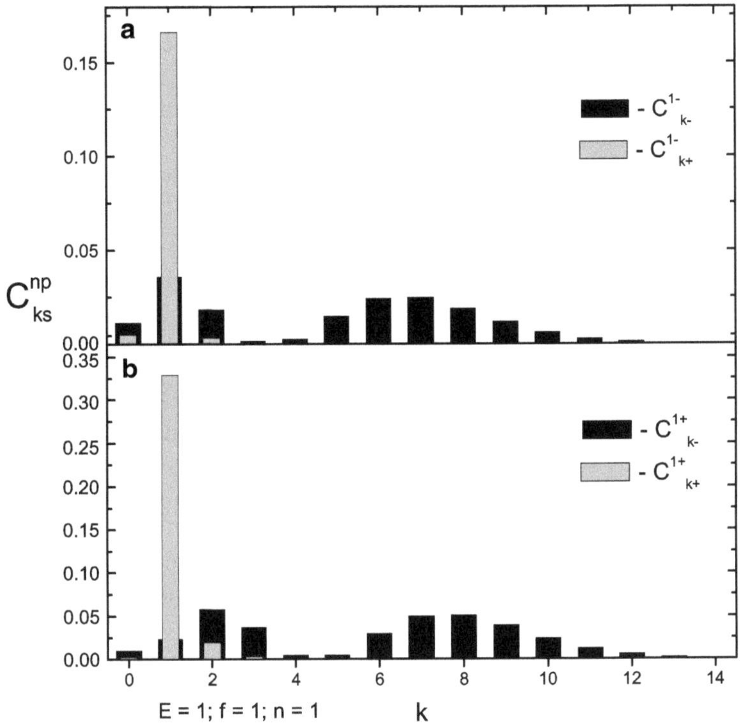

Fig. 7.2 Coefficients of the series on the basic set functions for the first excited even (**a**) and odd (**b**) states of the TLS

and quantum number variation. The solution to this problem can be obtained in the OM zeroth approximation, too.

In accordance with the OM algorithm one has to take into account only the diagonal part of the full Hamiltonian with respect to the considered basic set in order to find the OM zeroth-order approximation. To select this part from the matrix equations (7.68) let us expand the state vector $|\Psi_{np}\rangle$ in the eigenfunctions of the matrix σ_1:

$$|\Psi_{np}\rangle = |\varphi_{np+}\rangle\chi_+ + |\varphi_{np-}\rangle\chi_-, \tag{7.76}$$

and exclude the variables, related to the energy spin. Then the effective Schrödinger equation is:

$$\hat{H}_{eff}|\varphi_{np+}\rangle = \left[u^2 - 2uf + b^+b + (f-u)(b+b^+) + \frac{1}{2}pE\hat{S}\right]|\varphi_{np+}\rangle, \tag{7.77}$$

here $\hat{S}$ is the operator with matrix elements defined in the Eq. (7.71).

In accordance with the generalized form of the OM considered in Chap. 3 the diagonal form of the Hamiltonian can be extracted with respect to linear combinations of the vectors from a basic set which take into account the most essential properties of the exact wave functions. In the considered case such a peculiarity is defined by the degeneration (for $f = 0$ and $u = 0$) of the states with the quantum numbers n and $n + q$ ($q = p(-1)^n$) when $E = 1$, and can be described by means of the following correct linear combinations:

$$|\varphi^{(0)}_{np+}\rangle = A|n, u\rangle + B|n + q, u\rangle, \tag{7.78}$$

with the constant coefficients A and B.

Certainly, the exact eigenvalues do not depend on a choice of the parameter u, which defines only the wave functions representation. But just this artificial parameter ensures flexibility of the OM zeroth-order approximation for various coupling constants and there are several ways to choose the optimal value for this parameter. The optimal value u for effective Hamiltonian (7.77) is very simple $u = f$ and permits one to turn into zero the main part of the operator which defines the transitions between the nearest states. Applying the state vector (7.78) to the Eq. (7.77) leads to the following analytical formula for the stationary state energies in the OM zeroth approximation:

$$\mathcal{E}^{(0)}_{np} = n + \frac{1}{2}q - f^2 + \frac{1}{4}Eq(-1)^n(S_{nn}(f) + S_{n+q,n+q}(f)) -$$
$$-\frac{1}{2}q\sqrt{-\frac{1}{2}E(-1)^n(S_{nn}(f) - S_{n+q,n+q}(f))]^2 + E^2S^2_{n,n+q}(f)}. \tag{7.79}$$

Similarly to other applications of the OM the formula (7.79) proves to be uniformly suitable because it reproduces the exact asymptotic behavior of the function $\mathcal{E}_{np}$ in the limit cases $f \gg 1, f \ll 1$ and describes rather accurately the quantitative peculiarities of the energy spectrum for the intermediate values of the coupling constant (see Fig. 7.3).

Indeed, one can obtain the following formula for $f \ll 1$:

$$S_{nn}(f) \approx (-1)^n, \quad S_{n,n+q}(f) \approx (-1)^{n+\frac{1}{2}(1+q)} 2f\sqrt{n + \frac{1}{2}(1 + q)}, \tag{7.80}$$

and for $|1 - E| \leq f$:

$$\mathcal{E}^{(0)}_{np} \approx n + \frac{1}{2}q - \frac{1}{2}q\sqrt{(1 - E)^2 + 4f^2\left[n + \frac{1}{2}(1 + q)\right]} \tag{7.81}$$

which coincides with the results of RWA in Jaynes-Cummings model.

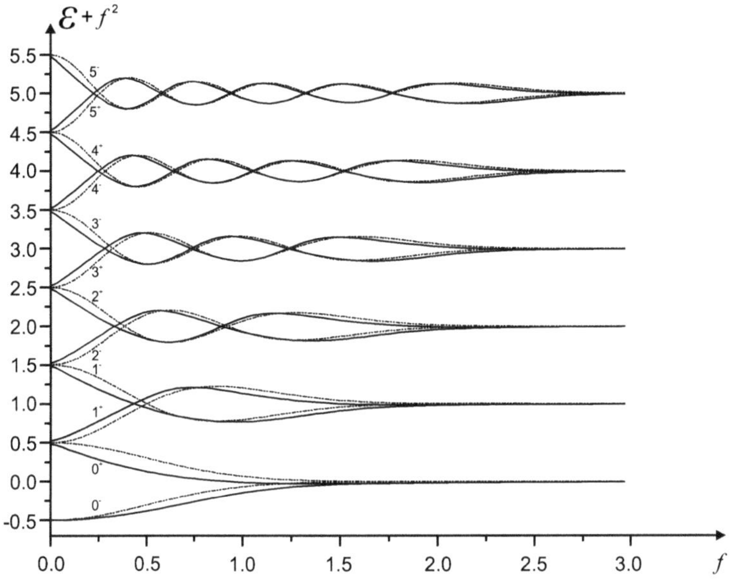

Fig. 7.3 Comparison of the exact eigenvalues (*solid lines*) with the OM zeroth approximation (*dotted lines*)

In the opposite limit case $f \gg 1$, the exponentially small terms in formula (7.79) can be omitted and it leads to the simple expression:

$$\mathcal{E}_{np}^{(0)} \approx n - f^2, \tag{7.82}$$

which is asymptotically exact.

A simpler formula may be used as the zeroth approximation takes into account the fact that the degeneration of levels is significant only in the domain of small f:

$$\mathcal{E}_{np}^{(0)} = n - f^2 + \frac{1}{2} qEe^{-2f^2} L_n(4f^2), \quad q = p(-1)^n. \tag{7.83}$$

As we could see above (Figs. 7.1 and 7.2) the characteristic feature of energy spectrum for the considered system is the intersection of the levels with the same quantum number n and different parity. The formula (7.79) [or (7.83)] shows that the OM zeroth-order approximation for spectrum has the same property (see Fig. 7.3).

The approximate estimation for the roots of the equation for f^2:

$$\mathcal{E}_{n,1}(f) = \mathcal{E}_{n,-1}(f) \tag{7.84}$$

with $\mathcal{E}$ from (7.83) can be found by means of the simple formula:

$$L_n(4f^2) = 0. \tag{7.85}$$

According to the exact results [46], the $L_n(4f^2)$ is the polynomial of the degree n with alternating terms in relation to f^2. As for example, the solutions of the Eq. (7.83) for $n = 1$ and $n = 2$ are written as:

$$f_1^2 = \frac{1}{4}, \quad n = 1; \qquad f_{1,2}^2 = \frac{1}{4}(2 \pm \sqrt{2}), \quad n = 2$$

and can be compared with analogous accurate values [46]:

$$f_1^2 = \frac{1}{4}(1 - \frac{E^2}{4}), \quad f_{1,2}^2 = \frac{1}{4}\left(2 - \frac{3}{16}E^2 \pm \sqrt{2 - \frac{E^2}{8} + \frac{E^4}{256}}\right).$$

Thus, the simple Eq. (7.83) permits one to find the characteristic points of the spectrum with accuracy $\sim 25\,\%$ in the most interesting resonant case.

The analytical expression for the system energy could be of special interest in case of large quantum numbers ($n \gg 1$) which are essential while considering the interaction of an atom with an intensive electromagnetic field. In this limit the matrix elements of the operator $\hat{S}$ are essentially simplified. Using the asymptotic formula (see [50]):

$$L_n^{\alpha}(x) = \frac{1}{\sqrt{\pi}} e^{\frac{1}{2}x} x^{-\frac{1}{2}\alpha - \frac{1}{4}} n^{\frac{1}{2}\alpha - \frac{1}{4}} \cos\left[2\sqrt{nx} - \frac{\alpha\pi}{2} - \frac{\pi}{4}\right] + O(n^{\frac{1}{2}\alpha - \frac{3}{4}}),$$

one may find:

$$S_{kn}(f) = \frac{(-1)^n}{\sqrt{2\pi f \sqrt{n}}} \cos\left(4f\sqrt{n} - \frac{(k-n)\pi}{2} - \frac{\pi}{4}\right), \quad n \gg 1, \quad (k-n) \ll n.$$

As a result, the formula (7.79) transforms as following:

$$\mathcal{E}_{np}^{(0)} = n - f^2 + \frac{qE}{2\sqrt{2\pi f \sqrt{n}}} \cos\left(4f\sqrt{n} - \frac{\pi}{4}\right), \tag{7.86}$$

and can be used in the limit when the field is described semiclassically.

The efficiency of the numerical solution is very important for the problem of the system time evolution when many stationary states have to be calculated. Therefore, we consider briefly the time evolution of some initial states of the system when the exact solution is used instead of RWA.

Let us suppose, for example, that at the moment $t = 0$ the atom occupies an excited state and the quantum field is in the vacuum state. It means that an initial wave packet in the considered system has the following form:

$$|\Psi(0)\rangle = |0\rangle\chi_{\uparrow}. \tag{7.87}$$

One can expand it in the exact stationary states (7.72) and calculate the probability $P_{\uparrow}(t)$ to find the atom in the excited state again at the moment t regardless of the field quantum number. It is well known that in the limits of the Jaynes-Cummings model the value $P_{\uparrow}(t)$ oscillates periodically at the Rabi frequency. The exact expression for $P_{\uparrow}(t)$ is the following:

$$\begin{aligned}
P_{\uparrow}(t) &= \frac{1}{2}\sum_{mp}\sum_{nq} A_{mp}A_{nq}\cos[(\mathcal{E}_{mp}-\mathcal{E}_{nq})t]\sum_{k}(C_{k+}^{mp}+C_{k-}^{mp})(C_{k+}^{nq}+C_{k-}^{nq}),\\
C_{k-}^{mp} &= \sum_{l} S_{kl}(f)C_{l+}^{mp},\\
A_{mp} &= \frac{1}{N\sqrt{2}}e^{-\frac{1}{2}f^2}\sum_{k}\frac{f^k}{\sqrt{k!}}(C_{k+}^{mp}+C_{k-}^{mp}),\\
N^2 &= \sum_{k}[(C_{k+}^{mp})^2+(C_{k-}^{mp})^2].
\end{aligned} \tag{7.88}$$

We calculated the required probability on the basis of the formula (7.88) and the numerical solution of the Eq. (7.72) for stationary states. Figure 7.4 shows the results of the calculation for various values of the coupling constant. The increase of the coupling constant leads to the modification of the oscillating regime of the function $P_{\uparrow}$ being periodic with period 2π to the specific case which corresponds to the quasi-periodic motion in the theory of instability [53] and arises as a result of a superposition of many motions with incommensurable frequencies.

The same tendency can be observed in Fig. 7.5 which shows the function $\langle P_{\uparrow}(n)\rangle$ averaged over the main period of oscillation that is:

$$\langle P_{\uparrow}(n)\rangle = \frac{1}{2\pi}\int_{2\pi n}^{2\pi(n+1)} dt \quad P_{\uparrow}(t).$$

One of the possible ways of chaotization in dynamical systems is connected with doubling of the oscillation frequency [53]. One can see (Figs. 7.6 and 7.7) that such kind of behavior appears in the considered system. In particular Fig. 7.6 shows the spectrum of the function $P_{\uparrow}(t)$ that is the function:

$$P_{\uparrow}(\omega) = \int_{-\infty}^{\infty} dt \quad e^{-i\omega t}P_{\uparrow}(t),$$

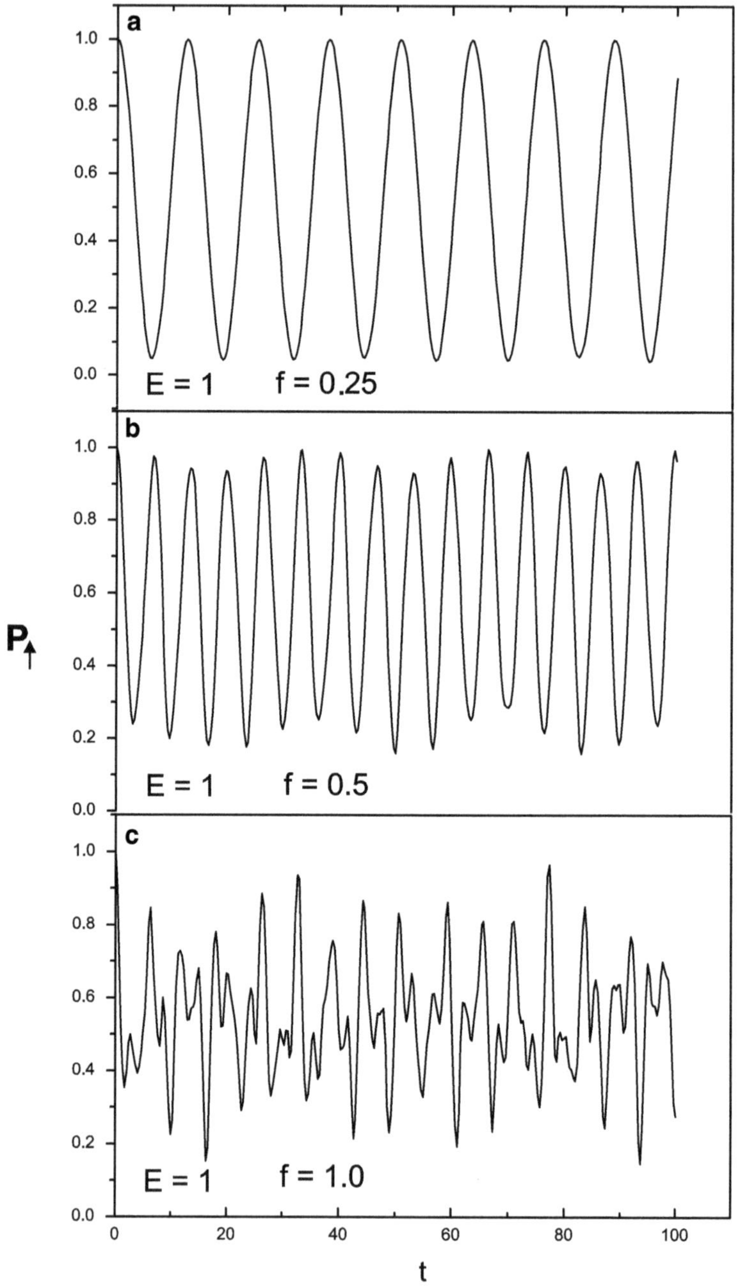

Fig. 7.4 Probability of the population of the atom excited state as a function of time

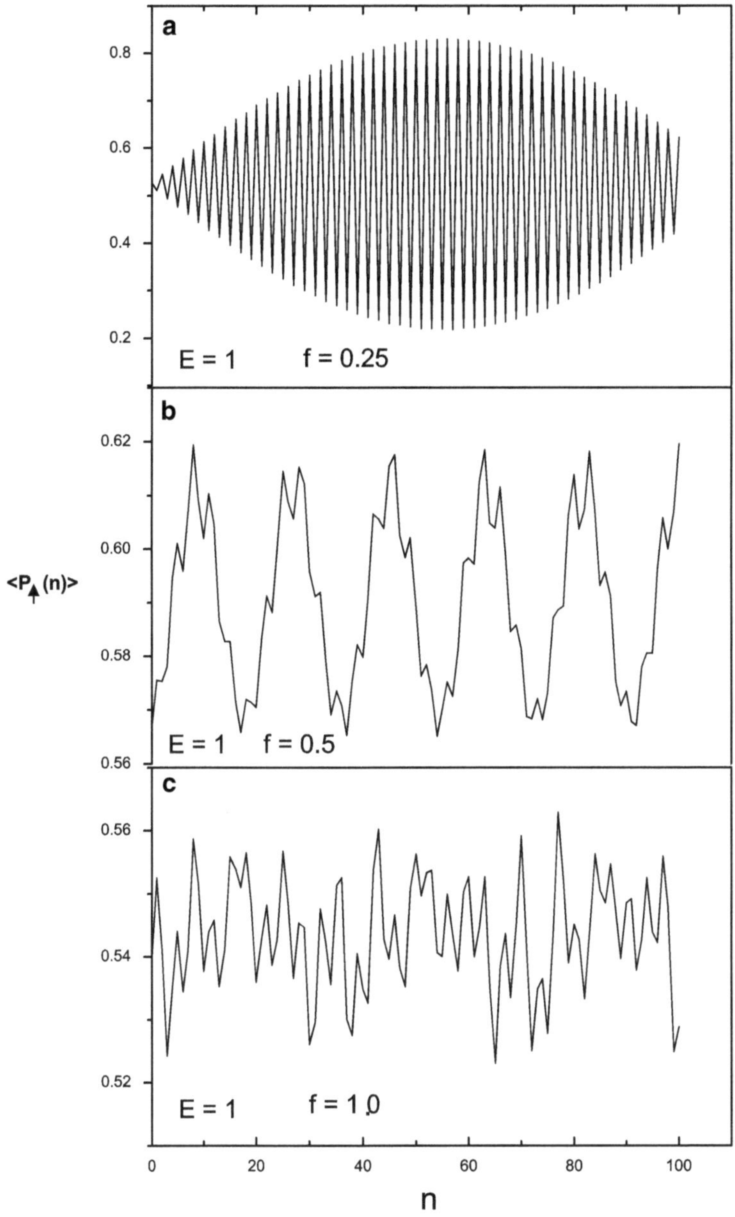

Fig. 7.5 Probability of the excited state population averaged over the main period

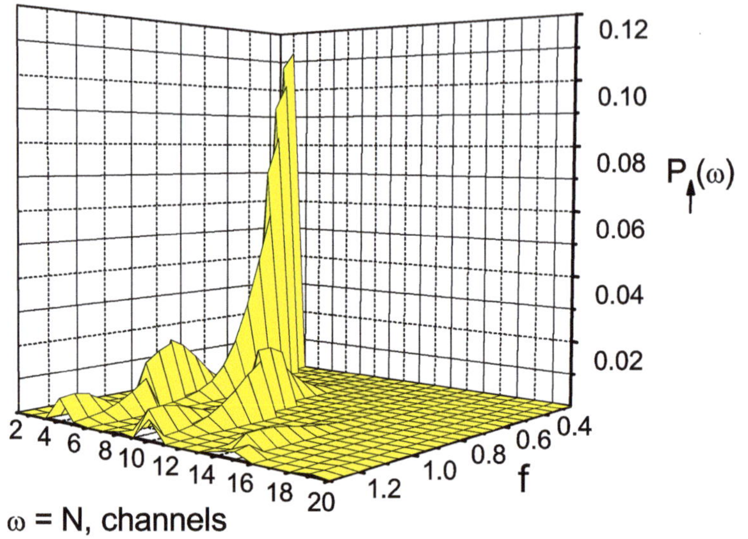

Fig. 7.6 Spectral function for the probability of the excited state population

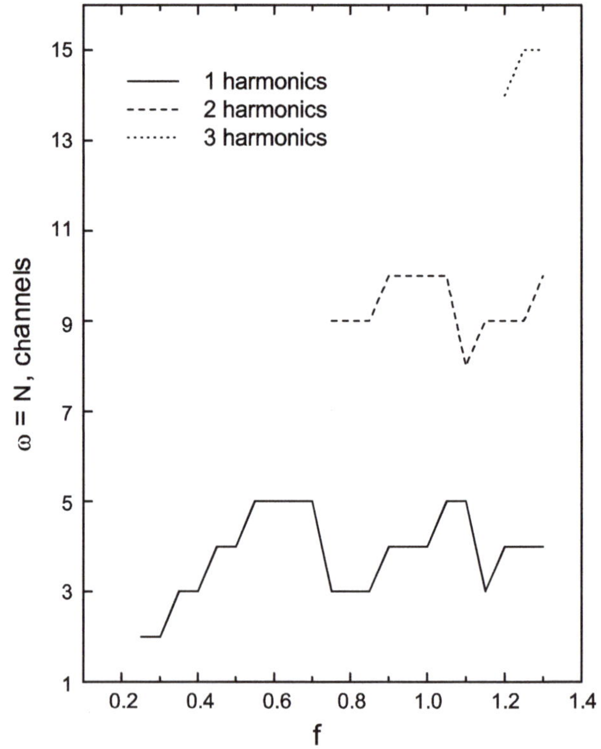

Fig. 7.7 Amplitudes of the main harmonics as the functions of coupling constant

depending on the coupling constant f. Figure 7.7 demonstrates a non-monotone dependence of the main harmonic amplitudes on the value f.

More detailed analysis of non-RWA effects in the evolution of TLS in a single-mode quantum field was considered recently in the papers [29–31].

References

1. L.I. Schiff, H. Snyder, Phys. Rev. **55**, 59 (1939)
2. Y. Yafet, R.W. Keyes, E.N. Adams, J. Phys. Chem. Solids **1**, 137 (1956)
3. H. Friedrich, D. Wintgen, Phys. Rep. **183**, 37 (1989)
4. L.D. Landau, E.M. Lifshitz, *Quantum Mechanics: Non-Relativistic* (Pergamon Press, Oxford, 1991)
5. A.R.P. Rau, L. Spruch, Astrophys. J. **207**, 671 (1976)
6. D. Delane, J.C. Gay, J. Phys. B Atom. Mol. Opt. Phys. **19**, L173 (1986)
7. J. Xi, X. He, B. Li, Phys. Rev. A **46**, 5806 (1992)
8. J.X. Zang, M.L. Rustgi, Phys. Rev. A **50**, 861 (1994)
9. C. Stubbins, K. Das, Y. Shiferaw, J. Phys. B Atom. Mol. Opt. Phys. **37**, 2201 (2004)
10. V.S. Popov, B.M. Karnakov, Phys. Usp. **57**, 257 (2014)
11. J.M. Lattimer, M. Prakash, Phys. Rep. **442**, 109 (2007)
12. O.I. Tolstikhin, T. Morishita, L.B. Madsen, Phys. Rev. A **84**, 053423 (2011)
13. B. Zhang, J. Yuan, Z. Zhao, Phys. Rev. Lett. **111**, 163001 (2013)
14. I.D. Feranchuk, L.I. Komarov, J. Phys. A Math. Gen. **17**, 3111 (1984)
15. L.I. Komarov, T.S. Romanova, J. Phys. B Atom. Mol. Phys. **18**, 859 (1985)
16. L.I. Komarov, T.S. Romanova, Z.A. Chan, Proc. Natl. Acad. Belarus Ser. Phys. Math. Sci. **1**, 90 (1987)
17. I.D. Feranchuk, X.H. Ly, Phys. Lett. A **137**, 385 (1989)
18. K. Ng. D. Yao, M.H. Nayfeh, Phys. A **35**, 2508 (1987)
19. P. Kustaanheimo, E. Stiefel, J. Reine Angew. Math. **1965**, 204 (1965)
20. L.I. Komarov, T.S. Romanova, Proc. Natl. Acad. Belarus Ser. Phys. Math. Sci. **2**, 98 (1982)
21. L.I. Komarov, T.S. Romanova, J. Phys. B Atom. Mol. Opt. Phys. **18**, 859 (1985)
22. H. Kleinert, *Group Dynamics of the Hydrogen Atom* (Gordon and Breach, New York, 1968)
23. H. Kleinert, Found. Phys. **23**, 769 (1993)
24. L. Allen, J. Eberly, *Optical Resonance and Two Level Atoms* (Wiley, New York, 1975)
25. H. Reik, H. Nusser, A. Ribeiro, J. Phys. A Math. Gen. **15**, 3491 (1982)
26. P. Meystre, T. Steimle, Phys. Lett. A **89**, 390 (1982)
27. R. Graham, M. Hoehnerbach, Phys. Lett. A **101**, 61 (1984)
28. I.D. Feranchuk, L.I. Komarov, A.P. Ulyanenkov, J. Phys. A Math. Gen. **29**, 4035 (1996)
29. I.D. Feranchuk, A.V. Leonov, Phys. Lett. A **373**, 4113 (2009)
30. I.D. Feranchuk, A.V. Leonov, Phys. Lett. A **373**, 517 (2009)
31. I.D. Feranchuk, A.V. Leonov, Phys. Lett. A **375**, 385 (2011)
32. V.H. Le, T. Viloria, Phys. Lett. A **171**, 23 (1992)
33. V.V. Kolosov, J. Phys. B Atom. Mol. Opt. Phys. **20**, 2539 (1987)
34. V.M. Vainberg, V.D. Mur, V.S. Popov, A.V. Sergeev, Sov. Phys. JETP Lett. **46**, 178 (1987)
35. E. Papp, Phys. Lett. A **132**, 127 (1988)
36. V.V. Kolosov, Opt. Spectroscop. **65**, 150 (1988)
37. V.D. Mur, V.S. Popov, Sov. Phys. JETP **67**, 2027 (1988)
38. I.D. Feranchuk, L.I. Komarov, Phys. Lett. A **88**, 211 (1982)
39. I.D. Feranchuk, L.I. Komarov, I.V. Nechipor, J. Phys. A Math. Gen. **20**, 3849 (1987)
40. D. Farrelly, W.P. Reinhardt, J. Phys. B Atom. Mol. **16**, 2103 (1983)
41. R.J. Damburg, V.G. Kolosov, J. Phys. B Atom. Mol. **9**, 3149 (1976)

42. C.Z. An, I.D. Feranchuk, L.I. Komarov, Phys. Lett. A **125**, 123 (1987)
43. V.B. Pavlov-Verevkin, B.I. Zhilinskii, Phys. Lett. A **75**, 279 (1980)
44. H. Friedrich, Phys. Rev. A **26**, 1827 (1982)
45. E.T. Jaynes, F.W. Cummings, Proc. IEEE **51**, 89 (1963)
46. M. Kus, M. Lewenstein, J. Phys. A Math. Gen. **19**, 305 (1986)
47. P. Lais, T. Steimle, Opt. Commun. **78**, 346 (1990)
48. E.K. Irish, Phys. Rev. Lett. **99**, 173601 (2007)
49. E.K. Irish, Phys. Rev. Lett. **99**, 259901 (2007)
50. P.M. Morse, H. Feshbach, *Methods of Theoretical Physics* (McGraw-Hill, New York, 1953)
51. H. Reik, M. Doucha, Phys. Rev. Lett. **57**, 787 (1986)
52. H. Reik, P. Lais, M. E. Stüzle, M. Doucha, J. Phys. A Math. Gen. **20**, 6327 (1987)
53. R. Richtmyer, *Principles of Advanced Mathematical Physics v.2* (Springer, New York, 1981)

Chapter 8
Many-Electron Atoms

The essential achievements in modern quantum theory are closely connected to the microscopic description of many-electron systems in quantum chemistry, biology and condensed matter physics by using density functional theory (DFT), introduced in pioneering works [1,2]. This approach is used as a basis for ab initio calculations for complex molecular systems and the details of the method are presented in numerous monographs and reviews [3–6]. The idea of DFT is a use of one-electron density $n_e(\boldsymbol{r})$ as a principle dynamic variable defining the state of the system, instead of many-particle wave function $\Psi(\xi_1, \xi_2, \ldots, \xi_N)$, which depends on the coordinate and spins of all electrons. The exact expression for the density in ground state is determined after the minimization of the density functional $F[n_e(\boldsymbol{r})]$, and thus the problem is reduced to the construction of the approximation for the functional at certain coordinates of the nuclei of the atoms of system (adiabatic approximation).

The most effective and simple way to obtain $F[n_e(\boldsymbol{r})]$ is based on the local density approximation (LDA) with the use of the model of Thomas-Fermi (MTF) or its modifications [7, 8]. The main advantages of this approach are the universal expressions for atomic values depending on nuclear charges Z, and the accurate asymptotic formulas for energy and other physical characteristics of atoms and ions. In the MTF concept, the atom is considered as electron gas with a small gradient of density, and thus one-electron basis is described by quasi-classic wave functions, which are close to plane waves [7]. These functions describe approximately the real localized wave functions of atomic electrons in the region $Z \gg 1$ only, that results in the disadvantages of MTF: unlimited increase of electron density near atomic nuclei, non-exponential decay at the infinity, the absence of shell effects in atomic characteristics, and asymptotic nature of the corrections to zeroth approximation. As a result, the LDA functional $F[n_e(\boldsymbol{r})]$ in the basis of MTF describes the distribution of electron density not accurately, especially for the atoms

I. Feranchuk et al., *Non-perturbative Description of Quantum Systems*,
Lecture Notes in Physics 894, DOI 10.1007/978-3-319-13006-4_8

with small number of electrons [4]. The correction for the density gradient by accurate quasi-classic approximation improves the results of the model [3], however, complicates essentially the calculations and is not uniformly suitable for arbitrary atomic charges.

The aim of this chapter is to illustrate the construction of LDA for DFT by operator method using the one-electron basis of either wave functions of harmonic oscillator (Sect. 8.1) or Coulomb wave functions (Sect. 8.3), which interpolate a real distribution of atomic electrons better than plane waves in MTF. This approach constructs the approximation for electron density of atom, which remains uniformly suitable at any distances from the nuclei and for atoms with any number of electrons. Within the framework of this approach, the shell effects are taken into account and the universal dependence of atomic characteristics on Z in the region $Z \gg 1$ is found.

Another development trend in quantum theory of many-electron systems involves the method of Hartree–Fock (HF), based on one-electron approximation for many-particle wave function [9, 10]. The wave function $\Psi(\xi_1, \xi_2, \ldots, \xi_N)$ of the system is represented as Slater determinant [11], composed of the one-electron functions (orbitals), which are found by numerical solution of complex system of integro-differential equations (see, for example, [12] and citations therein). In zeroth approximation of HF method, the contributions to the energy of the system are taken into account, which are determined by the Coulomb interaction between electrons and nuclei, the averaged potential of the repulsion between electrons, and quantum exchange interaction caused by the identity of electrons. The successive approximations take into account many-electron correlations by improving the wave functions with configuration interaction method [13]. A substantial difficulty for applications of HF method in quantum chemistry and condensed matter physics are cumbersome numerical calculations for the functions of zeroth approximation [12]. This fact complicates the analysis of qualitative properties of atoms in dependence on the nuclei charge Z, the construction of the excited states, the interaction of atoms with external fields and inter-atomic interaction in molecules. These problems are partly eliminated due to the use of Slater orbitals [11], which include the set of phenomenological parameters selected from best approximation of numerical wave functions of HF and depending on quantum number. The OM is also applicable in this case (Sect. 8.4), providing an effective charges model (ECM) [14, 15] for basic set of one-electron wave functions. Such approach guarantees the accuracy of zeroth approximation for atomic characteristics, which is comparable to the results after HF method. The operator method permits to construct the wave functions of zeroth approximation for excited states and to build a regular scheme for calculation of corrections to one-electron approximation.

8.1 Oscillator Model of Atom

We start here with the equations, which define the ground state of atom in OM zeroth approximation for one-electron wave functions of harmonic oscillator. These functions are not the best choice for the description of electron in Coulomb field (see Sect. 7.2), however, they are better than plane wave approximation of MTF and permit to account the effects of non-uniformity of electron density. The Schrödinger equation for the wave function of electron in self-consistent potential $v(\boldsymbol{r})$ is used, which includes both nucleus potential and averaged potential of electrons [9]. Without accounting of exchange and spin effects in Hartree approximation, this equation is written as:

$$
(\hat{H} - \epsilon_\lambda)|\psi_\lambda\rangle = \left(\frac{1}{2}p^2 + v(\boldsymbol{r}) - \epsilon_\lambda\right)|\psi_\lambda\rangle = 0;
$$
$$
\Psi(\xi_1, \xi_2, \ldots, \xi_N) = \prod_i \langle \xi_i | \psi_{\lambda_i} \rangle, \tag{8.1}
$$

where index λ includes four quantum numbers defining one-particle wave function $|\psi_\lambda\rangle$ and energy of electron ϵ_λ; $\boldsymbol{p}$ is a momentum operator. Without applying symmetrization of wave function, the numeration of electrons in atom is arbitrary, but due to Pauli principle the set of quantum numbers λ is in a single-valued correspondence with the index i, which defines the electron number in the selected sequence. In this chapter, the Coulomb units are used [9], where $\hbar = m = e = 1$, the distances are measured in the units of Bohr radius $a_B = \hbar^2/me^2 = 0.529$ Å, and energy in atomic units $\epsilon_B = me^4/\hbar^2 = 27.21\ eV$.

The mathematical derivation of Thomas-Fermi equation is based [7] on the quasi-classic approximation for Eq. (8.1). The closed equations for the potential are possible due to analytical expression for wave functions $|\psi_\lambda\rangle$ for arbitrary form of potential. The similar calculations can be performed for zeroth approximation of operator method, too. According to general receipt of OM (Chap. 3), the following canonic transformation for operators in the Eq. (8.1) is sufficient for the transition to oscillator basis:

$$
p_\mu = i\sqrt{\frac{\omega_\lambda}{2}}(a_\mu^+ - a_\mu); \quad x_\mu = \sqrt{\frac{1}{2\omega_\lambda}}(a_\mu^+ + a_\mu). \tag{8.2}
$$

Here $\mu = (x, y, z)$; a_μ, a_μ^+ are the operators of creation and annihilation, and the transformation parameters ω_λ will be defined later. These parameters depend on quantum numbers, which is specific for operator method. The operators a_μ, a_μ^+ are supplementary operators, which are used for determination of the state of separate electron in accordance with Eq. (8.1) and not for the state of the whole many-particle system. Therefore, these operators satisfy to commutative relations $[a_\mu, a_\nu^+] = \delta_{\mu\nu}$, corresponding to Bose-operators. The zeroth approximation of OM is determined by the part of one-electron Hamiltonian $\hat{H}$, which commutates with the operators of excitation numbers $a_\mu^+ a_\mu$, and has a following form:

$$\hat{H}_0 = \frac{\omega_\lambda}{4}(3 + 2\sum_\mu a_\mu^+ a_\mu) + \frac{1}{(2\pi)^3}\int d\boldsymbol{r} v(\boldsymbol{r}) \int d\boldsymbol{k} e^{-i\boldsymbol{k}\boldsymbol{r}} e^{-k^2/4\omega_\lambda} \sum_\mu \sum_{m=o}^{\infty} (-1)^m \frac{k_\mu^2}{4\omega_\lambda} \frac{(a_\mu^+)^m (a_\mu)^m}{(m!)^2}. \tag{8.3}$$

The eigenvectors of the operator $\hat{H}_0$ are defined by the quantum numbers $\lambda = (n_x, n_y, n_z)$, $|\psi_\lambda^0\rangle = |\{n_\mu\}\rangle$, and one-particle energy at arbitrary $v(\boldsymbol{r})$ is calculated from:

$$\epsilon_\lambda = \frac{\omega_\lambda}{4}\left(3 + 2\sum_\mu n_\mu\right) + \frac{1}{(2\pi)^3}\int d\boldsymbol{r} v(\boldsymbol{r}) \int d\boldsymbol{k} e^{-i\boldsymbol{k}\boldsymbol{r}} e^{-k^2/4\omega_\lambda} \prod_\mu L_{n_\mu}^0 (\frac{k_\mu^2}{4\omega_\lambda}), \tag{8.4}$$

where $L_n^m(x)$ are the Laguerre polynomials [16].

A further simplification of the model is attained assuming the self-consistent potential of the atom to be spherically symmetric, and due to this symmetry over all components, the parameters ω_λ depend on principle quantum number $n = n_x + n_y + n_z$ only, which describes the oscillatory shell of the atom. Under this assumption, all the electrons on the shell have equal energy, and the summation can be performed over all states at fixed n in formula (8.4) using the following relationship [16]:

$$\sum_{n_x=0}^{n} L_{n_x}^0 \left(\frac{k_x^2}{2\omega}\right) \sum_{n_y=0}^{n-n_x} L_{n_y}^0 \left(\frac{k_y^2}{2\omega}\right) L_{n-n_x-n_y}^0 \left(\frac{k_z^2}{2\omega}\right) = L_n^2 \left(\frac{k^2}{2\omega}\right). \tag{8.5}$$

As a result, the energy of the electron on the shell with a degeneracy (with two spin states) is:

$$g_n = (n+1)(n+2), \tag{8.6}$$

and is calculated from the formula:

$$\epsilon_n = \frac{\omega_n}{4}(2n+3) + \frac{2}{g_n}\frac{1}{(2\pi)^3}\int d\boldsymbol{r} v(\boldsymbol{r}) \int d\boldsymbol{k} e^{-i\boldsymbol{k}\boldsymbol{r}} e^{-k^2/4\omega_n} L_n^2 \left(\frac{k^2}{2\omega_n}\right). \tag{8.7}$$

The self-consistency of the model is achieved in a similar way as for MTF [7] by using Poisson equation for one-electron potential:

$$\Delta v(\boldsymbol{r}) = 4\pi Z\delta(\boldsymbol{r}) - 4\pi n_e(\boldsymbol{r}). \tag{8.8}$$

The solution for this equation for the nuclear charge Z is written as:

$$v(\boldsymbol{r}) = -\frac{Z}{r} + \frac{1}{\pi^2}\int d\boldsymbol{k}\frac{e^{-i\boldsymbol{k}\boldsymbol{r}}}{k^2}\sum_{n=0}^{n_m} e^{-k^2/4\omega_n} \times \sum_{n_x=0}^{n} L^0_{n_x}\left(\frac{k_x^2}{2\omega_n}\right)\sum_{n_y=0}^{n-n_x} L^0_{n_y}\left(\frac{k_y^2}{2\omega_n}\right) L^0_{n-n_x-n_y}\left(\frac{k_z^2}{2\omega_n}\right). \tag{8.9}$$

To calculate the electron density $n_e(\boldsymbol{r})$ for the system in a ground state, the summation in formula (8.9) is performed over the completed states, according to Pauli principle. The total number of states equals to number n_m of last complete shell, which is found from the normalization condition to the total number of electrons N_0 in atom or ion, and therefore $N_0 \neq Z$:

$$N_0 = 2\sum_{n=0}^{n_m}\sum_{n_x=0}^{n}\sum_{n_y}^{n-n_x} 1 = \sum_{n=0}^{n_m}(n+1)(n+2) = \frac{(n_m+1)}{3}(n_m^2+5n_m+6). \tag{8.10}$$

The formulas (8.7), (8.9) and (8.10) are useable for the atoms with completely filled shells. The real number of electrons in any atom equals:

$$N = N_0 + q, \tag{8.11}$$

where N_0 is a number of electrons on filled shells. For the exceeding equivalent electrons q on external uncomplete shell with a quantum number $n_m + 1$, the effective averaged energy from the formula below can be used:

$$\epsilon_{n_m,q} = \frac{q}{g_{n_m+1}}\epsilon_{n_m+1}. \tag{8.12}$$

By performing a partial summation of Legendre polynomials in (8.9) using expression (8.5) and integration over the angles, the spherically-symmetric self-consistent potential is written as:

$$v(r) = -\frac{Z}{r} + \frac{4}{\pi}\int_0^\infty dk\frac{\sin kr}{kr}\Phi(k);$$
$$\Phi(k) = \sum_{n=0}^{n_m} g_n e^{-\frac{k^2}{4\omega_n}} L_n^2\left(\frac{k^2}{2\omega_n}\right) + qe^{-\frac{k^2}{4\omega_{n_m+1}}} L^2_{n_m+1}\left(\frac{k^2}{2\omega_{n_m+1}}\right). \tag{8.13}$$

The radial distribution of the electron density is calculated by the formula:

$$n_e(r) = \frac{1}{\pi^2}\int_0^\infty dkk\frac{\sin kr}{r}\Phi(k), \tag{8.14}$$

and the expression for total energy of atom is found in a form:

$$E(N,Z,\{\omega_n\}) = \sum_{n=0}^{n_m} g_n\epsilon_n(\omega_n) + q\epsilon_{n_m+1} - \int d\mathbf{r} n_e(r)v_e(r);$$

$$v_e(r) = v(r) + \frac{Z}{r}, \tag{8.15}$$

where the last term in the Eq. (8.15) appears due to double accounting of the interaction of electrons in a sum of one-particle energies. According to general OM procedure, the optimal values for the sequence of parameters ω_n are determined by the extremum for the function (8.15) over these parameters:

$$\frac{\partial}{\partial\omega_n}E(N,Z,\{\omega_n\}) = 0, n = 0,1,2,\ldots,n_m+1. \tag{8.16}$$

Equation (8.16) are not generally equivalent to variational principle because of they can be used for both ground and excited states. The successive approximations of operator method over the operator $\hat{V} = \hat{H} - \hat{H}_0$ lead to the convergent and not asymptotic sequence (Chap. 3). The equations system for variables ω_n following from (8.16) contains the integrals with known polynomials, and for any real atom the number of the equations is quite limited ($n \leq 6$ for $Z = 112$). By substituting the solutions of these equations into (8.15) or by minimizing the value $E(N,Z,\{\omega_n\})$, the total energy of atom and other characteristics can be easily calculated as functions of Z, N.

The principle result of the model described above is the uniform suitability for all atomic characteristics calculated on the basis of this model and for any Z and N. Modern computing tools solve the Eq. (8.16) for quantum numbers with $n < 6$ with effectiveness comparable to one of analytical calculations. To illustrate this fact, the Table 8.1 shows the total energy of atoms calculated by OM zeroth approximation, the numerical method of Thomas-Fermi $E_{tot}^{(TF)}$, and Hartree–Fock method $E_{tot}^{(HF)}$ without relativistic corrections. The value $E_{tot}^{(OM)}$ is found from direct solution of the equations system for OM zeroth approximation (8.15), whereas the value $E_{tot}^{(\upsilon)}$ follows from the solution of these equations for continuous quantum numbers, which corresponds to the limit $Z \gg 1$ (see next section).

Table 8.1 The total energy of atoms calculated for oscillatory model

Z	20	40	70	112
ω_0	210.212	871.819	2712.76	7001.4
ω_1	21.7021	112.066	382.417	1033.07
ω_2	2.21197	28.1641	117.368	347.57
ω_3		2.51031	29.7238	111.928
ω_4			3.66659	36.9953
ω_5				4.06341
$E_{tot}^{(OM)}$	−601.025	−3095.55	−11659.3	−35542.5
$E_{tot}^{(v)}$	−820.233	−3991.12	−14450.1	−42769.3
$E_{tot}^{(TF)}$	−834.673	−4206.49	−15524.2	−46482.4
$E_{tot}^{(HF)}$	−676.758	−3538.97	−13391.5	−40937.8

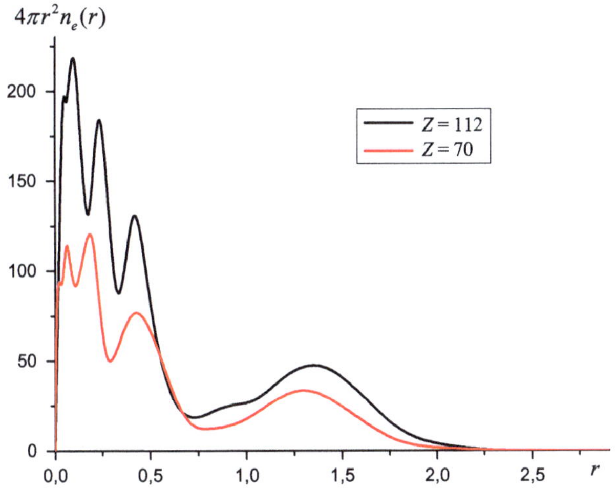

Fig. 8.1 A distribution of the radial electron density in neutral atoms with $Z = 70$ and $Z = 112$ using oscillatory model

The important advantage of the presented above atomic model in comparison to MTF one is the description of the electron density oscillations caused by shell structure of atoms. Figure 8.1 shows these oscillations calculated by OM zeroth approximation (8.14). The above results by OM zeroth approximation demonstrate the potential of this method and the uniform suitability of the obtained approximations for atomic characteristics providing the accuracy of $\sim$ 8–10 % in the whole range of physical parameters and describing correctly the qualitative properties of the system. The considerable improvement of the accuracy can be obtained using a basic set of the states of OM which takes into account a Coulomb nature of the potential (Sect. 8.3). The analytical model has also an advantage in easy development of the iteration technique for successive approximations of the operator method [17].

8.2 Continuous Oscillator Model in the Limit $Z \gg 1$

To apply the oscillatory atomic model in density function theory, the expressions obtained in previous section will be considered here in the limit $Z \gg 1$, which makes the equations universal similarly to the quasi-classic limit in statistical theory of atom. In this case, the state of most of the electrons corresponds to the quantum number $n \gg 1$, which makes possible a use of the asymptotic representation for the Laguerre polynomials [16]:

$$e^{-\frac{x}{2}} L_n^m(x) \Rightarrow \frac{\Gamma(n+m+1)}{n!(n+\frac{m+1}{2})^{m/2}} x^{-m/2} J_m\left(\sqrt{4x\left(n+\frac{m+1}{2}\right)}\right), \tag{8.17}$$

where Γ and J_m are Γ-function and Bessel function, respectively. Within this limit, the integrals over k in formulas (8.7)–(8.13) are expressed through elementary functions [16], for example, the one-particle energy (8.7) is:

$$\epsilon_n = \frac{\omega_n}{4}(2n+3) + \frac{16}{\pi}\left(\frac{\omega_n}{2n+3}\right)^{3/2}$$
$$\int_0^\infty drr^2 v(r)\sqrt{1-\frac{\omega_n r^2}{2n+3}}\,\Theta\left(1-\frac{\omega_n r^2}{2n+3}\right), \tag{8.18}$$

where Θ is a stepwise Heaviside function. A similar transformation can be done in other formulas, and new convenient parameters can be introduced instead of ω_n:

$$a_n^2 = 2\langle r^2\rangle_n = \frac{2n+3}{\omega_n}. \tag{8.19}$$

These values are directly connected to mean-square radii of the corresponding oscillator shells, and according to (8.18) define the borders of the electron density distribution within the shells in asymptotic limits, in which the quantum numbers n can be considered as continuous variables. As a result, the system of algebraic equations is transformed into integral equation (we consider here only neutral atoms $Z = N$), and new scale variables with universal dependence on Z can be introduced:

$$n + \frac{3}{2} = (3Z)^{1/3}\nu; \quad u_\nu = \frac{r(3Z)^{1/3}}{\alpha_\nu}; \quad \alpha_\nu = a_\nu(3Z)^{1/3}. \tag{8.20}$$

The atomic potential is then written as follows:

$$v(r) = -\frac{Z}{r}\left[1 - \frac{16}{\pi}\int_0^1 d\nu\nu^2 \int_0^{u_\nu} ds(1-s^2)^{\frac{3}{2}}\Theta(1-s)\right]. \tag{8.21}$$

and total energy of atom is:

$$E = 3^{1/3} Z^{7/3} [9 \int_0^1 d\nu \frac{\nu^4}{\alpha_\nu^2} - \frac{16}{\pi} \int_0^1 d\nu \frac{\nu^2}{\alpha_\nu}$$
$$+ \frac{24}{\pi^2} \int_0^1 d\nu \int_0^1 d\mu (\nu\mu)^2 \Big[\frac{1}{\alpha_\mu} \Theta\left(1 - \frac{\alpha_\nu}{\alpha_\mu}\right) F\left(\frac{1}{2}, -\frac{3}{2}; 3; \frac{\alpha_\nu^2}{\alpha_\mu^2}\right)$$
$$+ \frac{1}{\alpha_\nu} \Theta\left(1 - \frac{\alpha_\mu}{\alpha_\nu}\right) F\left(\frac{1}{2}, -\frac{3}{2}; 3; \frac{\alpha_\mu^2}{\alpha_\nu^2}\right) \Big], \tag{8.22}$$

where $F(a, b; c; x)$ is a hypergeometric function [16].

Thus, the atomic characteristics at large Z are described by universal functions depending on Z in a similar way as in the model of Thomas-Fermi. The coefficients in these functions are expressed through the sole function α_ν, which is the solution of dimensionless integral equation independent on Z. In fact, this equation plays the same role as MTF equation in the statistical theory of atom, it follows from the condition (8.16) as a result of the transfer to the continuous variable:

$$\frac{8}{9\pi}\alpha_\nu - \nu^2 = \frac{4}{3\pi}\alpha_\nu \int_0^1 d\mu \mu^2 \Big\{ A_{\nu\mu}^3 \Theta(1 - A_{\nu\mu}) F\left(\frac{3}{2}, -\frac{1}{2}; 4; A_{\nu\mu}^2\right)$$
$$+ \Theta(1 - A_{\mu\nu}) \Big[2F\left(\frac{1}{2}, -\frac{3}{2}; 3; A_{\mu\nu}^2\right) - A_{\mu\nu}^2 F\left(\frac{3}{2}, -\frac{1}{2}; 4; A_{\mu\nu}^2\right) \Big] \Big\}, \tag{8.23}$$

where

$$A_{\nu\mu} = A_{\mu\nu}^{-1} = \frac{\alpha_\nu}{\alpha_\mu}.$$

The electron density is an atomic characteristics, which is not universally depends on Z. This means the asymptotic limit $Z \to \infty$ is not uniformly available for density, and has various shapes for the center of atom, the external region and the region with maximal electron density. Mathematically it is manifested in the singularities of the integrals, which define the asymptotic expression for electron density:

$$n_e(r) = \frac{36}{\pi^2} Z^2 \int_\delta^{\nu_0} d\nu \nu^2 \alpha_\nu^{-3} \sqrt{1 - u_\nu^2} \Theta(1 - u_\nu). \tag{8.24}$$

At $Z \to \infty$, the value ν varies within the limits [0;1], which is utilized in formula (8.21), where the integrand has no singularities at the interval borders.

However, for any Z the integral limits δ and ν_0 are defined by the equations following from the normalization condition (8.10) and definition of scale variables (8.20), if the integral is replaced by the sum:

$$Z = \sum_{n=0}^{n_m}(n+1)(n+2) = \sum_{0}^{n_m}\left[\left(n+\frac{3}{2}\right)^2 - \frac{1}{4}\right] \approx 3Z\int_{\delta}^{\nu_0}\left[\nu^2 - \frac{\delta^2}{9}\right]d\nu;$$

$$\delta = \frac{3}{2}\frac{1}{(3Z)^{1/3}}; \quad \nu_0 = (3Z)^{-1/3}\left(n_m + \frac{3}{2}\right). \tag{8.25}$$

By using the Eq. (8.10) the expression for ν_0 is found:

$$\nu_0^3 + \delta\nu_0^2 + \frac{11}{9}\delta^2\nu_0 - \frac{5}{9}\delta^3 = 1;$$

$$\nu_0 \approx 1 - \frac{\delta}{3} - \frac{11}{27}\delta^2 + \dots \tag{8.26}$$

The behavior of the function α_ν on the borders of interval [0; 1] does not allow to set $\delta \to 0$ and $x_0 \to 1$ in the limit $Z \to \infty$, in opposite to the case of converging integrals for other atomic characteristics. This circumstance causes also the incorrect values of the electron density in asymptotic MTF theory at $r \to 0$ and $r \to \infty$. The universal function α_ν found from numerical solution of the Eq. (8.23), is shown in Fig. 8.2. In this figure, the discrete average radii a_n of the shell with number n are depicted by points, which are found from OM zeroth approximation (8.15) and (8.16) without transition to continuous limit. The function α_ν is an interpolation curve for discrete values a_n, which are located on this curve

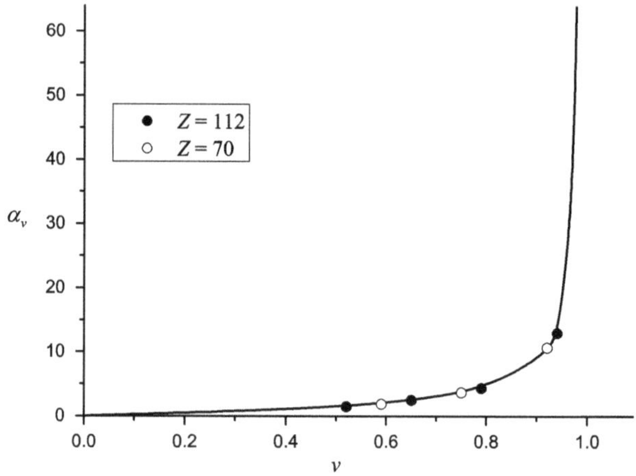

Fig. 8.2 Universal function α_ν for oscillator model of atom in continuous limit

in both asymptotic limit and for real atoms. Thus, the interpretation of the discrete values n as continuous is a good approximation for all values of nucleus charge Z.

The expressions above are valid for atoms with complete shells of three-dimensional spherical oscillator, which correspond to the charges Z_l following from the formulas (8.10):

$$Z_l = \frac{(l+1)}{3}(l^2+5l+6);\ l = 0;1;2;\ldots \tag{8.27}$$

For the physical characteristics $F(Z)$ of arbitrary atom with the nucleus charge in interval

$$Z_l \le Z \le Z_{(l+1)},$$

the following interpolation can be used:

$$F(Z) \approx F(Z_l) + \frac{Z - Z_l}{(l+2)(l+3)}[F(Z_{(l+1)}) - F(Z_l)]. \tag{8.28}$$

As follows from the Eq. (8.23), the universal function α_ν is monotonically increasing function on an interval [0;1], and the Eq. (8.23) can be re-written as:

$$\alpha_\nu = \frac{9\pi}{8}\nu^2\Bigg\{1 - \frac{3}{2}\int_\nu^1 d\mu\mu^2 A_{\nu\mu}^3 F\left(\frac{3}{2}, -\frac{1}{2}; 4; A_{\nu mu}^2\right) + \frac{3}{2}\int_0^\nu d\mu\mu^2\Bigg[A_{\mu\nu}^2 F\left(\frac{3}{2}, -\frac{1}{2}; 4; A_{\mu\nu}^2\right) - 2F\left(\frac{1}{2}, -\frac{3}{2}; 3; A_{\mu\nu}^2\right)\Bigg]\Bigg\}^{-1}, \tag{8.29}$$

that permits to investigate an asymptotic behavior of the function α_ν at both edges of the interval to make it useable for calculation of electron density from (8.24). In the limit of small ν:

$$\alpha_\nu \approx \frac{9\pi}{8}\nu^2, \quad \nu \ll 1. \tag{8.30}$$

At the opposite edge of the interval, the second integral in (8.29) tends to unity at $\nu \to 1$ due to normalization of electron density for neutral atom. Therefore $\alpha_\nu \sim (1-\nu)^{-1}$ at $\nu \to 1$, however, the coefficient is expressed through the integrals from hypergeometric functions, and it is easier to find this asymptotic numerically:

$$\alpha_\nu \approx \frac{0.57271}{1-\nu}, \quad \nu \to 1. \tag{8.31}$$

Now we can use Eq. (8.22) for calculation of the total energy of atom in the limit $Z \gg 1$:

$$E = -0,68480 Z^{7/3}.$$

The analogous coefficient in the exact asymptotic formula of MTF equals to −0.76874 [8]. The known qualitative restrictions of Thomas-Fermi model are the unlimited growth of electron density in the vicinity of nucleus and unlimited effective radius of atom. Equation (8.24) shows that the model based on the operator method leads to the limited value of $n_e(0)$ and certain value of atomic radius r_0, at which the electron density becomes zero, in opposite to MTF. At coordinates r lesser than radius of the first shell:

$$r < a_{n=0} = \frac{\alpha_\delta}{(3Z)^{1/3}} = \frac{27\pi}{32Z},$$

the electron density tends to constant value $n_e(0)$ computed from the formula:

$$n_e(0) = \frac{36}{\pi^2} Z^2 \int_\delta^{\nu_0} d\nu \frac{\nu^2}{\alpha_\nu^3}. \tag{8.32}$$

The main contribution to the integral is defined by the function at lower limit. Using an analytical approximation (8.30) for α_ν, the value n_e is:

$$n_e(0) \simeq \frac{36}{\pi^2} Z^2 (\frac{8}{9\pi})^3 \int_\delta^{\nu_0} \frac{d\nu}{\nu^4} = \frac{1}{\pi^5} (\frac{4}{3})^7 Z^3. \tag{8.33}$$

Due to uniformly suitable nature of OM, this function describes correctly the dependence $n_e(0)$ on the nucleus charge [18]. However, the numerical coefficient is underestimated in comparison with HF because the interpolation (8.17) has a poor accuracy at small n and the formula (8.15) has to be engaged with numerical summation of Laguerre polynomials (see Fig. 8.1). The external boundary of atom assumes the electron density (8.24) tends to zero when the coordinate r equals to the radius of external shell:

$$r_0 = \frac{\alpha_{\nu_0}}{(3Z)^{1/3}} \simeq \frac{0.57271}{(3Z)^{1/3}(1-\nu_0)}. \tag{8.34}$$

The function ν_0 tends to ($\nu_0 \simeq 1 - \frac{\delta}{3}$) at large Z, and the analytical approximation (8.31) can be used to calculate the atomic radius:

$$r_0 \simeq 3 \frac{0.57271}{(3Z)^{1/3}\delta} \approx 1.1454. \tag{8.35}$$

This result correlates with a qualitative estimate for atomic radius stating the r_0 depends weakly on the nucleus charge [19]. However, the absolute magnitude differs from the value found experimentally because of the wave functions of harmonic oscillator describe badly the asymptotic of Coulomb functions (Sect. 8.3).

Using the model of atom discussed in this section, the contribution of quantum effects into physical characteristics of atom can be also evaluated. This analysis is of certain interest for further applications of the model derived from the operator method. Besides the essential contribution into total energy of atom, which may improve the results of OM zeroth approximation, it can assist the calculation of complex atomic characteristics for exchange contribution to DFT functional, which determines the energy of electron subsystem in condensed matter. We start with the expression for exchange energy through the wave functions [20]:

$$E_{ex} = -\frac{1}{2\pi^2}\int \frac{d\boldsymbol{k}}{k^2}\sum_{\lambda}^{n_0}\sum_{\mu}^{n_0} |\langle\psi_\lambda|e^{i\boldsymbol{kr}}|\psi_\mu\rangle|^2. \tag{8.36}$$

The expression (8.36) is valid for atoms with filled shells and the summation is performed over the completed states of electrons. In the case of large Z, corresponding to continuous limit of the model, the influence of non-filled shells is relatively small and can be taken into account by interpolation formula (8.29). The representation of the exchange energy as a density functional based on the plane waves (uniform electron density) for the states $|\psi_\mu\rangle$ has been derived by Dirac [20]:

$$E_{ex}[n_e] = -\frac{3}{4}\left(\frac{3}{\pi}\right)^{1/3}\int n_e^{4/3} d\boldsymbol{r}. \tag{8.37}$$

This expression is used to account the exchange energy in DFT method within LDA approach [6]. Engaging the oscillatory model for atom, the numerical calculation of E_{ex} is pretty simple because of the matrix elements in (8.36) are expressed through the Laguerre polynomials. However, to be used in DFT method, this model has to implement a continuous approximation with continuous quantum number n and continuous function α_ν instead of set of values a_n. Then the most of calculations in (8.36) are performed analytically using operator technique and E_{ex} is represented as functional of α_ν. By using the following equation, the restrictions for summations can be avoided:

$$\int_0^{2\pi} d\varphi e^{i\varphi(n-m)} = 2\pi\delta_{m,n},$$

where $\delta_{m,n}$ is a Kronecker symbol. The expression (8.36) is then rewritten as:

$$E_{ex} = -\frac{1}{8\pi^4}\int \frac{d\boldsymbol{k}}{k^2}\sum_{n=0}^{n_0}\sum_{m=0}^{n_0}\int_0^{2\pi} d\varphi e^{-i\varphi n}\int_0^{2\pi} d\phi e^{-i\phi m} S;$$

$$S = \sum_{\lambda}\sum_{\mu}\langle n_\lambda|e^{i\varphi\hat{n}_\lambda}e^{i\boldsymbol{kr}}e^{i\phi\hat{m}_\mu}|m_\mu\rangle\langle m_\mu|e^{-i\boldsymbol{kr}}|n_\lambda\rangle. \tag{8.38}$$

where the operator of particles number $\hat{n}_\lambda$ is:

$$\hat{n}_\lambda = \hat{n}_x + \hat{n}_y + \hat{n}_z.$$

The eigenvalues of this operator define the number of the atomic shell and are limited by the values $n_0(Z)$. However, the summation over indices $n_{x,y,z}$ in formula (8.36) covers all values from zero to infinity, that means the summation over the entire set of vectors of intermediate states $|m_\mu\rangle$ results in unity operator and the function S is:

$$S = \sum_{\lambda}\langle n_\lambda|e^{i\varphi\hat{n}_\lambda}e^{i\boldsymbol{kr}}e^{i\phi\hat{m}_\mu}e^{-i\boldsymbol{kr}}|n_\lambda\rangle. \tag{8.39}$$

The calculation of the value S can be reduced to the summation of the diagonal elements (trace) of some operator, and this reduction is realized analytically because of all operators in the exponent contain only linear and quadratic forms of the creation and annihilation operators. These calculations, however, are cumbersome due to non-commutativity of operators of particle numbers $\hat{n}_\lambda$ and $\hat{m}_\mu$ because of they belong to different shells and therefore correspond to different frequencies. The principle term in the expression for E_{ex}, corresponding to continuous function, is calculated from the following estimate:

$$\frac{\partial\omega_n}{\partial n} \sim Z^{-1/3}\,\omega_n. \tag{8.40}$$

Thus, the variational parameter ω_n is independent on n within the used accuracy, which makes possible to refer both operators of particle numbers to the same frequency ω_n. The summation in the formula (8.39) is reduced to the product of three independent sums, and therefore it is enough to consider only one of them, for example, the sum S_x for x-component of the vector $\boldsymbol{r}$. Using the representation (8.2) for the coordinate, the function S_x in operator form is written as:

$$S_x = \sum_{l=0}^{\infty}\langle l|e^{i\varphi a^+a}e^{i(a+a^+)k_x/\sqrt{2\omega_n}}e^{i\phi a^+a}e^{-i(a+a^+)k_x/\sqrt{2\omega_n}}|l\rangle. \tag{8.41}$$

Here we omit the index x for operators of creation and annihilation. The formula (8.41) can be simplified by rearranging two last operators:

$$S_x = \sum_{l=0}^{\infty}\langle l|e^{i\varphi a^+a}e^{i\phi(a^++ik_x/\sqrt{2\omega_n})(a-ik_x/\sqrt{2\omega_n})}|l\rangle. \tag{8.42}$$

The expression (8.42) is a trace of the operator and it is independent on the choice of representation for state vectors $|l\rangle$. Therefore, to simplify the calculations and to reduce the operator to the diagonal form, the additional canonic transformation is used, which is equivalent to the change of representation for $|l\rangle$:

$$a = u + b; \quad a^+ = u^* + b^+, \tag{8.43}$$

where the classic parameter u will be defined later. As a next step, the first exponential operator in the expression (8.42) is reduced to the normal form, whereas the second to anti-normal form:

$$\begin{aligned} e^{i\varphi(u^*+b^+)(u+b)} &= e^{u(e^{i\varphi}-1)b^+} e^{i\varphi b^+ b} e^{u^*(e^{i\varphi}-1)b} e^{u^* u(e^{i\varphi}-1)}; \\ e^{i\phi(v^*+b^+)(v+b)} &= e^{v^*(1-e^{-i\phi})b} e^{i\phi b^+ b} e^{v(1-e^{-i\phi})b^+} e^{v^* v(1-e^{-i\phi})}; \\ v &= u - ik_x/\sqrt{2\omega_n}. \end{aligned} \tag{8.44}$$

We choose the parameter u in the following way:

$$\begin{aligned} u &= i\frac{k_x}{\sqrt{2\omega_n}}\frac{1-z_1}{1-zz_1}; \quad u^* = -i\frac{k_x}{\sqrt{2\omega_n}}\frac{1-z_1}{1-zz_1}; \\ z &= e^{i\varphi}; \qquad z_1 = e^{i\phi}, \end{aligned} \tag{8.45}$$

to eliminate the exponentials with linear operators. As a result, the function S_x is transformed to:

$$S_x = \sum_{l=0}^{\infty} \langle l| e^{i(\varphi+\phi)b^+ b} e^{-k_x^2/2\omega_n \frac{(1-z)(1-z_1)}{1-zz_1}} |l\rangle. \tag{8.46}$$

To calculate the integrals, we change to the complex integration variables z and z_1 instead of φ and ϕ in formula (8.38). The integration contours are deformed into the closed contours over the circle with radius $|z| = |z_1| = 1$. The functions $S_{x,y,z}$ are transformed after integration over $\boldsymbol{k}$, and the expression for the exchange energy is obtained:

$$\begin{aligned} E_{ex} = &\frac{1}{(2\pi)^2} \sum_{n=0}^{n_0} \sum_{m=0}^{n_0} \oint \frac{dz}{z^{n+1}} \oint \frac{dz_1}{z_1^{m+1}} \sqrt{\frac{2\omega_n}{\pi}} \\ &\times (1-zz_1)^{-5/2}(1-z)^{-1/2}(1-z_1)^{-1/2}. \end{aligned} \tag{8.47}$$

The summation over m and the calculation of contour integral in (8.47) leads to the following expression:

$$E_{ex} = -\sum_{n=0}^{n_0} \sum_{k=0}^{n} \sqrt{\frac{2\omega_n}{\pi}} A_k^{5/2} A_{n_0-k}^{3/2} A_{n-k}^{1/2}. \tag{8.48}$$

Here the values A_k^p are the coefficients at x^k in the expansion of binomial $(1-x)^{-p}$ in a power series. Finally, the last step is the transition to continuous limit over the variables n, k according to the definition (8.20) and use of asymptotic for the coefficients [21]:

$$A_k^p \simeq \frac{k^{p-1}}{\Gamma(p)}, \qquad k \gg 1.$$

As a result, the following expression for the functional of exchange energy is found:

$$E_{ex}[\alpha_\nu] = -\frac{2}{\pi} 3^{5/3} Z^{5/3} \int_0^1 d\nu \frac{\nu^{5/2}}{\alpha_\nu}. \tag{8.49}$$

The numerical value of the coefficient follows from the universal function α_ν (Fig. 8.2), which delivers:

$$E_{ex} = -0.518948 Z^{5/3}. \tag{8.50}$$

The expression obtained demonstrates a similar dependence on the nucleus charge as MTF model [8], however, with different coefficient:

$$E_{ex}^{MTF} = -0.269900 Z^{5/3}.$$

In the conclusion of this section, the functional of total energy in the DFT method unifies the formulas (8.22) and (8.50) if the exchange effect is taken into account:

$$\begin{aligned} E_{tot}[\alpha_\nu] = 3^{1/3} Z^{7/3} [9 \int_0^1 d\nu \frac{\nu^4}{\alpha_\nu^2} - \frac{16}{\pi} \int_0^1 d\nu \frac{\nu^2}{\alpha_\nu} \\ + \frac{24}{\pi^2} \int_0^1 d\nu \int_0^1 d\mu (\nu\mu)^2 \Big[\frac{1}{\alpha_\mu} \Theta \left(1 - \frac{\alpha_\nu}{\alpha_\mu}\right) F\left(\frac{1}{2}, -\frac{3}{2}; 3; \frac{\alpha_\nu^2}{\alpha_\mu^2}\right) \\ + \frac{1}{\alpha_\nu} \Theta \left(1 - \frac{\alpha_\mu}{\alpha_\nu}\right) F\left(\frac{1}{2}, -\frac{3}{2}; 3; \frac{\alpha_\mu^2}{\alpha_\nu^2}\right) \Big] - \frac{2}{\pi} 3^{5/3} Z^{5/3} \int_0^1 d\nu \frac{\nu^{5/2}}{\alpha_\nu}. \end{aligned} \tag{8.51}$$

Thus, the universal function α_ν, which expresses all atomic characteristics, is obtained from the minimum condition of full functional (8.51). At large yet finite Z this function depends on the nucleus charge and differs from the one displayed in Fig. 8.2.

8.3 Coulomb Based Atomic Model

According to general scheme of operator method (Sect. 3.1), the zeroth approximation is constructed on the basis of various basic sets. The approach described in previous section for oscillator atomic model can be generalized for the case of one-electron wave functions approximated by Coulomb basis, which takes into account the Coulomb nature of the nucleus potential. This application demonstrates the universality of the method and improves essentially the accuracy of OM zeroth approximation. The results of this section are published in the works [22, 23]. The electron density plays a crucial role in atomic-molecular calculations by DFT method:

$$\varrho(\mathbf{r}) = \langle \Psi | \sum_{j=1}^{N} \delta(\mathbf{r} - \mathbf{r}_j) | \Psi \rangle, \tag{8.52}$$

where $|\Psi\rangle$ is a state vector of the system and for many-electron atom $|\Psi\rangle$ is defined from the Schrödinger equation:

$$\left(\sum_{j=1}^{N} \frac{\mathbf{p}_j^2}{2} - \sum_{j=1}^{N} \frac{Z}{r_j} + \frac{1}{2} \sum_{i \neq j}^{N} \frac{1}{|\mathbf{r}_i - \mathbf{r}_j|} \right) |\Psi\rangle = E\, |\Psi\rangle. \tag{8.53}$$

Here N is a total number of electrons in atom; Z is a nucleus charge; $\mathbf{p}_j$ and $\mathbf{r}_j$ are the operators of momentum and location of jth electron; E is a total energy of the system. In the model of independent Hartree particles, the Coulomb wave functions of hydrogen-like atom $\psi_{nlm}(\mathbf{r})$ are selected as one-particle basis functions $\psi_i(\mathbf{r}_i)$ for the construction of the atomic wave function:

$$\Psi(\mathbf{r}_1, \mathbf{r}_2, \ldots, \mathbf{r}_N) = \psi_1(\mathbf{r}_1)\, \psi_2(\mathbf{r}_2) \ldots \psi_N(\mathbf{r}_N). \tag{8.54}$$

The electron density of nth atomic shell is written as:

$$\varrho_n(\mathbf{r}) = \sum_{l=0}^{n-1} \sum_{m=-l}^{l} \psi^*_{nlm}(\mathbf{r})\, \psi_{nlm}(\mathbf{r}), \tag{8.55}$$

where the summation over quantum numbers l and m corresponds to the summation over the degenerated states of hydrogen-like atom at fixed principle quantum number n. The sum (8.55) is calculated as analytical expression for electron density of the shell in asymptotic limit $n \to \infty$. This calculation can be performed algebraically if using the relationship between the solutions of Schrödinger equation for hydrogen-like atom with the solutions for isotropic harmonic oscillator in two-dimensional space with complex coordinates ξ_s ($s = 1, 2$), where Schrödinger equation has a following form:

$$H\tilde{\psi} = -\frac{1}{2}\frac{\partial^2\tilde{\psi}}{\partial\xi_s^*\partial\xi_s} + \frac{\omega^2}{2}\xi_s^*\xi_s\tilde{\psi} = Z\tilde{\psi}. \tag{8.56}$$

The coordinates ξ_s ($s = 1, 2$) are considered as components of spinor, and the asterisk corresponds to the conjuncted value; ω is a real positive number; the summation is performed over doubly repeated indices. Equation (8.56) is invariant relatively to transformation:

$$\xi_s \to \xi_s e^{i\phi}, \quad \xi_s^* \to \xi_s^* e^{-i\phi}, \tag{8.57}$$

that corresponds to nullification of the commutator of Hamiltonian H with operator:

$$Q = \xi_s^*\frac{\partial}{\partial\xi_s^*} - \xi_s\frac{\partial}{\partial\xi_s}. \tag{8.58}$$

The detailed description of the representation is given in Chap. 7 for application of OM to the quantum theory of hydrogen-like atom in external field. Below we provide a reader with the key characteristics of this representation required for the description of many-electron atoms. In the Eq. (8.56), we use new variables:

$$\begin{aligned} x_\lambda &= \xi_s^*(\sigma_\lambda)_{st}\xi_t \quad (\lambda = 1, 2, 3), \\ \Phi &= \operatorname{arctg}\left(\frac{\xi_1''}{\xi_1'}\right), \end{aligned} \tag{8.59}$$

where $(\sigma_\lambda)_{st}$ are the matrix elements of Pauli matrix σ_λ, and $\xi_s' = \operatorname{Re}\xi_s$, $\xi_s'' = \operatorname{Im}\xi_s$. The properties of the spinor ξ_s guarantee the real numbers x_λ to be the components of three-dimensional vector $\mathbf{r}$. Equation (8.56) in the variables (8.59) has the following form:

$$-\frac{1}{2}\left[r\Delta - \frac{1}{r + x_3}\left(x_1\frac{\partial}{\partial x_2} - x_2\frac{\partial}{\partial x_1}\right)\frac{\partial}{\partial\Phi} + \frac{1}{2(r + x_3)}\frac{\partial^2}{\partial\Phi^2}\right]\tilde{\psi} + \frac{\omega^2}{2}r\tilde{\psi} = Z\tilde{\psi}, \tag{8.60}$$

where $r = |\mathbf{r}|$, Δ is a Laplace operator. We consider here the physically meaningful solutions of the Eq. (8.60) independent on Φ and corresponding to zeroth eigenvalue of the operator Q, which in variables (8.59) has the following form:

$$Q = i\frac{\partial}{\partial\Phi}. \tag{8.61}$$

As follows from the Eq. (8.60), the physically meaningful solutions are the solutions of three-dimensional Schrödinger equation for the particle moving in Coulomb field and having the energy:

$$E = -\frac{\omega^2}{2}. \tag{8.62}$$

The relationship between the Schrödinger equation for hydrogen-like atom and Schrödinger equation for isotropic harmonic oscillator in two-dimensional complex space permits to present the solutions of these equations in a simple algebraic form. The scalar product of wave functions in ξ-space is:

$$\langle\tilde{\phi}|\tilde{\psi}\rangle = \int d^4\xi\, \tilde{\phi}^*(\xi_1^*, \xi_2^*, \xi_1, \xi_2)\, \tilde{\psi}(\xi_1^*, \xi_2^*, \xi_1, \xi_2), \tag{8.63}$$

where $d^4\xi = d\xi_1' d\xi_2' d\xi_1'' d\xi_2''$. Furthermore we introduce the annihilation operators a_s, b_s

$$a_s(\omega) = \left(\frac{\omega}{2}\right)^{1/2}\left(\xi_s + \frac{1}{\omega}\frac{\partial}{\partial\xi_s^*}\right),$$
$$b_s(\omega) = \left(\frac{\omega}{2}\right)^{1/2}\left(\xi_s^* + \frac{1}{\omega}\frac{\partial}{\partial\xi_s}\right), \tag{8.64}$$

and Hermitian conjugated with respect to the scalar product (8.63) creation operators a_s^+, b_s^+:

$$a_s^+(\omega) = \left(\frac{\omega}{2}\right)^{1/2}\left(\xi_s^* - \frac{1}{\omega}\frac{\partial}{\partial\xi_s}\right),$$
$$b_s^+(\omega) = \left(\frac{\omega}{2}\right)^{1/2}\left(\xi_s - \frac{1}{\omega}\frac{\partial}{\partial\xi_s^*}\right). \tag{8.65}$$

These operators satisfy the canonic transposition relationships:

$$\left[a_s, a_t^+\right] = \delta_{st}, \quad \left[b_s, b_t^+\right] = \delta_{st}, \tag{8.66}$$

and other commutators equal zero. By substituting the operators $\partial/\partial\xi_s$, ξ_s, $\partial/\partial\xi_s^*$, ξ_s^* expressed through operators (8.64) and (8.65) into (8.56), the following equation is obtained:

$$\frac{\omega}{2}\left(2 + a_s^+ a_s + b_s^+ b_s\right) = Z|\psi\rangle. \tag{8.67}$$

The solution of this equation follows from the eigenvectors of the creation operators. The ground state (vacuum) is defined by the equation:

$$a_s|0(\omega)\rangle = b_s|0(\omega)\rangle = 0. \tag{8.68}$$

Using the transposition formulas (8.66), the state vectors

$$|\tilde{\psi}_{n_1^a,n_1^b,n_2^a,n_2^b}(\omega)\rangle = \frac{(a_1^+)^{n_1^a}(b_1^+)^{n_1^b}(a_2^+)^{n_2^a}(b_2^+)^{n_2^b}}{(n_1^a!n_1^b!n_2^a!n_2^b!)^{1/2}}|0(\omega)\rangle, \tag{8.69}$$

with non-negative integer numbers n_1^a, n_1^b, n_2^a, n_2^b are the eigenvectors of commutating operators $a_1^+a_1$, $b_1^+b_1$, $a_2^+a_2$, $b_2^+b_2$, and these eigenvectors belong to the eigenvalues n_1^a, n_1^b, n_2^a, n_2^b. Thus, the states (8.66) are the solutions of the Eq. (8.67) in assumption of:

$$\omega = \frac{2Z}{2 + n_1^a + n_1^b + n_2^a + n_2^b}. \tag{8.70}$$

Equation (8.62) then lead to the expression:

$$E = -\frac{2Z^2}{(2 + n_1^a + n_1^b + n_2^a + n_2^b)^2}. \tag{8.71}$$

The physically meaningful solutions are the ones (8.69), which correspond to zeroth eigenvalue of the operator Q expressed through the operators (8.64) and (8.65) as follows:

$$Q = a_s^+a_s - b_s^+b_s. \tag{8.72}$$

This means that the only quantum numbers n_1^a, n_1^b, n_2^a, n_2^b satisfying the condition:

$$n_1^a + n_2^a = n_1^b + n_2^b, \tag{8.73}$$

are physically meaningful. The operator of the orbital momentum is defined through the operators (8.64) and (8.65):

$$l_\lambda = \frac{1}{2}(\sigma_\lambda)_{st}(a_s^+a_t - b_s^+b_t). \tag{8.74}$$

The connection between the quantum numbers (8.69) and spectroscopical classification of the states of hydrogen-like atom is established as follows: the energy is determined by principle quantum number $n = n + 1$ $(n = n_1^a + n_2^a = n_1^b + n_2^b)$, which takes the values $n = 1, 2, 3, \ldots$. The orbital quantum numbers l correspond to the states (without normalization multipliers):

$$l = 0: \ (a_s^+ b_s^+)^n |0\rangle, \quad n = 0, 1, 2, \ldots$$
$$l = 1: \ (\sigma_\lambda)_{st} a_s^+ b_t^+ (a_j^+ b_j^+)^n |0\rangle, \quad \lambda = 1, 2, 3, \quad n = 0, 1, 2, \ldots,$$
$$\ldots\ldots$$

which are the linear combinations of the states (8.69). The detailed calculations of matrix elements using state vectors (8.69) are presented in Chap. 7. Using the transformation (8.64) to express the Eq. (8.55) in the representation of two-dimensional complex oscillator, the sum (8.55) is written as:

$$\varrho_n(\mathbf{r}) = \sum_{l=0}^{n-1} \sum_{m=-l}^{l} \psi^*_{nlm}(\mathbf{r})\, \psi_{nlm}(\mathbf{r}) = \sum_{l=0}^{n-1} \sum_{m=-l}^{l} \int d^3\tilde{\mathbf{r}}\, \delta^{(3)}(\mathbf{r} - \tilde{\mathbf{r}})\, |\psi_{nlm}(\tilde{\mathbf{r}})|^2 =$$
$$= N^2 \sum_{n_1^a=0}^{n} \sum_{n_1^b=0}^{n} \int d^4\tilde{\xi}\, \delta^{(4)}(\xi - \tilde{\xi})\, |\tilde{\psi}_{n_1^a, n_1^b, n-n_1^a, n-n_1^b}(\tilde{\xi})|^2, \tag{8.75}$$

where the summation over n_1^a, n_1^b corresponds to the summation over the degenerated states, and the Eq. (8.73) and the relationships between integration measures and delta-functions in ξ- and r-spaces are utilized:

$$d^4\xi = \frac{1}{8(x_\lambda x_\lambda)^{1/2}} dx_1 dx_2 dx_3 d\Phi, \tag{8.76}$$

$$\int_0^{2\pi} d\tilde{\Phi}\, \delta^{(4)}(\xi - \tilde{\xi}) = 8\, r\, \delta^{(3)}(\mathbf{r} - \tilde{\mathbf{r}}), \tag{8.77}$$

$$\delta^{(4)}(\xi - \tilde{\xi}) = \int \frac{d^4k}{(2\pi)^4}\, e^{k\,(\xi - \tilde{\xi})}, \tag{8.78}$$

and $k\,(\xi - \tilde{\xi}) = k_1'\,(\xi_1' - \tilde{\xi}_1') + k_2'\,(\xi_2' - \tilde{\xi}_2') + k_1''\,(\xi_1'' - \tilde{\xi}_1'') + k_2''\,(\xi_2'' - \tilde{\xi}_2'')$.

Since the Coulomb wave functions $\tilde{\psi}(\xi)$ in the representation of two-dimensional complex space are normalized to unity in ξ-space relatively to scalar product (8.63), the normalization factor N in (8.75) is chosen as follows:

$$\int d^3\mathbf{r}\, |\psi_n(\mathbf{r})|^2 = \int d^3\mathbf{r} \int_0^{2\pi} \frac{d\Phi}{2\pi}\, |\psi_n(\mathbf{r})|^2 =$$
$$= N^2 \frac{4}{\pi} \int d^4\xi\; r\; |\tilde{\psi}_{n_1^a, n_1^b, n-n_1^a, n-n_1^b}(\xi)|^2 = N^2 \frac{4}{\pi} \frac{n+1}{\omega_n} = 1,$$
$$N^2 = \frac{\pi \omega_n}{4(n+1)}, \tag{8.79}$$

where $r = \xi_s^* \xi_s$, and ω_n is determined from the Eq. (8.70) at $n = n_1^a + n_2^a = n_1^b + n_2^b$. Using the definition of delta-function (8.78), the Eq. (8.75) is transformed to:

$$\varrho_n(\mathbf{r}) = \frac{\pi \omega_n}{4(n+1)} \int \frac{d^4k}{(2\pi)^4} e^{i k \xi} \times \sum_{n_1^a=0}^{n} \sum_{n_1^b=0}^{n} \langle \tilde{\psi}_{n_1^a, n_1^b, n-n_1^a, n-n_1^b}(\omega_n) | e^{-i k \tilde{\xi}} | \tilde{\psi}_{n_1^a, n_1^b, n-n_1^a, n-n_1^b}(\omega_n) \rangle . \tag{8.80}$$

The state vector $|\tilde{\psi}\rangle$ corresponds to the state of hydrogen-like atom with energy $E_n = -Z^2/2n^2$ and is a linear combination of Coulomb state vectors with different angular quantum numbers but with equal principle quantum number n. The method of the construction of state vectors with quantum numbers n, l, m in the basis (8.69) is explained in Chap. 7.

Using the definitions (8.68), (8.69), (8.64), and (8.65), the transposition formulas (8.66) and summing rule for Laguerre polynomials (8.974.4) from [16], the expression for sum (8.75) is obtained:

$$\varrho_n(\mathbf{r}) = \frac{\pi \omega_n}{4(n+1)} \int \frac{d^4k}{(2\pi)^4} e^{i k \xi} e^{-\frac{|k|^2}{8\omega_n}} \left[L_n^1 \left(\frac{|k|^2}{8\omega_n} \right) \right]^2 , \tag{8.81}$$

where $L_n^m(x)$ are Laguerre polynomials [16]. To integrate (8.81) over k, the following coordinate system is used:

$$\cos \phi_1 = \frac{k_1' \xi_1' + k_1'' \xi_1''}{|k_1|\,|\xi_1|}, \quad z \cos \alpha = |k_1|,$$
$$\cos \phi_2 = \frac{k_2' \xi_2' + k_2'' \xi_2''}{|k_2|\,|\xi_2|}, \quad z \sin \alpha = |k_2|, \tag{8.82}$$

with $0 \le \phi_1 \le 2\pi, 0 \le \phi_2 \le 2\pi$ and $0 \le z < \infty, 0 \le \alpha \le \pi/2$. By integrating over ϕ_1 and ϕ_2 in (8.81), using the integral representation of Bessel function (8.411.1) in [16], we arrive at equations:

$$\varrho_n(\mathbf{r}) = \frac{\pi \omega_n}{4(n+1)} \int_0^\infty \frac{dz}{(2\pi)^2} z^3 e^{-\frac{z^2}{8\omega_n}} \left[L_n^1 \left(\frac{z^2}{8\omega_n} \right) \right]^2 \times \int_0^{\pi/2} d\alpha \, \sin \alpha \, \cos \alpha \, J_0(z \sin \alpha \, |\xi_1|) J_0(z \cos \alpha \, |\xi_2|). \tag{8.83}$$

Finally, integrating over α with the use of formula (6.683.2) in [16], the following integral representation for the sum (8.75) is derived:

$$\varrho_n(\mathbf{r}) = \frac{\pi\omega_n}{4(n+1)} \int_0^\infty \frac{dz}{(2\pi)^2} z^3 e^{-\frac{z^2}{8\omega_n}} \left[L_n^1\left(\frac{z^2}{8\omega_n}\right)\right]^2 \frac{J_1(z\sqrt{r})}{z\sqrt{r}}, \quad (8.84)$$

where from (8.59) follows $r = |\xi_1|^2 + |\xi_2|^2$. Thus, the calculation of sum (8.75) is reduced to the calculation of the integral:

$$I = \int_0^\infty dz\, z^2 J_1(z\sqrt{r})\, e^{-\frac{z^2}{8\omega_n}} \left[L_n^1\left(\frac{z^2}{8\omega_n}\right)\right]^2, \quad (8.85)$$

which is calculated according to the technique proposed in [22, 23].

The Laguerre polynomials L_n^1 can be written in integral representation using the generating function (8.975.1) from [16]. The integration over z using formula (6.631.4) from [16] leads to:

$$I = \frac{16\,\omega_n^2\sqrt{r}}{(2\pi i)^2} \oint \frac{dz_1}{z_1^{n+1}} \frac{dz_2}{z_2^{n+1}} \frac{1}{(1-z_1z_2)^2} \exp\left[-2\omega_n r\, \frac{(1-z_1)(1-z_2)}{1-z_1z_2}\right]. \quad (8.86)$$

The use of generating function to perform inverse transformation leads to:

$$I = \frac{16\,\omega_n^2\sqrt{r}}{(2\pi i)^2} \sum_{p=0}^{\infty} \oint \frac{dz_1}{z_1^{n-p+1}} \oint \frac{dz_2}{z_2^{n-p+1}} L_p^1\left(2\omega_n r\, \frac{(1-z_1)}{z_1}\frac{(1-z_2)}{z_2}\right). \quad (8.87)$$

The Laguerre polynomial L_p^1 is represented through the sum of polynomials L_m using the formula (8.974.3) from [16] and L_m, and definition (8.970.1) from [16]. The expression is integrated over z_1 and z_2:

$$I = 16\,\omega_n^2\sqrt{r} \sum_{p=n}^{\infty} \frac{1}{[(p-n)!]^2} \sum_{l=0}^{p}\sum_{k=0}^{l} \frac{l!}{(n+k-p)!} \frac{(-2\omega_n r)^k}{(n+k-p)!(l-k)!}, \quad (8.88)$$

and is modified to the following formula for I:

$$\begin{aligned} I &= 16\,\omega_n^2\sqrt{r} \sum_{l=0}^{n} L_l(2\omega_n r)\left(1 + \frac{l+1}{(1!)^2}(-2\omega_n r) + \frac{(l+1)(l+2)}{(2!)^2}\right. \\ &\left.\times(-2\omega_n r)^2 + \ldots\right) = 16\,\omega_n^2\sqrt{r} \sum_{l=0}^{n} L_l(2\omega_n r)\, \Phi(l+1, 1, -2\omega_n r) \\ &= 16\,\omega_n^2\sqrt{r}\, e^{-2\omega_n r} \sum_{l=0}^{n} [L_l(2\omega_n r)]^2, \end{aligned} \quad (8.89)$$

where $\Phi(\alpha, \beta, z)$ is a degenerated hypergeometric function. The final expression for the integral (8.85) is derived by using the formula (8.974.1) from [16] for the sum of the products of Laguerre polynomials:

$$I = 8\omega_n(n+1)\sqrt{r}\,e^{-2\omega_n r}\left[\frac{dL_n(2\omega_n r)}{dr}L_{n+1}(2\omega_n r) - L_n(2\omega_n r)\frac{dL_{n+1}(2\omega_n r)}{dr}\right]. \tag{8.90}$$

Substituting (8.90) into (8.84), the sum (8.75) is written as:

$$\varrho_n(\mathbf{r}) = \frac{\omega_n^2}{2\pi}e^{-2\omega_n r} \times \left[\frac{dL_n(2\omega_n r)}{dr}L_{n+1}(2\omega_n r) - L_n(2\omega_n r)\frac{dL_{n+1}(2\omega_n r)}{dr}\right]. \tag{8.91}$$

To obtain the sum (8.91) in asymptotic limit $n \gg 1$, the integral representation (8.86) is used. Since the value $2\varrho_n$ has a meaning of the density of electrons number on the filled shell n of atom, the normalization condition for total number of electrons on shell is fulfilled:

$$\int d^3\mathbf{r}\, 2\varrho_n(\mathbf{r}) = 2\,(n+1)^2. \tag{8.92}$$

A total number of electrons in neutral atom with filled shells equals to $Z = \sum_n 2\,(n+1)^2$, and to switch to the limit $n \gg 1$ in (8.85), the asymptotic expression for Laguerre polynomials is used:

$$e^{-\frac{x}{2}}L_n^m(x) \Rightarrow \frac{\Gamma(n+m+1)}{n!(n+\frac{m+1}{2})^{m/2}}x^{-m/2}J_m\left(\sqrt{4x\left(n+\frac{m+1}{2}\right)}\right), \tag{8.93}$$

where Γ and J_m are Gamma and Bessel functions, respectively. Equation (8.85) is modified then to:

$$\varrho_n^{as}(\mathbf{r}) = \frac{\omega_n^2}{\pi}\int_0^\infty dz\,\frac{J_1(z\sqrt{r})}{\sqrt{r}}\,J_1\left(z\sqrt{\frac{n+1}{2\omega_n}}\right)^2. \tag{8.94}$$

Integrating over k with the use of formula (6.578.9) from [16], we obtain for ϱ_n in asymptotic limit $n \gg 1$ the following expression:

$$\varrho_n^{as}(\mathbf{r}) = \frac{2}{\pi^2}\frac{2(n+1)^2}{a_n^3}\sqrt{\frac{a_n}{r}-1}\,\Theta\left(1-\frac{r}{a_n}\right), \tag{8.95}$$

where $\Theta\left(1-\frac{r}{a_n}\right)$ is a Heaviside function, at $r > a_n$ the function $\Theta = 0$, and at $r < a_n$ the function $\Theta = 1$, $a_n = 2(n+1)/\omega_n$. The quantity ϱ_n^{as} satisfies the normalization condition (8.92). The expression (8.95) is an analogue of the electron density in MTF. Based on (8.95), the statistical model represents both electron density ϱ_n^{as} and electron potential φ_n^{as}:

$$\varphi_n^{as}(\mathbf{r}) = \int d^3\mathbf{r}' \, \frac{\varrho_n^{as}(\mathbf{r}')}{|\mathbf{r}-\mathbf{r}'|}, \tag{8.96}$$

in analytical form. Similarly to the oscillator model, the atom has a finite size because of at $r > a_\nu$ the electron density equals zero. Additionally, after the transformation

$$x = \frac{r}{a_\nu}, \quad \varrho^{as}(x) = \frac{a_n^3}{2\,(n+1)^2}\,\varrho_n^{as}(\mathbf{r}), \tag{8.97}$$

the electron density $\varrho^{as}(x)$, shown in Fig. 8.3, is the universal function for any shell due to the independence on n and Z.

The electron potential $\varphi^{as}(x)$, defined as:

$$x = \frac{r}{a_n}, \quad \varphi^{as}(x) = \frac{a_n}{2\,(n+1)^2}\,\varphi_n^{as}(\mathbf{r}), \tag{8.98}$$

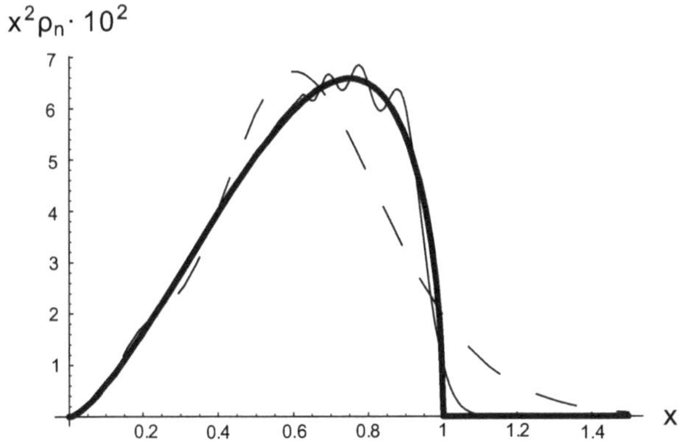

Fig. 8.3 The universal function of electron density $\varrho^{as}(x)$ (*thick line*), for third shell (*dashed line*), and for 31st shell (*thin line*)

is also a universal function, i.e. the potentials created by electron shells are similar each to the other. The potential $\varphi^{as}(x)$ is:

$$\varphi^{as}(x) = \Theta(1-x) + \frac{\Theta(x-1)}{2\,x} + \frac{\Theta(1-x)}{3\,\pi\,x}\{\sqrt{(1-x)\,x}\,[4\,(1-x)\,x-3] + \\ +(3-6\,x)\arcsin(\sqrt{x})\}\,. \tag{8.99}$$

Thus, this model preserves the universality of the atom description, which is equivalent to the universality of Thomas-Fermi model, because of $\varrho^{as}(x)$ is independent on the charge of the atomic nucleus Z.

The function ϱ_n^{as} is an exact limit of the function (8.91) at $\nu \rightarrow \infty$, as evident from the Fig. 8.3. The curves of the electron density (8.85) oscillate around asymptotic curve (8.97), however, with the increase of the shell number the amplitude decays rapidly and tends to zero at $\nu \rightarrow \infty$. As a matter of fact, both curves merge in asymptotic limit.

To construct the theory of atom taking into account the shell effects, the density of the electrons number in atom $\varrho = 2\sum_\nu \varrho_n$ can be used, because of it is regular in the origin of coordinates and has oscillating character in accordance with the shell structure of atom. Is a self-similarity preserved in such an atomic theory? To answer this question, we transform the electron density (8.91) according to (8.97):

$$\varrho_n(\mathbf{x}) = \frac{2\,(n+1)}{\pi}\,e^{-4\,(n+1)\,x}\left[\frac{dL_n\,(4\,(n+1)\,x)}{dx}L_{n+1}\,(4\,(n+1)\,x)\right. \\ \left. -\,L_n\,(4\,(n+1)\,x)\,\frac{dL_{n+1}\,(4\,(n+1)\,x)}{dx}\right], \tag{8.100}$$

and the corresponding transformations (8.98) for electron potential φ_n (8.96):

$$\varphi_n(\mathbf{x}) = \int d^3\mathbf{x}'\,\frac{\varrho_n(\mathbf{x}')}{|\mathbf{x}-\mathbf{x}'|}. \tag{8.101}$$

As follows from Fig. 8.4, the curve of electron density (8.100) is close to asymptotic curve at small values of ν, which corresponds to the shells of real atoms. The quantity $\varrho_n(x)$ is normalized to the same value for all shells.

In real applications, the interest is focused on physical values with electron density under integral sign, for example, electron potential (8.101), rather than electron density itself. The plots of potential (8.101) for the shells of real atoms and of asymptotic potential (8.99) are shown in Fig. 8.5. All curves practically coincide with the asymptotic line. For the quantitative assessment of the difference between the potentials of first four shells and asymptotic potential, Fig. 8.6 shows the deviations $\delta_n(x) = 100\,(\varphi_n(x) - \varphi^{as}(x))/\varphi^{as}(x)$. The largest deviation from the asymptotic has first shell $\delta_1 < 9\,\%$, the second shell is $\delta_2 < 3\,\%$, and for the rest

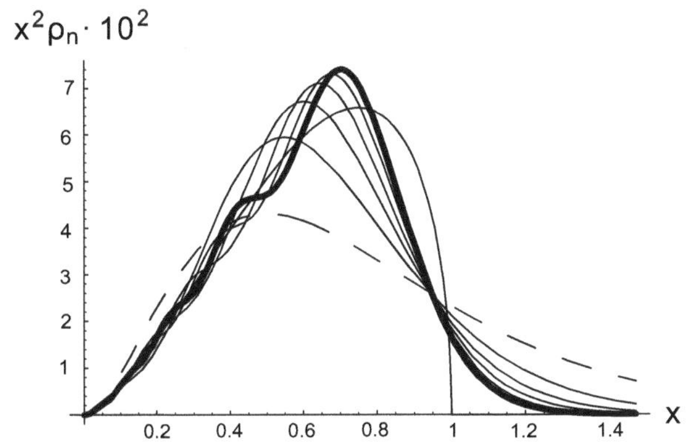

Fig. 8.4 The electron density ϱ_n for first six shells, the *dashed line* for first shell, the *thick line* for sixth shell

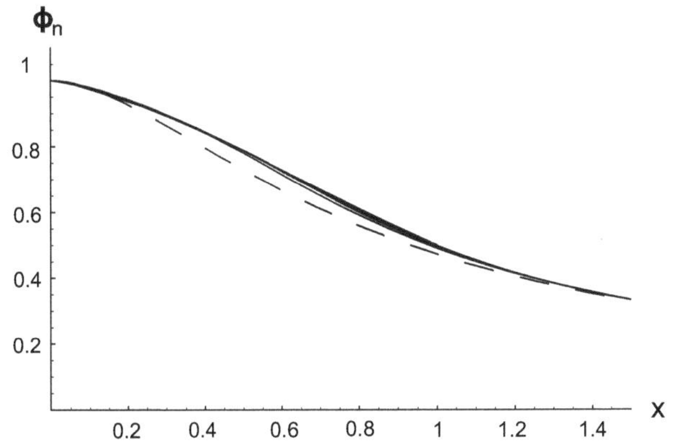

Fig. 8.5 Electron potential $\varphi_n(x)$ for first six shells and asymptotic potential $\varphi^{as}(x)$ (*thick line*); *dashed line* corresponds to the first shell

δ_n is lesser than one percent. This accuracy is typical for OM zeroth approximation, which describes the self-similarity of electron potential quite well. The electron potential $\varphi^{as}(x)$, i.e. the potential in asymptotic limit $y \to \infty$, plays a role of universal function.

The universal expressions for atomic characteristics have been also obtained besides MTF, in oscillator model of atom with eigenfunctions of three-dimensional symmetric oscillator as one-particle wave functions. To apply the model of two-dimensional complex oscillator for many-electron atom in approximation of independent particles, the one-particle equation for electron is written as:

$$\left(\frac{\mathbf{p}^2}{2} - \frac{Z}{r} + v(\mathbf{r}) - E_\lambda\right) |\psi_\lambda\rangle = 0. \tag{8.102}$$

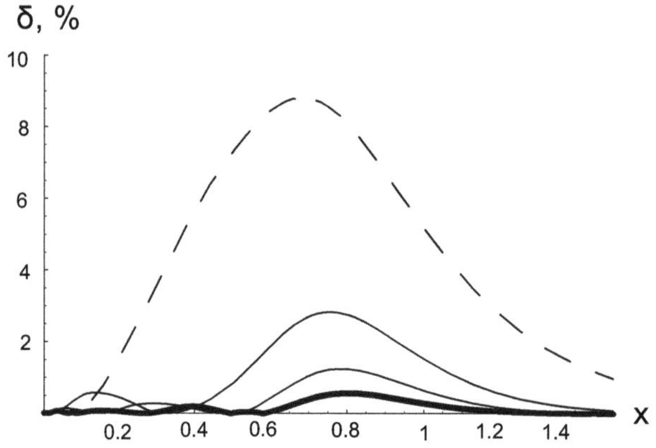

Fig. 8.6 The deviations $\delta_n(x)$ of the potentials of first four shells from the asymptotic potential; *dashed line* corresponds to first shell, *thick line* to sixth shell

The electron is assumed to be located in self-consistent potential $V(\mathbf{r}) = -\frac{Z}{r} + v(\mathbf{r})$, caused by other electrons and positively charged nucleus with the charge Z. In the Eq. (8.102), the parameter λ means the set of quantum numbers characterizing the state of electron in the atom; $\mathbf{p}$ is an operator of electron momentum; E_λ is the energy of one-electron state. To apply the OM to the solution of one-electron equation, the Eq. (8.102) is rewritten into secondary quantification form, using the transformations (8.64), (8.65):

$$\mathcal{L}|\psi_\lambda\rangle \equiv \left\{ \frac{\omega_\lambda}{4}\,(2 + N - M^+ - M) - Z + \int d^4\tilde{\xi}\;\tilde{\xi}^*\tilde{\xi}\;v(\tilde{\xi}^*,\tilde{\xi}) \int \frac{d^4k}{(2\pi)^4}\, e^{-i\,k\tilde{\xi}} \right.$$
$$\times\, e^{-\frac{k^2}{8\omega_\lambda}} \prod_{s=1}^{2} \sum_{n_s=0}^{\infty} \sum_{n_s'=0}^{\infty} \left(\frac{i\,k_s}{2\sqrt{2\omega_\lambda}}\right)^{n_s} \left(\frac{i\,k_s^*}{2\sqrt{2\omega_\lambda}}\right)^{n_s'} \frac{(a_s^+)^{n_s}\,(a_s)^{n_s'}}{n_s!\,n_s'!}$$
$$\left. \times \sum_{m_s=0}^{\infty} \sum_{m_s'=0}^{\infty} \left(\frac{i\,k_s}{2\sqrt{2\omega_\lambda}}\right)^{m_s} \left(\frac{i\,k_s^*}{2\sqrt{2\omega_\lambda}}\right)^{m_s'} \frac{(b_s^+)^{m_s}\,(b_s)^{m_s'}}{m_s!\,m_s'!} \right\} |\psi_\lambda\rangle$$
$$-\frac{1}{2\,\omega_\lambda}\,(2 + N + M^+ + M)\,E_\lambda\,|\psi_\lambda\rangle = 0. \tag{8.103}$$

The potential $v(\mathbf{r})$ is assumed to be spherically-symmetric, the vector $|\psi_\lambda\rangle$ belongs to Hilbert space of two-dimensional complex oscillator and is a state eigenvector of the operator Q (8.61) with a zero eigenvalue. According to procedure of OM applied to Eq. (8.103), the operator of zeroth approximation $\mathcal{L}_0$ commutating with the operator of particles number N is written as:

$$\mathcal{L}_0 = \left\{ \frac{\omega_\lambda}{4} (2+N) - Z + \int d^4\tilde{\xi}\ \tilde{\xi}^* \tilde{\xi}\ v(\tilde{\xi}^*, \tilde{\xi}) \int \frac{d^4 k}{(2\pi)^4} e^{-i k \tilde{\xi}} \right.$$
$$\times e^{-\frac{k^2}{8\omega_\lambda}} \prod_{s=1}^{2} \sum_{n_s=0}^{\infty} (-1)^{n_s} \left(\frac{|k_s|^2}{8\omega_\lambda} \right)^{n_s} \frac{(a_s^+)^{n_s} (a_s)^{n_s}}{(n_s!)^2} \sum_{m_s=0}^{\infty} (-1)^{m_s} \left(\frac{|k_s|^2}{8\omega_\lambda} \right)^{m_s}$$
$$\left. \times \frac{(b_s^+)^{m_s} (b_s)^{m_s}}{(m_s!)^2} \right\} - \frac{1}{2\,\omega_\lambda} (2+N)\, E_\lambda. \tag{8.104}$$

The eigenvectors of the operator $\mathcal{L}_0$ corresponding to zeroth eigenvalue of operator Q are determined from the condition $n_2^a = n - n_1^a$ and $n_2^b = n - n_1^b$. The index λ is then associated with three quantum numbers n, n_1^a, n_1^b and $|\psi_\lambda^{(0)}\rangle = |\psi_{n_1^a, n_1^b, n-n_1^a, n-n_1^b}\rangle$, where n, n_1^a, $n_1^b = 0, 1, 2, 3, 4 \ldots$. Hereinafter we make an assumption, which is justified below in comparison with the variational results for atom energy, obtained for Coulomb functions in coordinate representation. The parameter $\omega_\lambda = \omega_n$ is assumed to be dependent exclusively on principle quantum number $n+1$. The one-particle energy $E_{n, n_1^a, n_1^b, s}$ in zeroth approximation for the potential $v(r)$ is determined from the equation $\langle \psi_\lambda^{(0)} | \mathcal{L}_0 | \psi_\lambda^{(0)} \rangle = 0$ and equals to:

$$E^{(0)}_{n, n_1^a, n_1^b, s} = \frac{\omega_n^2}{2} - Z \frac{\omega_n}{n+1} + \int d^4\tilde{\xi}\ \tilde{\xi}^* \tilde{\xi}\ v(\tilde{\xi}^*, \tilde{\xi}) \int \frac{d^4 k}{(2\pi)^4} e^{-i k \tilde{\xi}} e^{-\frac{k^2}{8\omega_n}}$$
$$\times L_{n_1^a}\left(\frac{|k_1|^2}{2\sqrt{2\omega_n}} \right) L_{n_1^b}\left(\frac{|k_1|^2}{2\sqrt{2\omega_n}} \right) L_{n-n_1^a}\left(\frac{|k_2|^2}{2\sqrt{2\omega_n}} \right) L_{n-n_1^b}\left(\frac{|k_2|^2}{2\sqrt{2\omega_n}} \right), \tag{8.105}$$

where index $s = \pm\frac{1}{2}$ means the spin states of the electron. Because of the relativistic effects are not included into consideration, the one-particle energy is degenerated over s. Thus, according to Pauli principle the $(n+1)$th shell may accommodate $g_n = 2\sum_{n_1^a=0}^{n} \sum_{n_1^b=0}^{n} 1 = 2\,(n+1)^2$ electrons.

To make a model self-consistent, the potential $v(\mathbf{r})$ has to be determined from the Poisson equation, in a similar way as in Thomas-Fermi method:

$$\Delta v(\mathbf{r}) = -4\pi\, \varrho(\mathbf{r}), \tag{8.106}$$

where Δ is Laplace operator in three-dimensional space, and $\varrho(\mathbf{r})$ is electron density in the model of independent particles from the Eqs. (8.54) and (8.55). For neutral atoms with completely filled shells, the zeroth approximation of OM gives the electron density equal to:

$$\varrho(\mathbf{r}) = 2\sum_{n=0}^{n_0} \varrho_n(r), \tag{8.107}$$

where ϱ_n is determined from the Eq. (8.84), and n_0 from the equation $2\sum_{n=0}^{n_0}(n+1)^2 = Z$. The solution of the Poisson equation for potential $v(\mathbf{r})$ is found as:

$$v(r) = 2\sum_{n=0}^{n_0}\left(\frac{4\pi}{r}\int_0^r \mathrm{d}r'\, r'^2 \varrho_n(r') + 4\pi\int_r^\infty \mathrm{d}r'\, r' \varrho_n(r')\right). \tag{8.108}$$

As illustrated in Fig. 8.7, the above described model implements the shell structure of atom because of electron density has an oscillating nature in opposite to MTF, which provides the monotonic function.

The total energy of atom as a function of electrons number N_e, nucleus charge Z and parameters $\{\omega_n\}_{n=0}^{n_{max}}$ follows from the equation:

$$E\left(N_e, Z, \omega_1, \ldots, \omega_{n_{max}}\right) = \sum_{n, n_1^a, n_1^b, s} E^{(0)}_{n, n_1^a, n_1^b, s} - \frac{1}{2}\int \mathrm{d}^3\mathbf{r}\, \varrho(\mathbf{r})\, v(\mathbf{r}), \tag{8.109}$$

where summation is performed over one-particle states occupied by electrons under the condition $\sum_{n, n_1^a, n_1^b, s} 1 = N_e$. The parameters ω_n are determined from the extremum condition for energy (8.109) in accordance with OM procedure:

$$\frac{\partial E\left(N_e, Z, \omega_1, \ldots, \omega_{n_{max}}\right)}{\partial \omega_m} = 0, \qquad (m = 1, 2, \ldots, n_{max}). \tag{8.110}$$

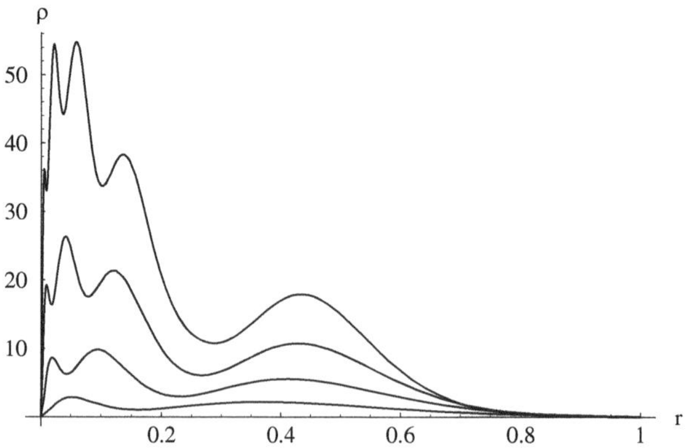

Fig. 8.7 The electron density $\rho(r)$ of the neutral atoms with completely filled shells for $Z = 10, 28, 60, 110$

This equation system is not fully equal to the variational condition for minimal energy, and can be applied for excited states, too. By resolving the algebraic system of Eq. (8.110), the total energy of atom and all related characteristics can be computed as functions of the total number of electrons N_e and nucleus charge Z.

To further discuss the presented model, we consider the limiting case $Z \to \infty$ and construct the analogue of quasi-classic limit in statistic MTF theory of atom. In the limit of large Z, the majority of electrons belongs to shells with $n \gg 1$, therefore we make use of asymptotic expressions (8.94), (8.96) for electron density ϱ_n^{as} and potential φ_n^{as} of the shell. For simplicity sake, the neutral atoms with completely filled shells are considered here, i.e. $N_e = Z$ and $(1 + n_0)(2 + n_0)(3 + 2\,n_0) = 3\,Z$. In asymptotic limit, the quantum number n is treated as continuous variable provided the following dimensionless variables are introduced:

$$x = \frac{n+1}{\left(\frac{3}{2}Z\right)^{1/3}}, \quad a(x) = a_n \left(\frac{3}{2}Z\right)^{1/3}, \quad u(x) = \frac{r}{a(x)} \left(\frac{3}{2}Z\right)^{1/3}, \tag{8.111}$$

The density of electrons (8.107) of a neutral atom with all filled shells in asymptotic limit equals:

$$\varrho(r) = \frac{18}{\pi^2} Z^2 \int_{\delta(Z)}^{x_0(Z)} \mathrm{d}x \, \frac{x^2}{a(x)^3} \sqrt{\frac{1}{u(x)} - 1}\, \Theta\left(1 - u(x)\right), \tag{8.112}$$

where the limits of integration are determined from the equations:

$$\delta = \left(\frac{2}{3\,Z}\right)^{1/3},$$
$$\frac{1}{2}(\delta + x_0)(2\delta + x_0)(3\delta + 2x_0) = 1. \tag{8.113}$$

Equation (8.113) has two complex and one real positive solution:

$$x_0 = -\frac{3\,\delta}{2} + \frac{3^{1/3}\,\delta^2 + \left(36 + \sqrt{3}\,\sqrt{432 - \delta^6}\right)^{2/3}}{2\,3^{2/3}\left(36 + \sqrt{3}\,\sqrt{432 - \delta^6}\right)^{1/3}} \approx 1 - \frac{3}{2}\,\delta + \frac{\delta^2}{12}. \tag{8.114}$$

It is evident that $\delta \to 0$ and $x_0 \to 1$ in asymptotic limit $Z \to \infty$, however, the behavior of the function $a(x)$ in points 0 and 1 does not allow to extend the integration limits from zero to unity in the expression for electron density (8.112).

All other atomic characteristics are expressed through the integrals converging on the interval [0, 1]. For example, the expression for electron potential $v(r)$ looks as follows:

$$v(r) = \frac{Z}{r} \int_0^1 \mathrm{d}x \, 6x^2 \, u(x) \, \varphi^{as} \left(u(x)\right), \tag{8.115}$$

where the function φ^{as} is determined from the Eq. (8.96). The total energy of atom as a functional of the function $a(x)$ is found from the Eq. (8.109):

$$E = 8 \left(\frac{3}{2} Z\right)^{7/3} \left\{ \int_0^1 \mathrm{d}x \left(\frac{x^4}{2\,a(x)^2} - \frac{x^2}{3\,a(x)} \right) + \right.$$
$$\left. + \frac{8}{\pi} \int_0^1 \mathrm{d}x \int_0^1 \mathrm{d}y \, \frac{x^2 y^2}{a(y)} \int_0^1 \mathrm{d}z \, z^2 \sqrt{\frac{1}{z} - 1} \, \varphi^{as} \left(z \frac{a(x)}{a(y)} \right) \right\} . \tag{8.116}$$

Thus, the atomic characteristics in the discussed model of atom at large Z are described by universal expressions, similar as MTF does. All values are expressed through single function $a(x)$, which is the solution of integral equation equivalent to the Eq. (8.110) and in explicit form follows from the condition of nullification of functional derivative of the functional of energy (8.116):

$$a(s) = 3\,s^2 - \frac{3\,a(s)^3}{s^2} \frac{\delta}{\delta a(s)} \left\{ \frac{8}{\pi} \int_0^1 \mathrm{d}x \int_0^1 \mathrm{d}y \, \frac{x^2 y^2}{a(y)} \int_0^1 \mathrm{d}z \, z^2 \times \right.$$
$$\left. \times \sqrt{\frac{1}{z} - 1} \, \varphi^{as} \left(z \frac{a(x)}{a(y)} \right) \right\} . \tag{8.117}$$

The dimensionless function $a(x)$ is an universal function for all atoms because of the charge of nucleus is not involved in the Eq. (8.117). Equation (8.117) is an analogue of MTF equation in statistical model of atom. The model proposed in this section is applicable for atom with arbitrary Z and N_e. Using modern computing algebraic resources, the total energy of atom E and shell parameters ω_n are easily found from the Eqs. (8.109) and (8.110). Table 8.2 shows the results for total energy of some atoms along with their accuracies in comparison to Hartree–Fock method, and accuracy does not exceed two percent. We have not accounted the exchange energy in calculations, and therefore the table presents total energy with subtracted self-interaction energy of electrons:

$$E_s = \frac{1}{2} \sum_i \int \mathrm{d}^3\mathbf{r} \int \mathrm{d}^3\mathbf{r}' \, \frac{|\psi_i(\mathbf{r}, \omega_i)|^2 \, |\psi_i(\mathbf{r}', \omega_j)|^2}{|\mathbf{r} - \mathbf{r}'|}, \tag{8.118}$$

Table 8.2 shows also the total energy accounting the interaction between electrons inside shells using the standard formula for exchange energy [20]. In this case, the

Table 8.2 Total energy and effective frequencies (charges) of neutral atoms with completely filled shells (Coulomb filling); δ is relative error of current results with respect to the results after Hartree–Fock method; a_{n_0} is effective radius of external shell; E_s is energy of self-interaction of electrons

Z	10	28	60	110	182	280
ω_0	9.63824	27.5657	59.5221	109.494	181.476	279.463
ω_1	2.87268	11.7705	27.5268	52.3491	88.2172	137.118
ω_2	–	4.32759	14.956	31.1974	54.8486	87.2306
ω_3	–	–	5.97149	18.6268	36.0454	60.0219
ω_4	–	–	–	7.77147	22.6507	41.549
ω_5	–	–	–	–	9.69789	26.9724
ω_6	–	–	–	–	–	11.7251
a_{n_0}	1.392	1.386	1.339	1.286	1.237	1.194
$-(E - E_s)$	125.905	1482.59	9157.45	38771.4	128138.2	355513.9
δ %	2.0	1.6	1.4	1.2	–	–
$-(E + E_{ex})$	129.374	1509.26	9262.34	–	–	–
δ %	0.7	0.2	0.2	–	–	–
$-E_{HF}$	128.547	1506.82	9283.70	39225.0	–	–
$-E_{TF}$	165.6	1830.1	10834.2	44568.6	144304.3	394289.2

self-interaction energy has not to be subtracted because of it is canceled by the same contribution in the exchange energy E_{ex}:

$$E_{ex} = -\frac{1}{2}\sum_{i,j}\int d^3\mathbf{r}\int d^3\mathbf{r}'\,\frac{\psi_i^*(\mathbf{r},\omega_i)}{|\mathbf{r}-\mathbf{r}'|}, \tag{8.119}$$

where ψ_i includes also the spinal part of the wave function of ith electron, the summation over electron assumes ith and jth electrons belong to the same shell, and other terms are neglected. This approximation is reasonable because of the exchange energy between shell electrons is essentially higher than exchange energy between electrons belonging to different shells. When summing up over all electrons, the formula (8.119) fails if one-electron wave functions are chosen as Coulomb functions with arbitrary parameter ω_i, because of these wave functions are not orthogonal for the electrons belonging to different shells. The functions ψ_i have been selected in a coordinate representation because of calculation of the exchange matrix elements is easier in coordinate representation than in the representation of two-dimensional complex oscillator. Here we have chosen the wave functions of the hydrogen-like atom for one-particle wave functions $|\psi_\lambda\rangle$:

$$\begin{aligned}|\psi_\lambda\rangle = Y_{lm}(\theta,\varphi)\,\frac{2\,(n\,\omega_{nl})^{3/2}}{n^{l+2}\,\Gamma(2l+2)}\sqrt{\frac{\Gamma(n+l+1)}{\Gamma(n-l)}}\,(2n\,\omega_{nl}\,r)^l\,e^{-\omega_{nl}\,r}\\ \times\Phi\left(-n+l+1,2l+2,2\omega_{nl}\,r\right),\end{aligned} \tag{8.120}$$

Table 8.3 Total energy and effective shell charges of neutral atoms with completely filled shells (experimentally observed filling)

Z	10	18	28	36	40	54
ω_{1s}	9.63122	17.5855	27.5573	35.5432	39.5364	53.5209
ω_{2s}	3.1717	7.0233	11.9444	15.8503	17.798	24.6776
ω_{2p}	2.97605	6.90317	11.8546	15.7909	17.7516	24.6649
ω_{3s}	–	2.84506	5.43641	7.74579	9.01713	13.3787
ω_{3p}	–	2.54858	5.23227	7.62518	8.91781	13.3193
ω_{3d}	–	–	4.34388	6.81171	8.18102	12.7479
ω_{4s}	–	–	1.84715	3.19168	4.30557	7.2486
ω_{4p}	–	–	–	2.84806	4.00172	7.00315
ω_{4d}	–	–	–	–	0.999796	6.1527
ω_{5s}	–	–	–	–	2.20132	3.36666
ω_{5p}	–	–	–	–	–	3.1042
$-(E + E_{ex})$	129.374	529.119	1514.24	2763.55	3552.21	7251.48
$-E_{HF}$	128.547	526.818	1506.82	2752.06	3538.97	7232.14
δ %	0.7	0.6	0.5	0.4	0.4	0.3

where $\Gamma(n)$ is Gamma function; $\Phi(a, b, x)$ is degenerated hypergeometric function; $Y_{lm}(\theta, \varphi)$ is spherical function; $\lambda = (n, l, m)$. The parameters ω_{nl} depend both on principle quantum number n and on orbital quantum number l.

Table 8.3 shows the energy values computed for several atoms using discussed model with experimentally observed $n + l$ filling of electrons in the shells, which is described approximately by Madelung-Klechkowski rule. Table 8.3 demonstrates a good approximation of HF functions by the wave functions of hydrogen-like atom, which follows from the agreement of calculated energies with ones resulted form HF method. The parameters ω_{nl} weakly depend on orbital quantum number l, therefore the domination of principle quantum number in the choice of approximation is reasonable. For example, the total energy of atom $Z = 40$ without contribution caused by dependence on l equals to 3550.32, which is close to HF energy 3552.21.

To accurately consider the asymptotic limit $Z \to \infty$ of the presented model, the Eq. (8.117) is written in the following form:

$$a(s) = \frac{3\,s^2}{1 - \frac{24}{\pi}\int_0^1 \mathrm{d}z\, z^2 \sqrt{\frac{1}{z} - 1}\int_0^1 \mathrm{d}x\, x^2\, \mathcal{F}\left(z, \frac{a(x)}{a(s)}\right)}, \tag{8.121}$$

$$\mathcal{F}\left(z, \frac{a(x)}{a(s)}\right) = \varphi^{as}\left(z\frac{a(x)}{a(s)}\right) + z\frac{a(s)}{a(x)}\frac{d}{dz}\varphi^{as}\left(z\frac{a(s)}{a(x)}\right) - z\frac{d}{dz}\varphi^{as}\left(z\frac{a(x)}{a(s)}\right).$$

The analysis of this equation indicates the behavior of function $a(s) \approx 3\,s^2$ in the limit $s \ll 1$, whereas $a(s) \approx 1/(1 - s)$ in the limit $s \to 1$, which means the existence of singularity in the point $s = 1$. The numerical solution of the integral equation (8.121) is cumbersome, however, the analytical one can be obtained from

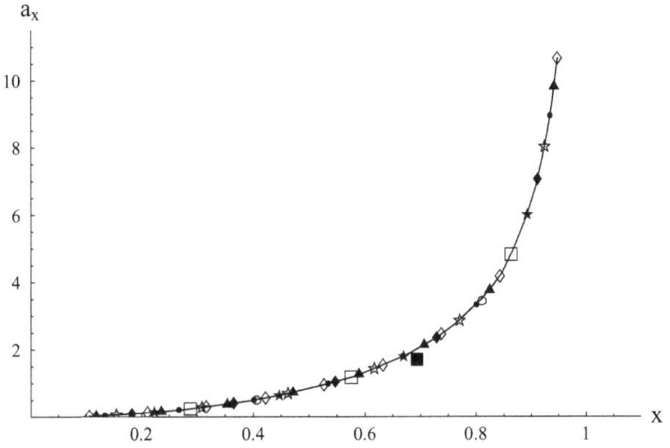

Fig. 8.8 The universal function for effective radii of shells $\tilde{a}(x)$ for different atoms $Z = 2, 10, 28, 60, 110, 182, 280, 408$

the asymptotics for $a(x)$ assuming the function $a(x)$ is an interpolative one for effective radii of atomic shells. Thus, by drawing all the points $\{x, \tilde{a}(x)\}$ on the graph ($\tilde{a}(x) = a_n(3Z/2)^{1/3}$, $x = (n+1)(2/3Z)^{1/3}$), where Z and a_n correspond to real atoms from Table 8.3, we obtain a single curve $\tilde{a}(x)$ shown in Fig. 8.8. The function $\tilde{a}(x)$ is approximated well by function $a(x)$, and $\tilde{a}(x) \approx a(x)$. As has been demonstrated in Sect. 8.2, the function $a(x)$ is interpolative for effective radii of real atoms, provided the one-particle functions are chosen as eigenfunctions of harmonic oscillator in oscillatory model of atom, and this feature remains valid for Coulomb model of atom.

To interpolate the points in Fig. 8.8 in zeroth approximation, the function $a(x) \approx a^{(0)}(x)$ is chosen in analytical form:

$$a^{(0)}(x) = 2a\left(\frac{1}{1-x^2} - 1\right) + (3 - 2a)\, x^2, \tag{8.122}$$

where a is a constant and the function $a^{(0)}(x)$ has proper asymptotics at zero and unity. The function $a(x)$ is specified as a sum of $a^{(0)}(x)$ and $a^{(1)}(x)$, which is interpolated by the following rational function:

$$a^{(1)}(x) = x^4\, \frac{c_1 + c_2\, x^4}{1 + b_1\, x^4 + b_2\, x^8}, \tag{8.123}$$

here the parameters a, c_1, c_2, b_1, b_2 are found from the fit of 35 points $\{x, \tilde{a}(x)\}$, corresponding to atoms with $Z = 10, 28, 60, 110, 182, 280$ and computed from the data presented in Table 8.2 using the relationship $a_n = 2(n+1)/\omega_n$ between ω_n and a_n. As a result, $a = 0.5261874$, $c_1 = 0.2634342$, $c_2 = -0.3480257$, $b_1 = -2.3631431$, $b_2 = 1.4504157$, and adding another eight points for $Z = 408$: $a = 0.4994434$, $c_1 = 0.3514095$, $c_2 = -0.43634995$, $b_1 = -2.2915775$, $b_2 = 1.3711871$. The accuracy of the interpolation obtained as $\tilde{a}(x) \approx a^{(0)}(x) + a^{(1)}(x)$, can be assessed by comparing the calculated total energy with the energy computed

Table 8.4 Total energy of atoms E^{int} from interpolative formula for parameters $\tilde{a}(x) \approx a^{(0)} + a^{(1)}$, and energy E obtained from exact minimization

Z	E^{int}	E	δ (%)
10	125.725	125.905	0.1
28	1482.11	1482.596	0.01
60	9156.12	9157.45	0.01
110	38767.61	38771.37	0.01

Table 8.5 The accuracy δ of the results by interpolation

δ, %				
0.49	2.36	0.97	1.06	0.26
0.61	1.42	0.75	0.04	0.20
0.79	1.28	0.27	1.40	0.20
1.06	1.55	0.46	0.70	0.03
1.51	1.21	1.60	0.73	0.07
1.12	4.60	0.37	0.35	0.02
1.26	1.58	0.44	1.92	0.01

by exact minimization. Table 8.4 displays the accuracy of this approximation better than hundredth of percent.

Table 8.5 shows the accuracy $\delta = 100\,|a^{(0)}+a^{(1)}-\tilde{a}(x)|/\tilde{a}(x)$ for the parameters in all 35 points. Thus, the accuracy of the interpolation is very good, and it has a simple analytical form and fits the asymptotics correctly.

Provided the expression for the function $a(x)$ is known, the formula for the atomic energy in the limit $Z \to \infty$ can be derived by substituting $a(x) \approx a^{(0)}+a^{(1)}$ into Eq. (8.116) and performing a numerical integration:

$$E = -0.76667\ Z^{7/3}. \tag{8.124}$$

The result demonstrates that the asymptotic value of E in the considered model is close to the energy of MTF $E = -0.76874\,Z^{7/3}$, which in its turn defines the exact asymptotic expression for total energy in the limit $Z \to \infty$. Thus, the approximation obtained for $a(x)$ is close to exact one, which corresponds to the minimum of the functional of energy (8.116) and is determined from the Eq. (8.121). The derived approximation does not violate the virial theorem: providing the mean kinetic energy E_k equals

$$E_k = 8\left(\frac{3}{2}\,Z\right)^{7/3} \int_0^1 \mathrm{d}x\, \frac{x^4}{2\,a(x)^2} = 0.762485\ Z^{7/3}, \tag{8.125}$$

the total energy of the system with Coulomb interaction is equal to the mean kinetic energy with the opposite sign $E = -E_k$. Comparing (8.122) and (8.124), the virial relation is evidently satisfied well with the accuracy 0.5 %. Thus, the approximate expression for $a(x)$, derived from the interpolation on the points corresponding to real atoms is a good interpolation for the solution of Eq. (8.121). As a consequence, the function $a(x)$ is an interpolation for effective radii of atoms with both large and small charges Z.

8.4 Effective Charges Model for Many-Electron Atom

In previous sections, the application of the operator method has been focused on calculation of the observable physical values of atom in ground state, for example, total energy or electron density. However, the numerous applications related to many-electron atoms in external fields, their interaction with electrons and photons require the wave functions of electron sub-system of atoms or ions in ground and excited states. This section deals with the application of OM to above mentioned problems and construction of analytical solutions using the representation of one-electron wave functions on the basis of effective charges model (ECM). This approach has been used in [14] to analytically derive the atomic scattering factors required for calculation of X-ray polarizabilities of crystals [24].

There are a lot of databases and papers where precise enough analytical approximation for one-electron wave functions has been derived (for example, [25–27]). In the most of these cases, the wave functions derived are a result of the numerical interpolation for solutions of the HF equations for free atoms in the ground states. This makes it difficult to use these functions for description of the excited states of the atoms and for calculation of the transition matrix elements which are necessary for description of interaction between atoms and external fields.

Here we use OM for the calculation of the atomic wave functions on the basis of the approximate solution of the Shrödinger equation that is valid for the entire range of Hamiltonian parameters. It was shown in the previous chapters that the essential feature of OM is its locality in the space of eigenstates of the investigated quantum system, i.e. a state vector in the zeroth-order approximation includes the variational parameters defined by the condition of the best description of exactly this state. This approach can be used both for the ground state and for the excited states of the system. Another important feature is the convergence of the successive approximation of OM in the entire range of changing of the Hamiltonian parameters. Both these features of the OM have permitted us to find the analytical expression for wave function of many-electron atoms using a comparatively small database of parameters with clear physical meaning.

Quantitative description of atomic systems is based mainly on the concept of independent movement of electrons in self-consistent field of HF potential [12]. One-electron state vectors are considered as a basis for construction of zeroth approximation for electron wave function in this common field. The OM allows one to extend the conception of a common self-consistent field by inclusion of an individual field for every electron in every state. It will be shown below that this helps to realize in full measure the idea of the independent movement of electrons, when the total energy of the atom is reduced to the sum of energies of individual electrons; this is in opposition to the total Hartree–Fock energy of atom, which does not satisfy this condition [28].

In accordance with the OM general rules (Chap. 3) its zeroth approximation is defined by the optimal choice of the wave function basis as distinct of the conventional PT which is defined by a choice of the zeroth order Hamiltonian. In the considered case the characteristics of the electron individual field are determined also by the selection of the basis of one-particle wave functions. For atoms the most natural basis is generated by functions describing the movement of each electron in its own Coulomb field defined by an effective charge. These individual effective charges are supposed to be different for different electron states and are considered as the OM variational parameters for the best zeroth approximation. The charges are transformed to the charge of atomic nucleus when the interaction between electrons is adiabatically cut off. The reason for such a parametrization can also be explained by the fact, that the Coulomb wave functions with a properly chosen effective charge approximates the Hartree–Fock functions very well [29].

An individualization of the self-consistent field for the electron in every quantum state does not contradict the permutation symmetry of the initial Hamiltonian of the whole atom. However, it is very important to use the basic set of orthonormalized one-electron wave functions for a correct calculation of form factors and higher-order approximations. The construction of an orthonormalized basis starts from the functions $R_{nl}(r, Z_{nl})$, which are the solutions of the radial Shrödinger equation for a particle in the Coulomb field of the charge Z_{nl} and spherical harmonics $Y_{lm}(\mathbf{n})$. Here $r = |\mathbf{r}|$; $\mathbf{n} = \mathbf{r}/r$ and n, l, m are the principal, orbital and azimuth quantum numbers [9], and the atomic units and traditional spectroscopic classification are used. Thus, a coordinate part of wave function of an electron in the first shell is chosen as the $1s$ state in the field with effective charge Z_{1s}:

$$^{1s}\psi(\mathbf{r}, Z_{1s}) = R_{10}(r, Z_{1s})\, Y_{00}(\mathbf{n}). \tag{8.126}$$

A general form for wave functions of the electrons in the second shell is

$$\begin{aligned} ^{2s}\psi(\mathbf{r}, Z_{2s}) &= (C_{20} R_{20}(r, Z_{2s}) + C_{21} R_{21}(r, Z_{2s}))Y_{00}(\mathbf{n}), \\ ^{2p}\psi_m(\mathbf{r}, Z_{2p}) &= R_{21}(r, Z_{2p})\, Y_{1m}(\mathbf{n}). \end{aligned} \tag{8.127}$$

Here the coefficients C_{20} and C_{21} are chosen from the conditions of the orthogonality of functions $^{1s}\psi$ and $^{2s}\psi$ and normalization of the function $^{2s}\psi$:

$$\begin{aligned} \int d\mathbf{r}\, ^{1s}\psi^*(\mathbf{r}, Z_{1s})\, ^{2s}\psi(\mathbf{r}, Z_{2s}) &= 0, \\ \int d\mathbf{r}\, |^{2s}\psi(\mathbf{r}, Z_{2s})|^2 &= 1. \end{aligned} \tag{8.128}$$

This procedure is repeated for the states of electrons in the third shell resulting in the relations:

$$\begin{aligned} ^{3s}\psi(\mathbf{r}, Z_{3s}) &= (C_{30}\, R_{30}(r, Z_{3s}) + C_{31}\, R_{31}(r, Z_{3s}) + C_{32}\, R_{32}(r, Z_{3s}))\, Y_{00}(\mathbf{n}), \\ ^{3p}\psi_m(\mathbf{r}, Z_{3p}) &= (D_{31}\, R_{31}(r, Z_{3p}) + D_{32}\, R_{32}(r, Z_{3p}))\, Y_{1m}(\mathbf{n}), \\ ^{3d}\psi_m(\mathbf{r}, Z_{3d}) &= R_{32}(r, Z_{3d})\, Y_{2m}(\mathbf{n}), \end{aligned} \tag{8.129}$$

where, again, the coefficients C_{30}, C_{31} and C_{32} are defined by the orthogonality of the function ${}^{3s}\psi(\mathbf{r}, Z_{3s})$ to the functions ${}^{1s}\psi(\mathbf{r}, Z_{1s})$ and ${}^{2s}\psi(\mathbf{r}, Z_{2s})$ and by normalization. The coefficients D_{31} and D_{32} can be found from the orthogonality of the functions ${}^{3p}\psi_m(\mathbf{r}, Z_{3p})$ and ${}^{2p}\psi_m(\mathbf{r}, Z_{2p})$ and their normalization.

The formulas above illustrate the procedure for construction of an orthonormalized basis from the Coulomb functions belonging to different effective charges. The set of radial functions with the same principal quantum number n is enough for a similar construction of electron wave functions in any shell. The effective charges are the only free variational parameters in the functions of the OM's zeroth approximation, contrary to the Slater orbitals [26], where essentially more parameters are introduced in order to interpolate the polynomial structure of radial functions.

The reasons for a considered way of orthonormal basis construction are also related to the influence of electron kinetic energy operator on these functions. For example,

$$-\frac{1}{2}\Delta\, {}^{1s}\psi(\mathbf{r}, Z_{1s}) = \left(-\frac{Z_{1s}^2}{2} + \frac{Z_{1s}}{r}\right) {}^{1s}\psi(\mathbf{r}, Z_{1s})$$

$$-\frac{1}{2}\Delta\, {}^{2s}\psi(\mathbf{r}, Z_{2s}) = (-\frac{Z_{2s}^2}{8} + \frac{Z_{2s}}{r})\, {}^{2s}\psi(\mathbf{r}, Z_{2s}$$

$$-\frac{C_{21}}{r^2} R_{21}(r, Z_{2s})\, Y_{00}(\mathbf{n}). \tag{8.130}$$

The components in the right hand sides of the equations $-(Z_{1s}^2/2)\, {}^{1s}\psi(\mathbf{r}, Z_{1s})$ and $-(Z_{2s}^2/8)\, {}^{2s}\psi(\mathbf{r}, Z_{2s})$ are eliminated when the total Hamiltonian acts on a many-particle wave function. In this case, the result is represented as the sum of one-particle energies of electrons. The considered here one-particle orthonormalized basis with effective charges of electrons as free parameters permits the construction of a wave function for the entire atom. For every concrete atomic state, the vector $|\Psi\rangle$ is presented as a sum of anti-symmetrized products of one-particle functions from the constructed basis in the form of Slater determinant (spin functions should be taken into account).

In the zeroth OM approximation, a diagonal matrix element $E = \langle\Psi|H|\Psi\rangle$ defines the energy of the atom with a nuclear charge Z and Hamiltonian:

$$H = -\frac{1}{2}\sum_i \Delta_i - \sum_i \frac{Z}{r_i} + \sum_{i<j} \frac{1}{|\mathbf{r}_i - \mathbf{r}_j|} \tag{8.131}$$

In accordance with OM, free parameters (effective charges in our case) can be chosen from the condition of independence of the diagonal matrix elements of the

Hamiltonian on a wave function representation. In our case, these conditions lead to the equations:

$$\frac{\partial E}{\partial Z_{1s}} = 0, \quad \frac{\partial E}{\partial Z_{2s}} = 0 \ldots \tag{8.132}$$

The Slater determinant is only considered for the mathematical foundation of the initial approximation. Certainly, the orthonormalization of the basis allows all the advantages of the secondary quantization formalism to be used in routine calculations. Let us restrict ourselves to completely filled nl states for illustrative purposes and introduce the notations for the following matrix elements:

$$e_{nl} = \frac{1}{2l+1} \int d\mathbf{r} \sum_m ({}^{nl}\psi^*_m(\mathbf{r}, Z_{nl})) \frac{1}{r} ({}^{nl}\psi_m(\mathbf{r}, Z_{nl})), \tag{8.133}$$

$$ec_{nl} = \frac{1}{2l+1} \int d\mathbf{r} \sum_m ({}^{nl}\psi^*_m(\mathbf{r}, Z_{nl})) \left(-\frac{1}{2}\Delta + \frac{Z^2_{nl}}{2n^2} - \frac{Z_{nl}}{r} \right) ({}^{nl}\psi_m(\mathbf{r}, Z_{nl})), \tag{8.134}$$

$$ee_{n_1 l_1 n_2 l_2} = \frac{1}{(2l_1+1)(2l_2+1)} \int d\mathbf{r}_1 \int d\mathbf{r}_2 \frac{1}{|\mathbf{r}_1 - \mathbf{r}_2|} \times \sum_{m_1, m_2} |{}^{n_1 l_1}\psi_{m_1}(\mathbf{r}_1, Z_{n_1 l_1})|^2 \; |{}^{n_2 l_2}\psi_{m_2}(\mathbf{r}_2, Z_{n_2 l_2})|^2, \tag{8.135}$$

$$\begin{aligned} ex_{n_1 l_1 n_2 l_2} = {} & \frac{1}{(2l_1+1)(2l_2+1)} \int d\mathbf{r}_1 \int d\mathbf{r}_2 \frac{1}{|\mathbf{r}_1 - \mathbf{r}_2|} \\ & \times \sum_{m_1, m_2} ({}^{n_1 l_1}\psi^*_{m_1}(\mathbf{r}_1, Z_{n_1 l_1})) \, ({}^{n_2 l_2}\psi^*_{m_2}(\mathbf{r}_2, Z_{n_1 l_1})) \\ & \times ({}^{n_2 l_2}\psi_{m_2}(\mathbf{r}_1, Z_{n_2 l_2})) \, ({}^{n_1 l_1}\psi_{m_1}(\mathbf{r}_2, Z_{n_1 l_1})). \end{aligned} \tag{8.136}$$

Then the ground state energy for any atom is performed as an algebraic expression. For example, for atom with four electrons and a nucleus charge Z the energy can be written as (population of the electron states is defined by the Pauli principle):

$$\begin{aligned} E &= E_0 + E_1; \\ E_0 &= -2\frac{Z^2_{1s}}{2} - 2\frac{Z^2_{2s}}{8}; \\ E_1 &= -2\,(Z - Z_{1s})\, e_{10} - 2\,(Z - Z_{2s})\, e_{20} - 2\, ec_{20} \\ & \quad + ee_{1010} + ee_{2020} + 4\, ee_{1020} - 2\, ex_{1020}. \end{aligned} \tag{8.137}$$

The analogous formula for the atom with ten electrons and nuclear charge Z is:

$$\begin{aligned} E &= E_0 + E_1; \\ E_0 &= -2\frac{Z_{1s}^2}{2} - 2\frac{Z_{2s}^2}{8} - 6\frac{Z_{2p}^2}{8}; \\ E_1 &= -2\,(Z - Z_{1s})\,e_{10} - 2\,(Z - Z_{2s})\,e_{20} - 6\,(Z - Z_{2p})\,e_{20} \\ &\quad -2\,ec_{20} + ee_{1010} + ee_{2020} + 4\,ee_{1020} + 12\,ee_{1021} + 12\,ee_{2021} \\ &\quad +15\,ee_{2121} - 2\,ex_{1020} - 6\,ex_{1021} - 6\,ex_{2021} - 6\,ex_{2121}. \end{aligned} \tag{8.138}$$

The numerical coefficients in the last formulas are determined from the calculation of the number of electrons in occupied states and number of interacting electron pairs, i.e. number of electron pairs with equally oriented spins contributing to the exchange interaction. According to the formulas (8.137)–(8.138), the energy of an atom can be written as a sum of two terms:

$$E = E_0 + E_1, \tag{8.139}$$

where E_0 represents a sum of one-particle energies of electrons, determined by the formulas following from the relations (8.130):

$$E_{1s} = -\frac{Z_{1s}^2}{2}, \quad E_{2s} = -\frac{Z_{2s}^2}{8}, \quad E_{2p} = -\frac{Z_{2p}^2}{8}, \quad \ldots \tag{8.140}$$

The value E_1 is the correction of the first order of OM, caused by an approximate presentation of the potential energy of the atom as the sum of individual potential energies of electrons.

The results of the OM zeroth approximation can be juxtaposed now with the HF results for some atomic characteristics. As for example, Table 8.6 shows the energies E_0 and E_1 for neutral Be, B, F, Ne, Na and Mg atoms, calculated assuming the definition of the effective charges as in Eq. (8.133). The effective charges for all neutral atoms and some ions were listed in the paper [14], the atomic units are used in the calculations and the total Hartree–Fock energy for nonrelativistic atoms

Table 8.6 Total energies of atoms calculated by OM (E) and by the HF method (E_{HF})

Atom (Z)	E_0	E_1	$E = E_0 + E_1$	E_{HF}
$Be(Z = 4)$	−15.4116	0.88164	− 14.5300	− 14.5730
$B(Z = 5)$	−25.8567	1.40611	−24.4506	−24.5291
$F(Z = 9)$	−104.454	4.57247	−99.8820	−99.4094
$Ne(Z = 10)$	−135.133	5.64220	−129.491	−128.547
$Na(Z = 11)$	−168.803	5.76189	−163.041	−161.859
$Mg(Z = 12)$	−207.919	7.06993	−200.849	−199.615

[26] is presented in the last column. The OM estimations for the total atomic energy are in a good agreement with the HF results. One of the advantages of the OM approach is the smallness of the nonadditive contribution E_1 in comparison with the total energy, this is in contrast to the HF method [28]. The simple structure of the wave functions can be also used for accounting of a self-consistent relativistic contribution to the atomic Hamiltonian [14].

Direct physical interpretation of the OM wave functions can be used to find the change of the atomic form factors in the external field. For example, the atoms C, Si and Ge in the magnetic field have in the ground state two external p-electrons, which form a state with the total spin $S = 1$ and orbital momentum $L = 1$. Taking into account the spin-orbital interaction, the ground state of these atoms corresponds to the zero eigenvalue of the total momentum $\boldsymbol{J} = \boldsymbol{L} + \boldsymbol{S}$ [9]. In a magnetic field Ξ, directed along the z axis, the Hamiltonian of the atom is changed by the value:

$$\delta H = A(\boldsymbol{LS}) + \mu_B \Xi (L_z + 2S_z), \tag{8.141}$$

where $A > 0$ is the constant of the spin-orbital interaction and μ_B is the Bohr magneton. The radial wave functions of the OM zeroth approximation $R_n1(r)$, $n = 2(C), 3(Si), 4(Ge)$ do not change but their spin-angular dependence defined by the eigenfunctions of the operator δH can be diagonalized by means of the eigenvectors of the full spin and the momentum operators $|M_L, M_S>$. The ground state vector of the atom in the field is transformed to the following linear combination:

$$\begin{aligned}
|\Phi_0> &= c_1|1,-1> + c_2|0,0> + c_3|-1,1>; \\
c_1 &= -\frac{1}{\sqrt{3}}(1 + \xi - \frac{2}{9}\xi^2), \quad c_2 = \frac{1}{\sqrt{3}}(1 - \frac{5}{9}\xi^2), \\
c_3 &= -\frac{1}{\sqrt{3}}(1 - \xi - \frac{2}{9}\xi^2), \\
\xi &= \frac{\mu_B \Xi}{A}.
\end{aligned} \tag{8.142}$$

These wave functions lead to the appearance of the non-spherical part of the electron density. As a result, the atomic scattering factor includes a term proportional to the second Legendre polynomial with an angle θ between the axis z and the vector $\boldsymbol{q}$. The amplitude of this contribution is defined by the integral of the radial electron density with Bessel function:

$$\begin{aligned}
f_m(\boldsymbol{q}) &= -\frac{10}{9}\xi^2 P_2(\cos\theta) \int_0^\infty dr r^2 R_{n1}^2(r) \sqrt{\frac{\pi}{2qr}} J_{5/2}(qr) \\
&= -\frac{10}{9}\xi^2 P_2(\cos\theta) f_m(s).
\end{aligned} \tag{8.143}$$

Thus, the use of OM for effective charges model delivers a simple and physically obvious algorithm for quantitative description of atomic characteristics in one-electron approximation with the accuracy satisfactory for numerous applications [24]. This approach is a basis for accounting of inter-electron correlations in successive approximations of operator method (Sect. 5.3).

References

1. P. Hohenberg, W. Kohn, Phys. Rev. **136**, B864 (1964)
2. W. Kohn, L.J. Sham, Phys. Rev. **140**, A1133 (1965)
3. W. Kohn, Rev. Mod. Phys. **71**, 1253 (1998)
4. R.O. Jones, O. Gunnarsson, Rev. Mod. Phys. **61**, 689 (1989)
5. R. Parr, W. Yang, *Density-Functional Theory of Atoms and Molecules* (Oxford University Press, New York, 1989)
6. W. Koch, M.C. Holthausen, *A Chemist's Guide to Density Functional Theory* (Wiley, New York, 2001)
7. B.-G. Englert, *Semiclassical Theory of Atoms* (Springer, Berlin, 1988)
8. I.K. Dmitrieva, G.I. Plindov, *Characteristics of Atoms and Ions in the Light of Statistical Theory* (in Russian) (Nauka and Tekhnika, Minsk, 1991)
9. L.D. Landau, E.M. Lifshitz, *Quantum Mechanics* (Fizmatgiz, Moscow, 2004)
10. A. Messiah, *Quantum Mechanics* (North-Holland, Amsterdam, 1976)
11. J.C. Slater, *Quantum Theory of Atomic Structure* (McGraw-Hill, New York, 1960)
12. C.F. Fischer, *The Hartree-Fock Method for Atoms. Numerical Approach* (Wiley, New York, 1977)
13. A. Szabo, N.S. Ostlund, *Modern Quantum Chemistry: Introduction to Advanced Electronic Structure Theory* (McGraw-Hill, New York, 1989)
14. I.D. Feranchuk, L.I. Gurskii, L.I. Komarov, O.M. Lugovskaya, F. Burgäzy, A.P. Ulyanenkov, Acta Crystallogr. A **58**, 370 (2002)
15. I.D. Feranchuk, V.V. Triguk, J. Appl. Spectrosc. **77**, 749 (2011)
16. I.S. Gradshtein, I.M. Ryzhik, *Tables of Integrals, Sums, Series and Products* (Fizmatgiz, Moscow, 1963)
17. I.D. Feranchuk, A.V. Leonov, Phys. Lett. A **375**, 385 (2011)
18. O.J. Heilmann, E.H. Lieb, Phys. Rev. A **82**, 3628 (1995)
19. J.C. Slater, *Quantum Theory of Molecules and Solids* (McGraw-Hill, New York, 1965)
20. P.A.M. Dirac, Proc. Camb. Philos. Soc. **26**, 376 (1930)
21. P.M. Morse, H. Feshbach, *Methods of Theoretical Physics* (McGraw-Hill, New York, 1953)
22. P.A. Khomyakov, *Development and Application of the Operator Method for Quantum Systems with many Degrees of Freedom* (in Russian), PhD Thesis, Belarusian University 2001
23. L.I. Komarov, P.A. Khomyakov, Nonlinear Phenom. Complex Syst. **4**, 341 (2001)
24. A. Benediktovitch, I. Feranchuk, A. Ulyanenkov, *Theoretical Concepts of X-Ray Nanoscale Analysis* (Springer, Heidelberg, 2014)
25. I. Clementi, D. Raimondi, J. Chem. Phys. **38**, 2686 (1963)
26. I. Clementi, C. Roetti, At. Data Nucl. Data Tables **14**, 177 (1974)
27. R.E. Stewart, J. Chem. Phys. **50**, 2485 (1969)
28. M.G. Veselov, L.N. Labzovskii, *The Theory of Atoms: Structure of Atomic Shells* (in Russian) (Nauka, Moscow, 1986)
29. M. Kregar, Phys. Scr. **29**, 438 (1984)

Chapter 9
Systems with Infinite Number of Degrees of Freedom

This chapter is dedicated to the applications of operator method for the analysis of propagation of electron in ionic crystal, called often as propagation of polaron of a large radius. This task is qualitatively distinguishable from the ones discussed in previous chapters by the fact that Hamiltonian includes the interaction of electron with the system possessing infinite degrees of freedom and describing the phonon field of the lattice vibration. A similar problem arises for the systems, which require the self-consistent description of objects with external media, and the significance of polaron covers much wider area than the description of the interaction between electrons and phonons initially constructed by Fröhlich [1]. The application of polaron in the condensed matter physics have been discussed in numerous monographs and reviews (see, for example, [2–5]). In this chapter we consider the application of the operator method to polaron problem.

The Hamiltonian of polaron is frequently considered not only for the description of the processes in solid state physics, but also as fundamental model to probe the non-perturbative methods of quantum field theory in the entire range of coupling constant α. Similar to other quantum systems, the polaron can be described both by Schrödinger equation and by calculating the functional integrals. The idea of the localized electron state in the field of optical phonons of ionic crystal, called polaron, has been induced from the variational solution of Schrödinger equation [6]. The variational estimate for the coupling energy $E_0(\alpha)$ of the ground state of polaron has been found from Schrödinger equation for large coupling constant $\alpha \gg 1$, which determines the interaction between electron and phonon field [6]. In the works [7–9], the adiabatic PT has been developed, which delivers the exact asymptotic description of ground state in the limit of strong coupling $\alpha \to \infty$. In Sect. 9.1, we demonstrate the capability of OM to find in analytical form a zeroth approximation for polaron in the limit of strong coupling, the successive approximations and the spectrum of the excited states for polaron as well as to explain the interaction of polaron with external field [10].

I. Feranchuk et al., *Non-perturbative Description of Quantum Systems*, Lecture Notes in Physics 894, DOI 10.1007/978-3-319-13006-4_9

The great advantage of variational approach introduced by Feynman for the path integral description of polaron is the approximation found for $E_0(\alpha)$, which is valid in the entire range of $0 < \alpha < \infty$, and the best estimate for coupling energy in the intermediate range of coupling force [11]. Recently the Monte-Carlo method [12] has been applied to calculate path integrals and to obtain accurate data for function $E_0(\alpha)$. The construction of uniformly available approximation for polaron, however, has a special importance for the description of spectrum of the excited states and for the systems with internal degrees of freedom. In Sect. 9.2, we apply the operator method for the description of one-dimensional polaron in the entire range of α, which makes possible to investigate the spectrum of polaron states [13].

There were many efforts spent to compute the ground state energy of three-dimensional polaron in the range of intermediate α on the basis of variational principle for Schrödinger equation. The trial functions used for investigation of the whole range of α in operator method, result in singularity of $E_0(\alpha)$ near the point $\alpha_c \simeq 7$ (see, for example, [14] and citations therein). This fact initiated the discussion about possible phase transition between two qualitative states of polaron near α_c. However, the number of publications (see review [15]) have reported the function $E_0(\alpha)$ to be analytical for all α and the phase transition does not exist. Section 9.3 shows the application of OM for the construction of uniformly available analytical approximation for polaron in the entire range of α [16, 17] and generalizes this approach for alternative interaction form between particle and quantum field [17].

9.1 OM for Strong Coupling Polaron

As has already been mentioned above, the majority of methods applied to complicated case of strong coupling polaron problem ($\alpha \gg 1$) are based on the variational principle in various forms. However, in contrast to the perturbation theory, the variational principle is not the regular calculation method, because it does not permit to increase directly the calculation precision, to find the spectrum of the excited states or to take into account the exact law of conservation of full momentum of the system. Bogolyubov [7] and Tyablikov [9] suggested a new method in the strong coupling polaron theory in which the polaron energy was expanded in a power series of the inverse value of the electron-phonon interaction constant α. The law of conservation of total momentum was taken into account exactly, however, the equations for zeroth-order approximation in this method were very complicated and the calculation of the next-order corrections required cumbersome numerical computations.

The regular method of the expansion of large-radius polaron energy into α^{-1} series is developed in present section by means of OM . The law of conservation of the momentum is taken into account exactly and, at the same time, the high-order corrections are calculated analytically. The high-order correction for the polaron effective mass and polaron polarizability in the external electric field are also calculated.

Let us consider the Hamiltonian of the large-radius polaron which was constructed by Fröhlich [1]:

$$\hat{H} = \frac{1}{2}\hat{p}^2 + 2^{5/4}\left(\frac{\pi\alpha}{\Omega}\right)^{1/2}\sum_k \frac{1}{k}\hat{q}_k \exp(i\boldsymbol{kr}) + \frac{1}{2}\sum_k (\hat{q}_k\hat{q}_{-k} + \hat{p}_k\hat{p}_{-k}), \quad (9.1)$$

where $\hat{p}$ is the particle momentum operator; $\hat{q}_k$ and $\hat{p}_k$ are the operators of the phonon coordinate and momentum, respectively; Ω is the normalization volume and the system of units is used where the electron mass m and phonon frequency are equal to unity. The Hamiltonian (9.1) commutates with the operator of the full momentum of the system:

$$\hat{P}_\mu = \hat{p}_\mu - i\sum_k k_\mu \hat{q}_k \hat{p}_k, \quad (9.2)$$

which is the integral of motion. As a consequence, the polaron state vector must be the common eigenvector for the Hamiltonian (9.1) and for the operator with components $\hat{P}_\mu$ ($\mu = 1, 2, 3$). The strong interaction of the particle with the phonon field leads to the appearance of the classical component in the field operators. In order to take into account this circumstance and to fulfil the exact law of conservation let us make a canonical transformation which was first introduced by Bogolyubov [7] and Tyablikov [9]:

$$\boldsymbol{r} = \boldsymbol{R} + \boldsymbol{\rho}; \quad \hat{q}_k = e^{-i\boldsymbol{kR}}(u_k + \hat{Q}_k), \quad (9.3)$$

where the new variable $\boldsymbol{R}$ is introduced in such a way that canonically conjugate operator $(-i\partial/\partial\boldsymbol{R})$ coincides with the operator of total momentum. To leave a total number of coordinates invariable let us put three additional conditions on new operators $\hat{Q}_k$:

$$\sum_k k_\mu v_k \hat{Q}_k = 0. \quad (9.4)$$

One can express the operator $\hat{p}_k$ through new variables using Eq. (9.3):

$$\hat{p}_k = -i\frac{\partial}{\partial q_k} = -i\frac{\partial R_\lambda}{\partial q_k}\left(\frac{\partial}{\partial R_\lambda} - \frac{\partial}{\partial \rho_\lambda}\right) + \sum_f \frac{\partial Q_f}{\partial q_k}\hat{P}_f;$$

$$\hat{P}_f = -i\frac{\partial}{\partial Q_f}; \quad \frac{\partial Q_f}{\partial q_k} = e^{i\boldsymbol{kR}}\delta_{kf} + i\frac{\partial R_\lambda}{\partial q_k} f_\lambda(u_f + \hat{Q}_f). \quad (9.5)$$

The conditions (9.4) lead to the equation for the derivative:

$$-i\frac{\partial R_\lambda}{\partial q_k} = e^{ikR}k_\lambda v_k + i\sum_f f_\lambda f_\mu \hat{Q}_f\frac{\partial R_\lambda}{\partial q_k}. \tag{9.6}$$

Here the function v_k satisfies the condition of orthogonality:

$$\sum_k k_\lambda k_\mu u_k v_k = \delta_{\lambda\mu}. \tag{9.7}$$

In order to realize further calculation at the operator level and not to use the explicit form of the wave function, let us introduce a new operator notation for $\boldsymbol{\rho}$ and $(-i\partial/\partial\boldsymbol{\rho})$:

$$\rho_\lambda = \frac{1}{\sqrt{2\omega}}(a_\lambda + a_\lambda^+); \quad -i\partial/\partial\rho_\lambda = -i\sqrt{\frac{\omega}{2}}(a_\lambda - a_\lambda^+), \tag{9.8}$$

where a_λ and a_λ^+ are the creation and annihilation operators which satisfy the standard commutative conditions:

$$a_\lambda a_\mu^+ - a_\mu^+ a_\lambda = \delta_{\lambda\mu}, \tag{9.9}$$

and the parameter ω will be defined below. The formulae (9.3)–(9.8) express the Hamiltonian of the system through new variables. The main idea of the method considered below consists of expanding of the Hamiltonian in a power series of the operators $a, a^+, \hat{Q}_k, \hat{P}_k$. Then only the quantities of the second order inclusively remain in Hamiltonian and the quadratic quantic is diagonalized exactly. The rest of the Hamiltonian is taken into account by means of a standard perturbation theory.

To construct the above mentioned series, the Eq. (9.6) has to be solved by iteration method with an accuracy up to the first order on the operator $\hat{Q}_k$. As a result, one can find the following expression for the operator $\hat{p}_k$ through new variables:

$$\hat{p}_k = e^{ikR}v_k(k_\lambda - \sum_f k_\mu f_\lambda f_\mu v_f \hat{Q}_f)(\partial/\partial R_\lambda - \partial/\partial\rho_\lambda) + e^{ikR}\hat{P}_k';$$

$$\hat{P}_k' = \hat{P}_k - k_\lambda v_k \sum_f f_\lambda u_f \hat{P}_f; \quad \sum_f u_f \hat{P}_f' = 0. \tag{9.10}$$

The coordinate $\boldsymbol{R}$ enters in $\hat{H}$ only in the form of the canonically conjugate operator $-i\partial/\partial\boldsymbol{R}$ which coincides with the momentum of the system. Therefore the polaron wave function is assumed in the following form:

$$|\Psi\rangle = e^{i\boldsymbol{PR}}|\psi\rangle, \tag{9.11}$$

and the operator $-i\partial/\partial \boldsymbol{R}$ in the Hamiltonian may be changed by constant vector $\boldsymbol{P}$. Let us also note that the operator $\exp(i\boldsymbol{k}\boldsymbol{\rho})$ can be expanded into power series of a_λ and a_λ^+ by means of the following operator identity:

$$e^{i(k_\lambda/\sqrt{2\omega})(a_\lambda+a_\lambda^+)} = e^{-k^2/4\omega} e^{ik_\lambda a_\lambda^+/\sqrt{2\omega}} e^{ik_\lambda a_\lambda/\sqrt{2\omega}}. \tag{9.12}$$

As a result, all the quantities in this series will be presented in a normal form. After simple but unwieldy transformations, the Hamiltonian (9.6) may be written with an accuracy of the second order of the field operators as follows:

$$\hat{H} = \hat{H}_0 + \hat{H}_1,$$

where

$$\begin{aligned}\hat{H}_0 = &\frac{3}{4}\omega + \frac{\omega}{2}(2a_\lambda^+ a_\lambda - a_\lambda^+ a_\lambda^+ - a_\lambda a_\lambda) + \frac{1}{2}\sum_{\boldsymbol{k}}(u_{\boldsymbol{k}} u_{-\boldsymbol{k}} + 2u_{\boldsymbol{k}}\hat{Q}_{-\boldsymbol{k}} + \\ &+\hat{Q}_{\boldsymbol{k}}\hat{Q}_{-\boldsymbol{k}} + \hat{P}'_{\boldsymbol{k}}\hat{P}'_{-\boldsymbol{k}}) + \frac{1}{2}\sum_{\boldsymbol{k}}(k_\lambda k_\mu v_{\boldsymbol{k}} v_{-\boldsymbol{k}} P_\lambda P_\mu + 2k_\lambda v_{\boldsymbol{k}} \hat{P}'_{-\boldsymbol{k}} P_\lambda) + \\ &+2^{5/4}\left(\frac{\pi\alpha}{\Omega}\right)^{1/2}\sum_{\boldsymbol{k}}\frac{1}{k}(u_{\boldsymbol{k}} + \hat{Q}_{\boldsymbol{k}})e^{-k^2/4\omega}\left[1 + i\frac{k_\lambda}{\sqrt{2\omega}}(a_\lambda + a_\lambda^+)\right] - \\ &-2^{5/4}\left(\frac{\pi\alpha}{\Omega}\right)^{1/2}\sum_{\boldsymbol{k}}\frac{1}{k}e^{-k^2/4\omega}\frac{u_{\boldsymbol{k}}}{4\omega}k_\lambda k_\mu(2a_\lambda^+ a_\mu + a_\lambda^+ a_\mu^+ + a_\lambda a_\mu),\end{aligned} \tag{9.13}$$

and the perturbation operator $\hat{H}_1$ is defined by the following expression:

$$\begin{aligned}\hat{H}_1 = &2^{5/4}\left(\frac{\pi\alpha}{\Omega}\right)^{1/2}\sum_{\boldsymbol{k}}\frac{1}{k}e^{-k^2/4\omega}\left\{(u_{\boldsymbol{k}} + \hat{Q}_{\boldsymbol{k}})\left[e^{ik_\lambda a_\lambda^+/\sqrt{2\omega}} e^{ik_\lambda a_\lambda/\sqrt{2\omega}} - \right.\right. \\ &\left.\left. -1 - i\frac{k_\lambda}{\sqrt{2\omega}}(a_\lambda + a_\lambda^+)\right] + \frac{u_{\boldsymbol{k}}}{4\omega}k_\lambda k_\mu(2a_\lambda^+ a_\mu + a_\lambda^+ a_\mu^+ + a_\lambda a_\mu)\right\}.\end{aligned} \tag{9.14}$$

The operators $\sim P_\lambda P_\mu$ are small values when $\alpha \gg 1$ but they should be included in zeroth-order Hamiltonian $\hat{H}_0$ in order to find the effective mass of the polaron. As soon as the polaron binding energy is calculated, the condition $P_\lambda = 0$ can be utilized. Then those terms in Hamiltonian which consist of the operator $\hat{P}'_{\boldsymbol{k}}$ in a linear form become equal to zero if the numbers $v_{\boldsymbol{k}}$ are chosen as:

$$v_{\boldsymbol{k}} = Au_{-\boldsymbol{k}}.$$

As follows from the definition of the operator $\hat{P}'_k$ and from the orthogonality condition (9.10):

$$\sum_k u_k \hat{P}'_k = 0.$$

The value A can be found by means of Eq. (9.7):

$$A \sum_k k_\lambda k_\mu u_k u_{-k} = 1. \tag{9.15}$$

The terms in the Hamiltonian (9.13) which are proportional to $\hat{Q}_k$ become equal to zero under the condition:

$$u_k = -2^{5/4} \left(\frac{\pi\alpha}{\Omega}\right)^{1/2} \frac{1}{k} e^{-k^2/4\omega}. \tag{9.16}$$

Taking into account the condition (9.4), one can see that the terms in formula (9.13) which consist of a_λ and $\hat{Q}_k$ products also become equal to zero if the value u_k is chosen according to formula (9.16). Let us now introduce the following presentation for the operators $\hat{Q}_k$ and $\hat{P}'_k$ through new creation and annihilation operators which are connected by additional conditions:

$$\hat{Q}_k = \frac{1}{\sqrt{2}}(\tilde{b}_k + \tilde{b}^+_{-k}); \quad \hat{P}'_k = \frac{1}{\sqrt{2}}(\tilde{b}^+_k - \tilde{b}_{-k});$$
$$\tilde{b}_k = b_k - k_\lambda v_k \sum_f f_\lambda u_f b_f; \quad [b_k b^+_f] = \delta \boldsymbol{k} \boldsymbol{f}. \tag{9.17}$$

As a result of above mentioned transformations and after the integration in formula (9.13), the Hamiltonian $\hat{H}_0$ is expressed as follows:

$$\hat{H}_0 = \frac{3}{4}\omega + \frac{\omega}{4}(2a^+_\lambda a_\lambda - a^+_\lambda a^+_\lambda - a_\lambda a_\lambda) - \alpha \left(\frac{\omega}{\pi}\right)^{1/2} +$$
$$+ \frac{\alpha}{6}\left(\frac{\omega}{\pi}\right)^{1/2} (2a^+_\lambda a_\mu + a^+_\lambda a^+_\mu + a_\lambda a_\mu) - \frac{3}{2} + \sum_k \tilde{b}^+_k \tilde{b}_k. \tag{9.18}$$

The parameter ω, which was introduced in formula (9.8), is chosen from the condition that non-diagonal terms in the Hamiltonian $\hat{H}_0$ are equal to zero:

$$\omega = \frac{4\alpha^2}{9\pi};$$
$$\hat{H}_0 = -\frac{\alpha^2}{9\pi} - \frac{3}{2} + \frac{4\alpha^2}{9\pi} a^+_\lambda a_\lambda + \sum_k \tilde{b}^+_k \tilde{b}_k, \tag{9.19}$$

and the polaron ground state vector is defined by the following equations:

$$a_\lambda|\psi_0>=b_{\boldsymbol{k}}|\psi_0>=0. \qquad (9.20)$$

The OM enables the calculation of the correction to the energy of polaron ground state. This correction is defined by the operator $\hat{H}_1$ in formula (9.14) and it may be analytically calculated by means of canonical perturbation theory because the spectrum of the excitation of zeroth-order Hamiltonian $\hat{H}_0$ in formula (9.18) is already known. The first-order correction is equal to zero identically, and the second-order correction is defined by the following expression:

$$\Delta E^{(2)}=<\psi_0|\hat{H}_1[E^{(0)}-\hat{H}_0]^{-1}\hat{H}_1|\psi_0>. \qquad (9.21)$$

Let us use the following integral representation for the operator $[E^{(0)}-\hat{H}_0]^{-1}$:

$$(E^{(0)}-\hat{H}_0)^{-1}=-\int_0^\infty dx \exp\left[-x\left(\omega\sum_{\lambda=1}^{3}\hat{n}_\lambda+\sum_{\boldsymbol{k}}\hat{N}_{\boldsymbol{k}}\right)\right], \qquad (9.22)$$

where

$$E^{(0)}=-\frac{\alpha^2}{3\pi}-\frac{3}{2};\quad \hat{n}_\lambda=a_\lambda^+a_\lambda;\quad \hat{N}_{\boldsymbol{k}}=\tilde{b}_{\boldsymbol{k}}^+\tilde{b}_{\boldsymbol{k}}. \qquad (9.23)$$

Here the particle number operators describe the excitations of the system in zeroth approximation. In order to calculate the value (9.21) it is enough to transform the operators in formula (9.21) to the normal form. This operation may be performed on the basis of the following identities:

$$\begin{aligned}&\hat{c}\exp[x(\hat{n}+1)]\hat{c};\quad \hat{n}=\hat{c}^+\hat{c};\quad [\hat{c}\hat{c}^+]=1;\\ &\exp(\beta\hat{c})\exp(x\hat{n})=\exp(x\hat{n})\exp(e^x\beta\hat{c});\quad [\tilde{b}_{\boldsymbol{k}}\tilde{b}_{\boldsymbol{f}}^+]=\delta_{\boldsymbol{k}\boldsymbol{f}}-(\boldsymbol{k}\boldsymbol{f})u_{\boldsymbol{k}}v_{\boldsymbol{f}}.\end{aligned} \qquad (9.24)$$

Thus, the value $\Delta E^{(2)}$ can be written as:

$$\begin{aligned}\Delta E^{(2)}&=\Delta E_1^{(2)}+\Delta E_2^{(2)}+\Delta E_3^{(2)};\\ \Delta E_1^{(2)}&=-\frac{\alpha^2}{3\pi}\left[12-4\pi-12\ln(2+\sqrt{3})+24\ln 2-\frac{1}{4}\right];\\ \Delta E_2^{(2)}&=\alpha\sqrt{\frac{\omega}{\pi}}\{1-\int_0^\infty dx\frac{\exp(-x)}{[1-\exp(-\omega x)]^{1/2}}\};\\ \Delta E_3^{(2)}&=\frac{12}{\alpha\sqrt{\pi\omega}}\int_0^\infty dx\exp[x(\omega-1)]\Big\{[4-\exp(-2x\omega)]^{1/2}-\\ &\quad-\exp(\omega x)\sin^{-1}\left[\frac{1}{2}\exp(-\omega x)\right]\Big\}.\end{aligned} \qquad (9.25)$$

The function $\Delta E^{(2)}$ can be also used in the case when the coupling constant $\alpha \to 0$. Furthermore, the value $\Delta E^{(0)} + \Delta E^{(2)}$ coincides with the perturbation theory result, i.e. $(-\alpha)$ when $\alpha \to 0$. In the limit $\alpha \gg 1$, the formula (9.25) leads to the following expression for the polaron energy in the ground state:

$$E_0 = E^{(0)} + \Delta E^{(2)} = -1.016\frac{\alpha^2}{3\pi} - 2.767. \tag{9.26}$$

This expression may be compared with the formula found by Feynman [11] on the basis of the variational method:

$$E_F = -\frac{\alpha^2}{3\pi} - 2.83, \tag{9.27}$$

and in strong-coupling region $\alpha > 8$, where both formulae may be applied, one can find that our value $E_0 < E_F$.

The operator method allows to find directly physical characteristics of the polaron. In particular, in order to calculate the polaron effective mass we leave in Hamiltonian (9.13) those items which depend on the total momentum, and then it is sufficient to calculate the energy of the ground state with an accuracy of P^2:

$$E(P) \approx E_0 + \frac{P^2}{2m^*}. \tag{9.28}$$

where m^* is an effective mass of the polaron. In zeroth-order approximation ($\alpha \gg 1$) the same expression has been obtained by Bogolyubov [7] and Tyablikov [9]:

$$m^* = \frac{1}{3}\sum_{\boldsymbol{k}} k^2 u_k^2 \tag{9.29}$$

with values u_k defined according to formula (9.16), which results in:

$$m^* = \frac{16\alpha^4}{81\pi^2}. \tag{9.30}$$

In order to find next correction to m^* in the α^{-1} series the fact is used that the excitation energy of polaron changes when the total momentum is not equal to zero. We introduce here two frequencies,which correspond to the cases when particle coordinate and momentum are expressed through the creation and annihilation operators:

$$\rho_{1,2} = \frac{1}{\sqrt{2\omega_\perp}}(a_{1,2}^+ + a_{1,2}); \quad \rho_3 = \frac{1}{\sqrt{2\omega_\parallel}}(a_3^+ + a_3), \tag{9.31}$$

with Z-axis direction coincides with the direction of total momentum.

The procedure for further calculation is the same as described above for calculation of polaron energy. The expressions for frequencies $\omega_\perp$ and $\omega_\parallel$ is found from zeroth-order Hamiltonian in the following form:

$$\omega_\perp = \omega + \frac{2}{3}\epsilon; \quad \omega_\parallel = \omega + \epsilon; \quad \epsilon = -\frac{L(M+3N)}{N(M+N) - 2M^2};$$
$$L = \omega^{3/2}/5\alpha^3\sqrt{\pi}; \quad M = \alpha/20\sqrt{\pi\omega}; \quad N = -1 - \alpha/60\sqrt{\pi\omega}. \tag{9.32}$$

A change of the excitation energy of the polaron leads to appearance of the terms proportional to P^2 in the second-order corrections. Taking into account these corrections, the following expression for the effective mass of the polaron are found:

$$m^* = 0.9984\frac{16\alpha^4}{81\pi^2} + 0.458\alpha^2. \tag{9.33}$$

The OM enables the calculation of the response of quantum systems to external field. As for example, we calculate here the polaron polarizability β in a weak electric field. The polarizability of a system is defined if the excited states are taken into account and that is why the variational calculation of the polarizability is very difficult. In order to calculate the polarizability on the basis of OM, it is sufficient to introduce in Hamiltonian the following operator:

$$\hat{H}_{int} = -e(\boldsymbol{\rho}\boldsymbol{E}), \tag{9.34}$$

which describes the interaction of the polaron and the field $\boldsymbol{E}$, and to use the following representation for the operators connected with the particle:

$$\rho_\lambda = \frac{1}{\sqrt{2\omega_1}}(a_\lambda + a_\lambda^+ + f_\lambda); \quad -i\partial/\partial\rho_\lambda = -i\sqrt{\frac{\omega_1}{2}}(a_\lambda - a_\lambda^+), \tag{9.35}$$

where $\boldsymbol{f}$ is a constant vector. Then the polaron energy and the parameter ω_1 are calculated with an accuracy of E^2. The vector $\boldsymbol{f}$ is defined from the condition that term $\sim \boldsymbol{E}$ in the polaron energy is equal to zero.Finally, the calculation of β is reduced to the finite integrals, and in the limit $\alpha \gg 1$ the following expression is found:

$$\omega_1 = \omega; \quad \beta = \frac{e^2}{2\omega^2}\left[\left(2 - \frac{4}{9\pi} - \lambda_1\right) - \frac{\lambda_2}{\alpha^2}\right];$$
$$\lambda_1 \approx 0.043; \quad \lambda_2 \approx 4.422. \tag{9.36}$$

To conclude, the regular strong coupling method considered in the present section may be used not only in the polaron theory but in any problem where the structure of a physical particle is formed as the result of the strong interaction of a bare particle and the quantum field.

9.2 One-Dimensional Polaron

In previous section, the dependence of polaron energy on the coupling constant has been considered in the limit of strong interaction of particle with field, which results in classic component of the field with amplitude depending on particle wave function. In the present section, the relation of this result to ones obtained for weak coupling by canonic PT and the construction of UAA for polaron energy for entire range of coupling constant are considered for one-dimensional polaron. This case preserves all qualitative features of general polaron problem, however, makes it possible to perform all calculations analytically [13, 18, 19].

The Hamiltonian of optical polaron of a large radius, being an analogue of the operator (9.1), is expressed in the following form [20, 21]:

$$\hat{H} = \frac{\hat{p}^2}{2} + \sum_k \omega_k a_k^+ a_k + \sum_k \left(V_k^* \, a_k^+ e^{-ikx} + V_k \, a_k e^{ikx}\right), \tag{9.37}$$

where ω_k is frequency of phonon with momentum k, for optical polaron the quantity $\omega_k = \omega$ is independent on k and $V_k = 2^{1/4}(\alpha/L)^{1/2}$; $\hat{p} = -i\partial/\partial x$, x are operators of momentum and position of electron; a_k^+, a_k are phonon creation and annihilation operators with momentum k; L is normalization length; α is a constant of electron-phonon interaction. The units are used which imply $\hbar$, ω and electron mass equal unity. The large radius polaron assumes the crystal is a uniform dielectric media, and thus the sum $\sum_k$ is replaced by the integral $L \int dk/2\pi$. The stationary Schrödinger equation for (9.37) is written as:

$$\hat{H}|\Psi\rangle = E|\Psi\rangle. \tag{9.38}$$

First of all, we consider the limit of weak coupling $\alpha \ll 1$ for optical polaron with Hamiltonian (9.37). The perturbation theory over the parameter α is applicable for this case, which is built by preserving a total momentum (9.2) and using the unitary transformation Lee-Low-Pines [22] in the Eq. (9.38):

$$\begin{aligned} \hat{H}' &= U^{-1}\hat{H}U, \\ |\Psi\rangle &= U|\Psi'\rangle, \\ U &= \exp\left[i\,(P - \sum_k k\, a_k^+ a_k)\, x\right], \end{aligned} \tag{9.39}$$

where P is a c-number defining total momentum of the system. After the unitary transformation (9.39), the Eq. (9.38) is written in the following form:

$$(\hat{H}_0 + \hat{H}_1)|\Psi'\rangle = E|\Psi'\rangle,$$

$$\hat{H}_0 = \frac{1}{2}(P - \sum_k k\, a_k^+ a_k)^2 + \sum_k a_k^+ a_k, \tag{9.40}$$

$$\hat{H}_1 = \sum_k \left(V_k^*\, a_k^+ + V_k\, a_k\right),$$

where $\hat{H}_0$ is unperturbed Hamiltonian, and $\hat{H}_1$ is a perturbation operator. The modified Hamiltonian is independent on the particle coordinate, and the state vector $|\Psi\rangle$ is an eigenvector for the operator of total momentum at arbitrary $|\Psi'\rangle$. We use here the canonic form of RSPT [23] for the solution of Schrödinger equation (9.38) in Fock basis, which corresponds to the set of eigenstates of the operator of particle number for phonons $\hat{n}_k = a_k^+ a_k$. The diagonal operator $\hat{H}_0$ is a Hamiltonian of zeroth approximation.

The zeroth approximation for the state vector $|\Psi'\rangle$ is defined as vacuum state $|0\rangle$ of the phonon field, and the state vector $|0\rangle$ is determined from the equation:

$$a_k|0\rangle = 0; \quad \langle 0|a_k^+ = 0. \tag{9.41}$$

Since the state vector of the vacuum of phonon field (9.41) is an eigenvector of the operator of zeroth approximation $\hat{H}_0$, the polaron energy in zeroth approximation equals to:

$$E_0^{(0)} = \langle 0|H_0|0\rangle = \frac{P^2}{2}. \tag{9.42}$$

As follows from the Eq. (9.42), in zeroth approximation of weak coupling the polaron is an electron propagating freely in the crystal and not interacting with the oscillations of crystallographic lattice (phonons). To take into account the interaction of electron with the lattice, the corrections of higher order have to be calculated with $\hat{H}_1$ as a perturbation operator to zeroth approximation Hamiltonian $\hat{H}_0$. The correction of the first order for polaron energy $E_1^{(0)} = \langle 0|H_1|0\rangle$ equals zeroth due to relationship (9.41). All other corrections of odd order equal zero $E_0^{(1)} = E_0^{(3)} = \ldots = E_0^{(n)} = 0$ ($n = 1, 3, 5, \ldots$), too. Therefore, only the corrections of even order are considered hereinafter $E_0^{(n)}$ ($n = 0, 2, 4, 6, \ldots$). For example, the second-order correction equals [21]:

$$E_0^{(2)} = \langle 0|\hat{H}_1 \frac{1}{E_0^{(0)} - \hat{H}_0} \hat{H}_1|0\rangle = $$

$$= \frac{2^{1/2}\alpha}{2\pi} \int_{-\infty}^{\infty} dk_1 \frac{1}{E_0^{(0)} - \frac{1}{2}(P - k_1)^2 - 1}. \tag{9.43}$$

A special diagram technique is established to analytically calculate the successive approximations of RSPT over the operator $\hat{H}_1$, which helps to derive the terms of the expansion over α and to investigate the convergence of the series [18, 19]. The calculus is reduced to the calculation of one-dimensional integrals (9.43) and here we present only the selected results of this technique. The energy of polaron with the accuracy up to the sixth order is:

$$E_0 = E_0^{(0)} + E_0^{(2)} + E_0^{(4)} + \dots, \tag{9.44}$$

where the corrections for the energy are:

$$\begin{aligned} E_0^{(0)}(P) &= \frac{P^2}{2}, \\ E_0^{(2)}(P) &= -\frac{2^{1/2}}{(2-P^2)^{1/2}}\alpha = -\alpha - \frac{P^2}{4}\alpha + o(P^4), \\ E_0^{(4)}(P) &= -\left[\frac{P^2(P^2-4)+6}{(2-P^2)^{3/2}(4-P^2)^{1/2}} - \frac{P^2(P^2-3)+4}{(2-P^2)^2}\right]\alpha^2 \\ &= -\left(\frac{3\sqrt{2}}{4} - 1\right)\alpha^2 + \frac{P^2}{32}(8-5\sqrt{2})\alpha^2 + o(P^4). \end{aligned} \tag{9.45}$$

Using the Eqs. (9.44)–(9.45), the ground state energy of slow polaron ($P \ll 1$) is written as:

$$E_0 \approx \frac{P^2}{2m^*} - \alpha - 0.0607\,\alpha^2 - 0.0084\,\alpha^3 - 0.0015\,\alpha^4 - \dots, \tag{9.46}$$

where the effective mass of electron follows from the equation:

$$\frac{1}{m^*} = \left.\frac{\partial^2 E_0}{\partial P^2}\right|_{P=0}, \tag{9.47}$$

which results in the expansion:

$$\begin{aligned} m^* &= 1 + \frac{\alpha}{2} + \frac{5-2\sqrt{2}}{8\sqrt{2}}\alpha^2 + \\ &+ \frac{128094744 + 90576657\sqrt{2} - 73960128\sqrt{3} - 52297711\sqrt{6}}{576\,(-960449 - 679140\sqrt{2} + 554400\sqrt{3} + 392020\sqrt{6})}\alpha^3 \\ &\simeq 1 + 0.5\,\alpha + 0.1919\,\alpha^2 + 0.0691\,\alpha^3 + \dots . \end{aligned} \tag{9.48}$$

As follows from the Eq. (9.45), there are exist singularities in energy at certain values of total momentum of the system P, when $P = \pm\sqrt{n}$, $n = 2, 4, 6, 8, \dots$. The existence of these singularities is related to the infinite degeneration of the energy levels of unperturbed Hamiltonian H_0. For example, the energy $P^2/2$ in the state $|0\rangle$ and the energy $(P-k)^2/2+1$ in the state $|1_k\rangle$ are equal when $P \geq \sqrt{2}$, $k = P \pm (P^2-2)^{1/2}$. Therefore, the secular equation has to be resolved for the case $P \approx \sqrt{2}$, and the perturbation theory above is only applicable for the case $P \ll \sqrt{2}$ (slow polaron). The investigation of the convergence of PT series [18] based on the corrections for energy $E_0(\alpha) \equiv E_0(\alpha)|_{P=0}$ of the ground state of polaron, concludes that the series (9.46) has a limited radius of convergence over the coupling constant $\alpha < \alpha_c \sim 4-5$. This fact points to degeneration of the solution of Eq. (9.38) along with the solution corresponding to the strong coupling case $\alpha \gg 1$, presented in Sect. 9.1. To obtain UAA for the function $E_0(\alpha)$ for the entire range of coupling constant, the region near degeneracy point $\alpha \approx \alpha_c$ has to be investigated out of PT scope. For this purpose, the Eq. (9.38) is considered in a strong coupling regime on the basis of adiabatic PT [7,9] (Sect. 9.1) applied to one-dimensional case. Neglecting the dependence on the total momentum, the solution is easily obtained by using the unitary transformation of Hamiltonian (9.37), which distinguishes a classic component u_k of phonon field typical for strong coupling case:

$$\hat{H}' = S^{-1}\hat{H}S,$$
$$|\Psi\rangle = S|\Psi'\rangle,$$
$$S = \exp\left[\sum_k (u_k a_k^+ - \frac{1}{2}u_k^* u_k)\right], \tag{9.49}$$

where u_k is an arbitrary complex number determined later from the condition of energy minimum for polaron. After the transformation (9.49), the Hamiltonian (9.37) and Schrödinger equation are written as:

$$\hat{H}' = \frac{\hat{p}^2}{2} + v(x) + \sum_k a_k^+ a_k + \sum_k \left(u_k^* a_k + u_k a_k^+\right) + $$
$$+ \sum_k V_k \left(e^{i\,kx} a_k + e^{-i\,kx} a_k^+\right);$$
$$H'|\Psi'\rangle = E'|\Psi'\rangle, \tag{9.50}$$

where the potential $v(x)$ and eigenvalues of energy E' are:

$$v(x) = \sum_k V_k \left(u_k e^{i\,kx} + u_k^* e^{-i\,kx}\right),$$
$$E = E' + \sum_k u_k^* u_k. \tag{9.51}$$

We select the trial state vector for the Eq. (9.49) in the form:

$$|\Psi'\rangle = \psi(x)|0\rangle, \tag{9.52}$$

where $\psi(x)$ is a normalized function, and $|0\rangle$ is a vacuum state of the phonon field, which is defined from the equation $a_k|0\rangle = 0$. By varying E over the parameters u_k and function ψ with normalization condition:

$$\int_{-\infty}^{\infty} \mathrm{d}x\ |\psi(x)|^2 = 1, \tag{9.53}$$

the following system of equations is derived for ψ and u_k:

$$\left[\frac{p^2}{2} + v(x)\right]\psi(x) = E'\psi(x),$$
$$u_k = -V_k \int_{-\infty}^{\infty} \mathrm{d}x\ e^{-i\,kx}|\psi(x)|^2. \tag{9.54}$$

Using the definition of the potential in (9.51), the system of linear equations (9.54) is transformed to non-linear equation for the function $\psi(x)$:

$$\frac{\hat{p}^2}{2}\psi(x) - 2^{3/2}\,\alpha\psi(x)^3 = E'\psi(x). \tag{9.55}$$

Equation (9.55) has an exact solution corresponding to coupled state with energy E_0':

$$\psi_0(x) = \sqrt{\frac{\alpha}{2^{1/2}}}\frac{1}{\mathrm{ch}(2^{1/2}\alpha\, x)},$$
$$E_0' = -\alpha^2. \tag{9.56}$$

At fixed potential, the Schrödinger equation for the excited states of continuous spectrum corresponding to momentum p is written as:

$$\frac{p^2}{2}\psi(x) - 2^{3/2}\,\alpha\psi_0(x)^2\,\psi(x) = E'\psi(x) \tag{9.57}$$

and has exact solutions E_p':

$$\psi_p(x) = \frac{1}{\sqrt{L(p^2+2\alpha^2)}}e^{i\,px}\left[p + 2^{1/2}i\,\alpha\mathrm{th}(2^{1/2}x)\right],$$
$$E_p' = \frac{p^2}{2}, \tag{9.58}$$

where $0 \le p < +\infty$. This spectrum describes Frank-Condon excited states of polaron, using the terminology of molecules theory [24] in adiabatic approximation.

For normalized state vector Ψ and energy of ground state E_0, we obtain the following exact analytical expressions in the limit of strong coupling:

$$|\Psi_0\rangle = \psi_0(x) \exp\left[\sum_k (u_k a_k^+ - \frac{1}{2} u_k^* u_k)\right] |0\rangle,$$

$$E_0 = -\frac{\alpha^2}{3}, \qquad \alpha \longrightarrow \infty. \tag{9.59}$$

The value of the energy in ground state in strong coupling limit calculated in Feynman model of polaron with one-dimensional integral equals:

$$E_0^F = -\frac{\alpha^2}{\pi}, \tag{9.60}$$

with an accuracy $\approx 4.5\%$ in comparison to exact value (9.59).

The formal intersection point for the solutions for weak (9.46) and strong (9.59) coupling corresponds to the value $\alpha_0 \approx 3.68$, which correlates to above estimate for the radius of convergence of PT series. To perform more detailed analysis, the structure of the excited states near the ground one has to be closely considered.

There is a classification of the excited states of polaron system [25]: (i) scattering states, one or more real phonons are excited; (ii) the states, characterizing the internal excitations of electron subsystem of polaron [26], and there is no excitation of real phonons (RES states); (iii) Frank-Condon states [26], characterizing the polaron states in fixed potential corresponding to the ground state of polaron. The second and third types of excitations are the resonances of metastable polaron states. There are only exist the experimental observations of scattering states for free polarons [25].

The internal excited states play important role in bipolarons [25]. Frank-Condon states have a limited life time and turn into RES states, which are also metastable but with longer life time in case of intermediate and strong electron-phonon coupling. In the region of weak coupling, the RES states are unstable because of the mutual ground state of phonon and polaron is energetically expedient $E_{RES} > E_0 + \hbar\omega$ [26]. The operator calculations show that there exists a RES state $E_{RES} < E_0 + \hbar\omega$, which causes the quasi-crossing of the excited state with ground state near the point $\alpha \sim 5$ [13]. To build a first excited state, the following trial vector is chosen:

$$|\Psi_1\rangle = \int_{-\infty}^{\infty} \mathrm{d}k\, C_k\, a_k^+ \,|0\rangle, \tag{9.61}$$

where C_k is a trial function of phonon momentum k normalized to unity:

$$\int_{-\infty}^{\infty} \mathrm{d}k\, |C_k|^2 = 1. \tag{9.62}$$

The energy of zeroth approximation for the first excited state of one-dimensional optical polaron described by Hamiltonian (9.37) is then defined by the expression:

$$E_1^{(0)}\left[C_k\right] = \int_{-\infty}^{\infty} \mathrm{d}k\, C_k^2 + \int_{-\infty}^{\infty} \mathrm{d}k\, \frac{k^2}{2}\, C_k^2. \tag{9.63}$$

When considering the expression (9.63) as functional $E_1^{(0)}[C_k]$, and accounting the normalization condition (9.62), the extremum for energy $E_1^{(0)} = 1$ (equivalent to the condition of independence of eigenvalues on the selected basis) is realized at $C_k = f(k,\delta)|_{\delta\to 0}$, where $f(k,\delta)$ is normalized to unity function, which tends to delta function in the limit $\delta \to 0$. This situation conforms to the polaron in ground state and phonon with momentum $k = 0$, which is the nearest excited scattering state to the ground one.

With the increase of the coupling constant, the energy of this excited state is defined as follows (including the RSPT correction of the second order α):

$$E_1\left[C_k\right] = \int_{-\infty}^{\infty} \mathrm{d}k\, C_k^2 + \int_{-\infty}^{\infty} \mathrm{d}k\, \frac{k^2}{2}\, C_k^2 + \frac{2^{1/2}\alpha}{2\pi} \times$$
$$\times \int_{-\infty}^{\infty} \mathrm{d}k_1 \int_{-\infty}^{\infty} \mathrm{d}k_2\, C_{k_1} C_{k_2} \left(\frac{1}{E_1^{(0)} - \frac{1}{2}\left(k_1 + k_2\right)^2} - \frac{1}{E_1^{(0)}} \right). \tag{9.64}$$

The coefficients C_k are determined from the extremum of functional $E_1[C_k]$, with the normalization condition (9.62), and the following integral equation for the function C_k is obtained:

$$\left(1 + \frac{k^2}{2} - E_1^{(0)}\right) C_k + \frac{2^{1/2}\alpha}{2\pi} \int_{-\infty}^{\infty} \mathrm{d}k_1 C_{k_1}$$
$$\times \left(\frac{1}{E_1^{(0)} - \frac{1}{2}\left(k + k_1\right)^2} - \frac{1}{E_1^{(0)}} \right) = 0. \tag{9.65}$$

Using the Fourier transformation $C_k = \int_{-\infty}^{\infty} \mathrm{d}k\, \Psi(x)\, e^{-i\,k\,x}$ in the Eq. (9.65), we arrive at:

$$-\frac{1}{2}\frac{d^2\Psi(x)}{dx^2} + \sqrt{2}\alpha\, \delta(x)\, \Psi(x) - \alpha\, e^{-\sqrt{2}|x|}\Psi(x) = -\epsilon\, \Psi(x), \tag{9.66}$$

where $\epsilon = 1 - \alpha - E_1$ is the difference between the energy of first scattering state $E_1^{sc} = 1 - \alpha$ and the energy of the first excited state of the system, calculated in the second order of perturbation theory. Equation (9.66) after the substitution $z = -\sqrt{2}\, e^{-\frac{x}{\sqrt{2}}}$ becomes a standard equation for Bessel function. The solution of the Eq. (9.66), satisfying the condition of the quadratic integrity of wave function has a form:

$$\Psi(x) = N\, J_{2\sqrt{\epsilon}}\left(2\sqrt{\alpha}\, e^{-\sqrt{2}|x|}\right), \tag{9.67}$$

where N is normalization factor and $\epsilon > 0$. The eigenvalues are found from the continuity of the first derivative:

$$\Psi'(0-) - \Psi'(0+) = 2\sqrt{2}\alpha\, \Psi(0), \tag{9.68}$$

which experiences the discontinuity in zero point due to delta-shaped potential. Using the explicit form of the wave function (9.67), the continuity condition is reduced to the equation:

$$J_{2\sqrt{\epsilon}+1}(2\sqrt{\alpha}) - J_{2\sqrt{\epsilon}-1}(2\sqrt{\alpha}) = 2\sqrt{\alpha}\, J_{2\sqrt{\epsilon}}(2\sqrt{\alpha}), \tag{9.69}$$

which has a solution at $\alpha > 0$. Table 9.1 shows the calculated values of energy of the first excited state $E_1 = 1 - \alpha - \epsilon$ within the interval $0 < \alpha \leq 6$. For all values of the coupling constant $\alpha > 0$, the energy $E_1 < E_1^{sc}$, that points out to the stability of the constructed excited polaron state $|\Psi_1\rangle$ and prevent a system from the decay into phonon and polaron in ground state. Thus, the state with all phonon modes excited and with the distribution of the probability density $|C_k|^2$ is more stable than the state with a single excited zeroth mode. Table 9.1 contains also the energies E_1 for the first RES state obtained in [26]. Nevertheless, the state $|\Psi_1\rangle$ is metastable,

Table 9.1 The energies E_0, E_1 and E_1^{sc}, calculated by the second-order perturbation theory

α	E_0	E_1	E_1^{sc}	E_1 from [26]	$E_0 - E_1$
0.1	-0.1	0.89998	0.9	0.93333	0.99998
0.5	-0.5	0.49544	0.5	0.66667	0.99544
0.8	-0.8	0.18307	0.2	0.46666	0.98307
1.0	-1.0	−0.02999	0.0	0.33333	0.97001
1.5	-1.5	−0.57831	−0.5	0.00000	0.92169
2.5	-2.5	−1.73094	−1.5	−0.66667	0.76905
3.0	-3.0	−2.32981	−2.0	−1.00000	0.67018
3.5	-3.5	−2.94103	−2.5	−1.33333	0.55897
4.0	-4.0	−3.56308	−1.0	−1.66667	0.43691
4.5	-4.5	−4.19479	−3.5	−2.00000	0.30521
5.0	-5.0	−4.83516	−4.0	−2.33333	0.16484
5.557	-5.557	−5.55771	−4.557	−2.70467	−0.00072
6.0	-6.0	−6.13870	−5.0	−3.00000	−0.13870

Fig. 9.1 The normalized to unity function C_k for $\alpha = 2.5$ (*thick line*) and $\alpha = 1.0$ (*thin line*)

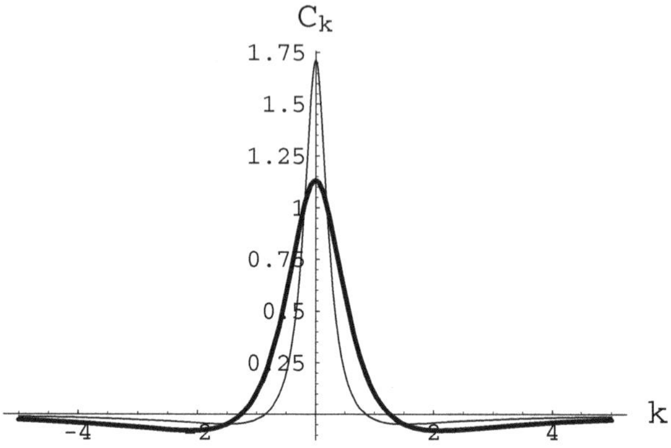

because of the transition into ground state $|\Psi_1\rangle \to |\Psi_0\rangle$ is possible due to non-zero matrix element of the interaction operator. The state $|\Psi_1\rangle$ can be interpreted as coupled with electron one-phonon state, because of the function C_k is quadratically integrable (see Fig. 9.1), that distinguishes qualitatively this state from the ground and scattering states.

Figure 9.1 demonstrates that the function C_k tends to delta function in zero point at $\alpha \to 0$, because of the case $\alpha = 0$ corresponds to free electron with momentum $p = 0$ and phonon with momentum $k = 0$ and energy $E_1^{sc} = 1$. Table 9.1 shows that the energy $E_1 \to 1$ at $\alpha \to 0$, and energies E_1 and E_0 coincide in $\alpha = 5.557$, i.e. there exists an intersection of energy levels of the first excited RES state with the ground state in the first-order approximation. This intersection is, however, a quasi-crossing because of the matrix element of interaction operator for the transition between these two states is non-zero [27], and the linear combination of the states $|\Psi_0\rangle$ and $|\Psi_1\rangle$ can be constructed in such a way that the energy levels $|\Psi_\pm\rangle = C_0^\pm|\Psi_0\rangle \pm C_1^\pm|\Psi_1\rangle$ are close each to other near $\alpha \sim 5.557$ but not really intersect.

Thus, the RES energy in the second order of PT has an intersection with ground state near the point $\alpha \sim 5.557$, that correlates well with the existence of singularity in $E_0(\alpha)$ near the real axis. The value of the convergence radius of PT series obtained from the asymptotic expansion (9.45) is located between $\alpha_c \sim 5.07$ and $\alpha_c \sim 5.57$ [18]. To clarify the question of the quasi-intersection between the ground and excited states of polaron energy, the many-phonon states $C_{k_1,k_2,\ldots}$ have to be taken into account, which may also have the intersections with ground and first RES one-phonon states. This kind of OM analysis requires the choice of more general than (9.61) form of variational function for excited state, which is considered in next section for three-dimensional and one-dimensional polarons.

9.3 UAA for Three-Dimensional Polaron Energy

In this section, we make a re-use of the Fröhlich Hamiltonian for three-dimensional optical polaron of large radius (9.1), represented through the operators of creation and annihilation of phonons:

$$\hat{H} = -\frac{1}{2}\Delta + \sum_k a_k^+ a_k + 2^{3/4}\left(\frac{\pi\alpha}{\Omega}\right)^{1/2} \sum_k \frac{1}{k}\left(e^{ikr} a_k + e^{-ikr} a_k^+\right). \tag{9.70}$$

In general case, to reconstruct the spectrum of polaron states the solution of Schrödinger equation has to be found:

$$\hat{H}|\Psi_P\rangle = E(\alpha, \mathbf{P})|\Psi_P\rangle, \tag{9.71}$$

as well as the equation, corresponding to the conservation law for total momentum and following from the translational invariance:

$$\hat{\mathbf{P}}|\Psi_P\rangle = \Big(-i\nabla + \sum_k \mathbf{k} a_k^+ a_k\Big)|\Psi_P\rangle = \mathbf{P}|\Psi_P\rangle. \tag{9.72}$$

As has been illustrated in previous section for one-dimensional polaron, there are two forms of variational solution of these equations corresponding to asymptotic limits of strong and weak coupling. These states can be used as basics for the construction of UAA for state vector of system in the intermediate range. For three-dimensional polaron, the limit of weak coupling $\alpha \ll 1$ corresponds to the state vector found by the unitary transformation operator Lee-Low-Pines [22] and initial Hamiltonian (9.71):

$$\begin{aligned} \hat{H}_P &= \hat{R}\hat{H}\hat{R}^{-1}; \\ |\Psi_P\rangle &= \hat{R}|\Psi\rangle; \\ \hat{R} &= \exp\left[i(\mathbf{P} - \sum_k \mathbf{k} a_k^+ a_k)\mathbf{r}\right]. \end{aligned} \tag{9.73}$$

The operator $\hat{R}$ is the projection operator, which provides the state vector satisfying the Eq. (9.73) at arbitrary vector $|\Psi\rangle$. To take into account the influence of many-phonon transitions, the state vector $|\Psi\rangle$ is chosen in a general trial form, which admits the separation of classic component, in opposite to (9.63):

$$|\Psi\rangle = \frac{1}{\sqrt{\Omega}} e^{\sum_k [v_k(P) a_k^+ - v_k^*(P) a_k]}|0\rangle;$$

$$|\Psi_P^{(D)}(\mathbf{r})\rangle = \hat{R}|\Psi\rangle = \frac{1}{\sqrt{\Omega}} e^{\left(i\mathbf{Pr}+\sum_k\left(e^{-ikr}v_k^*(\mathbf{P})a_k^+ - v_k(\mathbf{P})e^{ikr}a_k\right)\right)}|0\rangle;$$
$$a_k|0\rangle = 0. \tag{9.74}$$

This polaron state vector defines the delocalized state (D), and the wave function of electron corresponds to continuous spectrum and the electron density:

$$\rho(\boldsymbol{r}) = | < \boldsymbol{r}|\Psi_P^{(D)} > |^2$$

and is independent on coordinate. The state vector (9.74) is an exact eigenfunction of the total momentum operator, and to make it also the best approximation for the solution of Schrödinger equation (9.71), the fittable parameters $v_k(\mathbf{P})$ have to be chosen from the minimum of energy calculated as an average from the total Hamiltonian (9.70) over this state. The variation of the average over $v_k(\mathbf{P})$ leads to the following result [17]:

$$v_k(\mathbf{P}) = -2^{7/4}\left(\frac{\pi\alpha}{\Omega}\right)^{1/2}\frac{1}{k[k^2 - 2(\mathbf{P}-\mathbf{Q})\mathbf{k} + 2]},$$
$$\mathbf{Q} = \sum_k \mathbf{k} v_{\mathbf{k}}^2(\mathbf{P}). \tag{9.75}$$

For simplicity sake, all further calculations are presented for the polaron settle on standing $\mathbf{P} = 0$, when the coupling energy of polaron is defined by formula [3]:

$$E_D(\alpha) = H_{DD} = \int d\mathbf{r}\langle\Psi_0^{(D)}(\mathbf{r})|\hat{H}|\Psi_0^{(D)}(\mathbf{r})\rangle = -\alpha. \tag{9.76}$$

The second basis state is defined by the variational wave function, which has been introduced for the first time in [6] and supposes the hypothesis about trapping of the slowly moving electron in ionic crystal into potential well created due to polarization of ionic lattice by the same electron. As shown in Sect. 9.1 and [7, 9], the method exists which provides the translational invariance of this state, and the state is an asymptotically exact solution of the Schrödinger equation in the limit $\alpha \to \infty$. According to [3], the translational invariant state vector describing self-localized polaron is written as:

$$|\Psi(\mathbf{r},\mathbf{R})\rangle = \phi(\mathbf{r}-\mathbf{R})e^{\left(\sum_k\left(e^{-ik\mathbf{R}}u_k^*a_k^+ - e^{ik\mathbf{R}}u_k a_k\right)\right)}|0\rangle;$$
$$\int d\mathbf{r}|\phi(\mathbf{r})|^2 = 1. \tag{9.77}$$

Here $\mathbf{R}$ is an arbitrary point in space, in the vicinity of this point the electron with the wave function $\phi(\mathbf{r}-\mathbf{R})$ is localized. The classic component of the phonon field u_k is determined from the minimum of energy and is [6]:

$$u_k = -2^{3/4}\left(\frac{\pi\alpha}{\Omega}\right)^{1/2}\frac{1}{k}\int d\mathbf{r}|\phi(\mathbf{r})|^2 e^{-ikr}. \tag{9.78}$$

In turn, the wave function of electron in this state satisfies the following non-linear equation with eigenvalue E_P, derived by Pekar:

$$\left[-\frac{1}{2}\Delta - 2^{1/2}\alpha\int d\mathbf{r}_1\frac{|\phi(\mathbf{r}_1)|^2}{|\mathbf{r}_1-\mathbf{r}|}+\right.$$
$$\left.+\frac{\alpha}{\sqrt{2}}\int d\mathbf{r}_1\int d\mathbf{r}_2\frac{|\phi(\mathbf{r}_1)|^2|\phi(\mathbf{r}_2)|^2}{|\mathbf{r}_1-\mathbf{r}_2|} - E_P\right]\phi(\mathbf{r}) = 0. \tag{9.79}$$

The exact solution of this equation is found numerically (see, for example, [28] and citations therein). However, for analytical simulations the wave function $\phi_0(\mathbf{r})$ can be used, which has been introduced by Pekar as a good approximation [3]:

$$\phi_0(\mathbf{r}) = \sqrt{\frac{8}{14+42c+45c^2}}(1+br+cb^2r^2)e^{-br}, \tag{9.80}$$

with optimal variational parameters

$$b = 0.931307\alpha, \quad c = 0.451668,$$

at which the coupling energy of polaron is:

$$E_P = -0.108504\alpha^2,$$

that is close to exact result $-0.108513\alpha^2$.

The collection of the state vectors (9.80) corresponding to various $\boldsymbol{R}$ describes the localized states (L) of electron. Each of these states is related to local violation of translational symmetry in the sense discussed in [29]. However, due to this symmetry the energy of the system is independent on localization point $\mathbf{R}$, that is secured by the selection of correct linear combination of degenerated states $|\Psi(\mathbf{r},\mathbf{R})\rangle$, being the eigenvector for the operator of total momentum:

$$|\Psi_P^{(L)}(\mathbf{r})\rangle = \frac{1}{N_P\sqrt{\Omega}}\int d\mathbf{R}e^{i\boldsymbol{PR}}|\Psi(\mathbf{r},\mathbf{R})\rangle, \tag{9.81}$$

with normalization constant:

$$N_P^2 = \int d\mathbf{R}\int d\mathbf{r}\phi(\mathbf{r})\phi(\mathbf{r}+\mathbf{R})\exp\left[\sum_k u_k^2(e^{-i\boldsymbol{kR}}-1)\right].$$

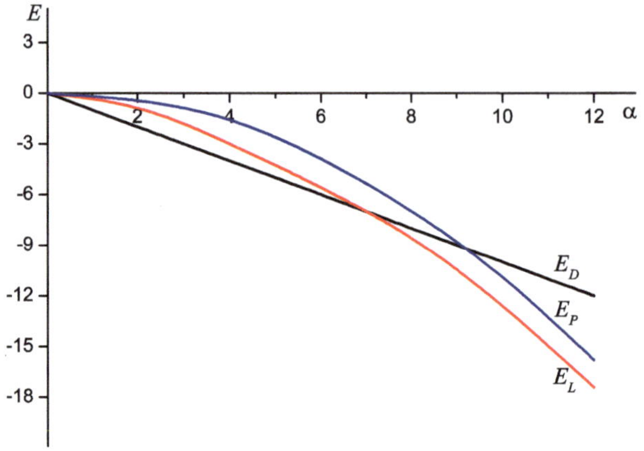

Fig. 9.2 The polaron energies calculated by Pekar (E_P), in delocalized (E_D) and localized (E_L) states

The state vector (9.81) combined with the expression (9.80) results in the following estimate for coupling energy of localized polaron ar arbitrary α:

$$E_L(\alpha) = H_{LL} = \int d\mathbf{r}\, \langle \Psi_0^{(L)}(\mathbf{r})|\hat{H}|\Psi_0^{(L)}(\mathbf{r})\rangle. \tag{9.82}$$

Figure 9.2 compares the functions $E_D(\alpha)$ and $E_L(\alpha)$ with the values $E_P(\alpha)$ computed on the basis of state vector (9.77), which does not account the translational symmetry of the system.

As follows from Fig. 9.2, there is no restriction on coupling constant α for both continuous functions $E_D(\alpha)$ and $E_L(\alpha)$, describing two qualitatively different states. However, there is an intersection of these energy terms near the point $\alpha \simeq \alpha_c \simeq 7$, that corresponds to the degeneration of states. A similar degeneration is known in the electronic band theory in crystals as the intersection of resonant bands (see, for example, [30]). In opposite to real intersection of energetic levels with different symmetries discussed in Sect. 3.5 for two-level system, here we observe a quasi-intersection caused by approximate nature of the mathematical description of these states. This degeneracy is eliminated if a proper linear combination of both states is chosen for variational wave function [27]. In accordance with OM principles, this combination is modeled by ansatz built on the basis of functions (9.74) and (9.81):

$$|\Phi_P(\mathbf{r})\rangle = C_1|\Psi_P^{(D)}(\mathbf{r})\rangle + C_2|\Psi_P^{(L)}(\mathbf{r})\rangle. \tag{9.83}$$

This ansatz uses only linear variational parameters $C_{1,2}$, whereas the parameters for basis states are fixed. In typical variational calculus of polaron energy for intermediate values of coupling constant, the complex variational functions have been used being connected to either localized or delocalized polaron states independently, and this fact results in the fictitious phase transition in the point α_c [3]. The simple trial function (9.83), which is not reduced to any of these states and provides the continuity of energetic terms in the point α_c, has been proposed in [16, 17].

As prescribed by OM algorithm, and due to the fact that both basis vectors are exact eigenfunctions of the operator of total momentum, the variational function (9.83) is used for Schrödinger equation (9.71). The condition for existence of non-trivial solution for coefficients $C_{1,2}$ results in the following formula for eigenvalues for adiabatic terms of polaron:

$$\begin{aligned} E_{\pm}(\alpha) &= \frac{1}{2(1-S^2)}\left(H_{DD}+H_{LL}-2SH_{LD}\pm E_R\right) \\ E_R &= [(H_{DD}+H_{LL}-2SH_{LD})^2 - \\ &\quad 4(1-S^2)(H_{DD}H_{LL}-H_{LD}H_{LD})]^{1/2}. \end{aligned} \tag{9.84}$$

Here H_{DL}, H_{DD} and H_{LL} are matrix elements of total Hamiltonian (9.1) between the basis states, which are generally not orthogonal, and $S \neq 0$ is an overlapping integral. We skip all intermediate calculations and present the resulting analytical formulas at $\mathbf{P} = 0$. The simplest expression is:

$$H_{DD} = -\alpha, \tag{9.85}$$

and the energy of localized state is more cumbersome:

$$\begin{aligned} H_{LL} = \frac{1}{N^2}\int d\mathbf{R}\int d\mathbf{r}\,\phi_0(\mathbf{r}+\mathbf{R})\exp\left(\sum_k u_k^2 e^{-ikR}\right) \\ \left\{-\frac{1}{2}\Delta + \sum_k u_k^2 e^{-ikR} + \right. \\ \left. 2^{3/4}\left(\frac{\pi\alpha}{\Omega}\right)^{1/2}\sum_k \frac{u_k}{k}\left(e^{ikr}+e^{-ik(r+R)}\right)\right\}\phi_0(\mathbf{r}) \end{aligned} \tag{9.86}$$

where the normalization constant is:

$$N^2 = \int d\mathbf{R}\int d\mathbf{r}\,\phi_0(\mathbf{r})\phi_0(\mathbf{r}+\mathbf{R})\exp\left(\sum_k u_k^2 e^{-ikR}\right).$$

The matrix element of transition between basis states is written:

$$H_{LD} = \frac{e^{-\alpha/4}}{N}\int d\mathbf{r}\,\phi_0(\mathbf{r})\left\{-\alpha + \frac{1}{2}\left(\sum_k \mathbf{k}u_k v_k\right)^2\right\}\exp\left(\sum_k u_k v_k e^{ikr}\right). \tag{9.87}$$

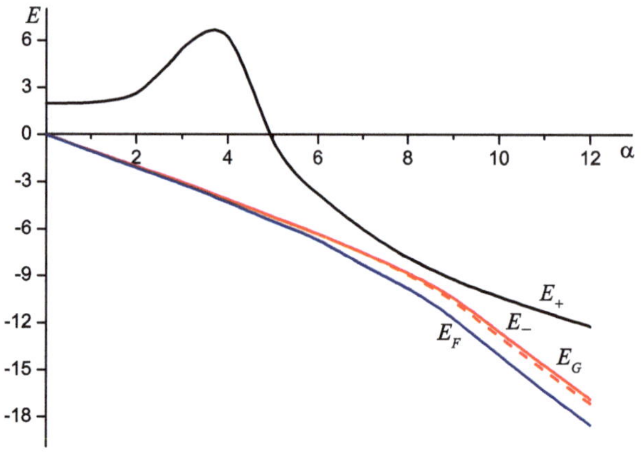

Fig. 9.3 Energetic terms of polaron: after Feynman (E_F) and on the basis of formula (9.84) ($E_{\pm}$); E_G is an analogous of E_- at Gaussian approximation for wave function of electron

Finally, the overlapping integral of vectors is:

$$S = \frac{e^{-\alpha/4}}{N} \int d\mathbf{r}\, \phi_0(\mathbf{r}) \exp\left(\sum_k u_k v_k e^{ikr}\right). \tag{9.88}$$

The formulas (9.84)–(9.88) calculate the adiabatic terms of polaron $E_{\pm}(\alpha)$ in the entire range of coupling constant, that is demonstrated in Fig. 9.3. The estimate after Feynman $E_F(\alpha)$ [11] is depicted in figure, too. The functions E_- and E_F are almost coinciding continuous function, and E_- is located below E_F in asymptotic region $\alpha \gg 1$, which is not shown in figure. This is due to more accurate choice of trial function (9.77) comparing to [11], where the path integral for harmonic oscillator is used for description of localized state. The magnitude of term E_- depends weakly on the form of electron wave function $\phi_0(\mathbf{r})$, as follows from the curve E_G corresponding to Gaussian approximation for this function.

The visible difference between adiabatic term after OM and Feynman approximation in the domain of intermediate α is due to more accurate accounting of the non-dependent on α contribution to the coupling energy of polaron in functional approach. As proved in Sect. 9.1, the second-order OM approximation delivers more precise results and the term becomes $E_- < E_F$. This statement is illustrated in Fig. 9.4, where the results for one-dimensional polaron with Hamiltonian (9.37) are presented. The role of constant term is suppressed here and OM calculates polaron energy more accurately than method of functional integration.

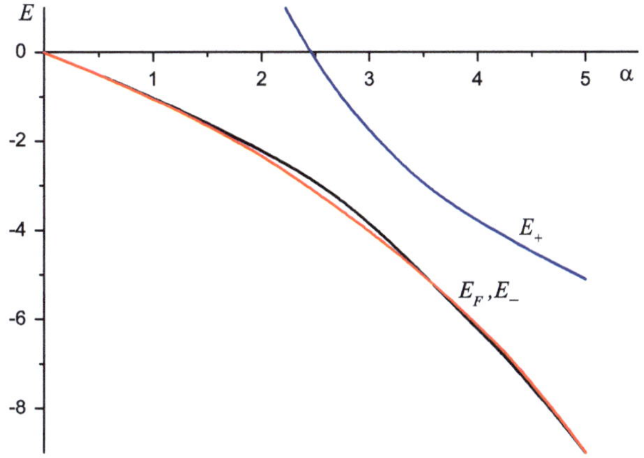

Fig. 9.4 The same functions as in Fig. 9.3, for one-dimensional polaron

9.4 Particle-Field Interaction Model with a Divergent Perturbation Theory

The polaron problem is an unique physical model in quantum field theory due to the fact that all PT diagrams, including energetic ones, result in convergent integrals and therefore the PT series is not requiring additional mass re-normalization [19]. However, in real models used in quantum field theory, the diagrams of standard PT tend to infinite values. The integrals diverge both in the region of large momenta of intermediate states (ultraviolet divergence) and in the region of small momenta (infrared divergence) [31, 32]. Here we discuss the results of the application of operator method to localized states in the model of interaction of particle with quantum field, which is more general than polaron. The Hamiltonian describing the interaction of non-relativistic particle with scalar quantum field is used for this purpose [17]:

$$\hat{H} = -\frac{1}{2}\Delta + \sum_{k} \omega_k a_k^+ a_k + \\ + \frac{g}{\sqrt{\Omega}} \sum_{k} \frac{1}{\sqrt{2\omega_k}} \left(e^{i\boldsymbol{kr}} a_k + e^{-i\boldsymbol{kr}} a_k^+ \right), \ \omega_k = k = |\mathbf{k}|, \tag{9.89}$$

where g is a coupling constant for the particle and the field. An analogous operator describes the interaction of electron with the field of acoustic phonons in the framework of continuous crystal model [30].

The self-energy diagram corresponding to the second-order PT (9.43), being integrated over three-dimensional space of vectors $\boldsymbol{k}$ tends to infinity because of the absent multiplier $1/k$ in the interaction operator. As a result, there is no presence of delocalized state of particle equivalent to (9.74). However, the localized state similar to (9.81) is possible, if the OM approach is used instead of PT. We use here the following ansatz for the vector of localized state:

$$|\Psi(\mathbf{r},\mathbf{R})\rangle = \phi(\mathbf{r}-\mathbf{R})\exp\left(\sum_k \left(u_k^* e^{-ikR} a_k^+ - \frac{1}{2}u_k^2\right)\right)|0\rangle. \tag{9.90}$$

In accordance with OM procedure, the parameters of the classic component of field u_k and the wave function of electron have to be found from the extremum of the average of total Hamiltonian (9.89) over the state vector, i.e. nullification of the following variational derivatives:

$$\frac{\delta}{\delta u_k}[\langle\Psi(\mathbf{r},\mathbf{R})|\hat{H}|\Psi(\mathbf{r},\mathbf{R})\rangle] = \frac{\delta}{\delta\phi(\mathbf{r}-\mathbf{R})}[\langle\Psi(\mathbf{r},\mathbf{R})|\hat{H}|\Psi(\mathbf{r},\mathbf{R})\rangle] = 0. \tag{9.91}$$

From this expression, the classic component of the field equals:

$$u_k = -\frac{g}{\sqrt{2\Omega\omega_k^3}}\int d\mathbf{r}\phi^2(\mathbf{r})e^{-ik\mathbf{r}}, \tag{9.92}$$

and wave function $\phi(\mathbf{r})$ satisfies to non-linear equation analogous to (9.79), which corresponds to optical polaron. The total energy of the state is found from (9.91) and (9.92):

$$E_0 = \int d\mathbf{r}\phi(\mathbf{r})\left(-\frac{1}{2}\Delta\right)\phi(\mathbf{r}) - \sum_k \omega_k u_k^2, \tag{9.93}$$

and does not undergo an ultraviolet divergence due to particle form-factor conditioned by the wave function of localized state. The value (9.93) is always finite at arbitrary value of the coupling constant g. At the same time, the infrared divergence remains, which causes the state vector (9.90) to contain an infinite expression $\sim \sum_k u_k^2$, which defines the average number of the excited quanta of field and implements the integral diverging at $k \to 0$. The situation changes drastically if the translational symmetry is restored in the state vector (9.92) in a way similar to formula (9.81):

$$|\Psi_P^{(L)}(\mathbf{r}) >= \frac{1}{N_P\sqrt{\Omega}}\int d\mathbf{R}\,\phi(\mathbf{r}-\mathbf{R})\exp(i\mathbf{PR} +$$
$$\sum_k (u_k e^{-ikR} - \frac{1}{2}u_k^2))|0> . \tag{9.94}$$

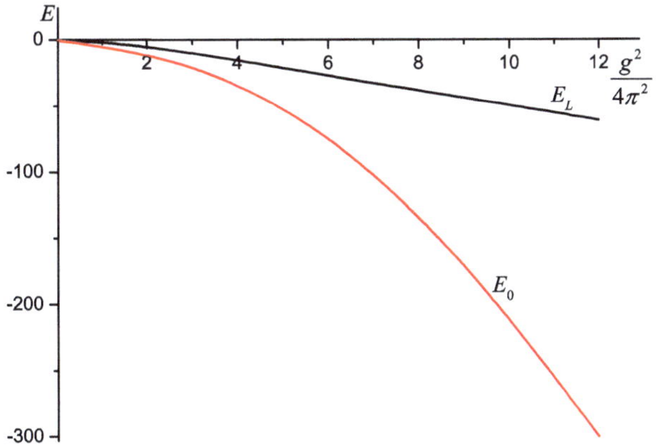

Fig. 9.5 The intrinsic energy E_L of "dressed" particle for the model with Hamiltonian (9.89); E_0 is the energy (9.93) of the state without translational symmetry

The normalizing integral has a following form (we assume here $\mathbf{P} = 0$):

$$N_0^2 = \int d\mathbf{R} \int d\mathbf{r}\, \phi(\mathbf{r})\phi(\mathbf{r}+\mathbf{R}) \exp\left(\sum_k u_k^2 (e^{-ikR} - 1)\right), \tag{9.95}$$

and contains the convergent integrals only. Figure 9.5 shows the coupling energy of "dressed" particle:

$$E_L(g) = \int d\mathbf{r} < \Psi_0^{(L)}(\mathbf{r})|\hat{H}|\Psi_0^{(L)}(\mathbf{r}) > .$$

The figure demonstrates that the intrinsic energy of "dressed" in phonon field particle is well-defined in OM and remains finite in the entire range of the coupling constant. The physical characteristics of "dressed" particle are determined by localized in a space wave function and therefore cannot be computed using standard perturbation theory, which implements the plane wave basis for zeroth approximation. A similar problem appears in the case of moving quantum particle in potential field, if the coupled localized states are found by canonic PT.

References

1. H. Fröhlich, Adv. Phys. **3**, 325 (1950)
2. Y.A. Firsov, *Polarons, Ed. (in Russian)* (Nauka, Moscow, 1975)
3. T.K. Mitra, A. Chatterjee, Mikhopadhyay, Phys. Rep. **153**, 91 (1987)
4. A.S. Alexandrov, N. Mott, *Polarons and Bipolarons* (World Scientific, Singapore, 1996)
5. J.T. Devreese, A.S. Alexandrov, Rep. Progr. Phys. **72**, 066501 (2009)
6. S.I. Pekar, Zh. Exper. Theor. Fiziki (in Russian) **16**, 341 (1946)

7. N.N. Bogoliubov, Matemat. Zh (in Russian) **2**, 3 (1950)
8. N.N. Bogolubov, *Selected Works, v.2, p.499 (in russian)* (Navukova Dumka, Kiev, 1972)
9. S.V. Tyablikov, Zh. Exper. Theor. Fiziki (in Russian) **21**, 377 (1952)
10. I.D. Feranchuk, L.I. Komarov, J. Phys. C Solid State Phys. **15**, 1965 (1982)
11. R.P. Feynman, Phys. Rev. **97**, 660 (1955)
12. A.S. Mishchenko, N.V. Prokof'ev, A. Sakamoto, B.V. Svistunov, Phys. Rev. B **62**, 6317 (2000)
13. I.D. Feranchuk, P.A. Khomyakov, Nonlinear Phenom. Complex Syst. **4**, 347 (2001)
14. I.D. Feranchuk, S.I. Fisher, L.I. Komarov, J. Phys. C Solid State Phys. **18**, 5083 (1985)
15. B. Gerlach, H. Löwen, Rev. Mod. Phys. **63**, 63 (1991)
16. G.D. Fillipis, V. Gataudella, V.M. Ramaglia, C.A. Perroni, D. Bersioux, Eur. Phys. J. B **36**, 65 (2003)
17. I.D. Feranchuk, L.I. Komarov, arxiv.org. **cond-mat.**, 0510510 (2005)
18. P.A. Khomyakov, *Development and Application of the Operator Method for Quantum Systems with many Degrees of Freedom (in Russian)*, PhD Thesis, Belarusian University, 2001
19. P.A. Khomyakov, Phys. Rev. B **63**, 153405 (2001)
20. M.H. Degani, G.A. Farias, Phys. Rev. B **42**, 11950 (1990)
21. F.M. Peeters, M.A. Smondyrev, Phys. Rev. B **43**, 4920 (1991)
22. T.D. Lee, F.M. Low, D. Pines, Phys. Rev. **90**, 297 (1953)
23. P.M. Morse, H. Feshbach, *Methods of Theoretical Physics* (McGraw-Hill, New York, 1953)
24. K.P. Huber, G. Gerzberg, *Molecular Spectra and Molecular Structure. vol. 4 Constants of Diatomic Molecules* (Van Rostrand Reinhold, New York, 1979)
25. J.T. Devreese, *Encyclopedia of Applied Physics*, vol. 14 (Wiley-VCH, New York, 1996), p. 383
26. R. Evrard, Phys. Lett. **14**, 295 (1965)
27. L.D. Landau, E.M. Lifshitz, *Quantum Mechanics* (Fizmatgiz, Moscow, 2004)
28. A.-T. Le, L.I. Komarov, J. Phys. C Cond. Matt. **11**, 11679 (1998)
29. N.N. Bogoliubov, Preprint JINR (in Russian) **R-1451**, 1 (1963)
30. N.W. Ashcroft, N.D. Mermin, *Solid State Physics* (Holt and Rinehart and Winsto, New York, 1976)
31. A.I. Akhiezer, V.B. Beresteckij, *Quantum Electrodynamics*, 3rd edn. (Nauka, Moscow, 1969)
32. L.D. Landau, E.M. Lifshitz, *Quantum Electrodynamics*, 1st edn. (Pergamon, New York, 1971)

Index

Acceleration of the convergence, 30
Accuracy of the approximation, 4
Adiabatic approximation, 21, 187
  for CQAO, 197
Adiabatic expansion, 2
Adiabatic terms of polaron, 353
Algorithm stability, 3
Analytical continuation, 29
Analytical solution(s), 233
Analytical solution of TF equation, 117
Annihilation operators, 9, 31
Approximate factorization, 96, 115
Approximate methods, 2
Approximating Hamiltonian, 3
Artificial parameter, 32
Asymptotic behavior, 16
Asymptotic curve for density, 312
Asymptotic expansion, 35
Asymptotic methods of averaging, 51
Asymptotic operator, 230, 231
Asymptotic series, 2, 4, 11
Atomic radius, 298
Average atom distance, 177
Averaged exponential operator, 93

Bogolyubov-Feynman inequality, 119
Bogolyubov-Tyablikov transformation, 333
Borel's summation, 21
Borel's transform, 22
Born-Oppenheimer, 21, 187
Bound states, 92
Bra-vector, 6
Brillouin-Wigner, 7

Canonical transformation, 31, 33, 89
Choice of the functions, 81
Classical component, 333
  of the field, 273
Classical law of motion, 57
Classical limit, 51
Classical nonlinear oscillations, 51
Classical periodic fields, 103
Classical trajectory, 55
Closed algebra of operators, 245
Coefficient of linear expansion, 177
Coherent state, 73
Combined parity, 272
Complete set, 6
Complex coordinates, 218
Complex eigenvalues, 65
Complex parameter, 67
Complex plane, 66
Complex-valued parameter, 68
Conditions of completeness, 38
Conditions of orthogonality, 38
Continued-fraction algorithm, 272
Continuous oscillator model, 294
Continuous variables for atom, 317
Convergence, 42
  radius, 10
  rate, 42, 224
  of successive approximations, 59
Conversion of OM, 48
Coordinate representation, 6
Coordinate rotation, 66
Coordinate scale transformation, 32
Correlational corrections, 187
Coulomb basis model, 303
Coulomb units, 289

I. Feranchuk et al., *Non-perturbative Description of Quantum Systems*,
Lecture Notes in Physics 894, DOI 10.1007/978-3-319-13006-4

Coulomb wave functions, 307
Coupled anharmonic oscillator, 5
Coupled harmonic oscillators, 19, 188
Coupled quantum anharmonic oscillators, 160, 192
Creation operators, 9, 31
Cumulant(s), 130
Cumulant expansion, 130
  for CQAO, 163

Decreasing sequence, 4
Degeneration, 7
  of CQAO, 194
Delocalized state, 350
Delta-expansion method, 33
Density functional theory, 287
Density matrix, 166
  of QAO, 172
Diagonal matrix elements, 39
Dimensionless parameter, 2
Dimensionless physical parameter, 6
Dipole transition probability, 46
Discrete spectrum, 31
Double well potential, 72

Effective charges model, 288, 323
Effective inverse temperature, 130, 131
Effective masses, 243
Effective parameter, 14
Eigenfunction(s), 2
Eigenfunction calculation, 45
Eigenvalues, 2
Electromagnetic interactions, 1
Electron density, 292, 317
  in atom, 311
Electron potential in atom, 311
Energy levels of CQAO, 162, 193
Ensemble of rotators, 139
Exact law of conservation, 333
Exchange energy, 299
  functional, 302
Exchange matrix elements, 319
Excitation number operator, 10
Exciton excited states, 240
Exciton ground state, 236
Exciton in a magnetic field, 216
Expansion parameter, 12
Exponential operator for PF, 129

Formal small parameter, 40
Four-dimensional harmonic oscillator, 252
Fröhlich Hamiltonian, 333
Free energy of QAO, 136
Frequency transformation operator, 46
Full orthonormal set, 32
Functional dependence, 35
Functional integrals, 1

Generalized QAO, 82
Geometric progression, 42
Ground state of QAO, 18

Hamiltonian, 2
Hartree-Fock method, 187, 288
He excited states, 208
He Rydberg states, 213
High polynomial potential, 82
High precision calculation, 224
Highly excited states, 51
Hill determinant, 104
Hydrogen atom in electric field, 259
Hydrogen atom in magnetic field, 266
Hydrogen-like atom, 254

Initial condition, 50
Integrals of motion, 72
Intersection point, 345
Intersections of the levels, 275
Inverse problem, 48
Inverse temperature, 129
Iteration scheme, 38, 43, 221, 265
Iteration scheme for integrals, 120

Jaynes-Cummings model, 271

Ket-vector, 6
Kustaanheimo-Stiefel, 251

Lattice model, 3
Lee-Low-Pines transformation, 340, 349
Levi-Civita transformation, 216, 217
Local density approximation, 287
Localized polaron, 350, 352

Madelung-Klechkowski rule, 320
Many-electron systems, 287
Mathieu equation, 95
Mathieu functions, 95
Mean-square radius, 294
Modification of iteration scheme, 78, 102

Morse potential, 92
Multi-dimensional systems, 187

Negatively charged exciton, 243
Non-adiabatic effects, 187
Non-degenerated, 7
Non-hermitian Hamiltonian, 66
Nonlinear algebraic equations, 39
Non-linear scaling, 5
Nonlinear oscillations, 51
Non-perturbative methods, 3
Non-polynomial Hamiltonian, 44
Non-polynomial potential, 109
Non-quadratic Hamiltonian, 9
Non-symmetric potential, 70, 88
Normal form, 28, 31
Normalization condition, 8, 38, 310
Normalization for periodic functions, 104
Number of the diagrams, 63
Numerical solution, 1

One-dimensional polaron, 340
One-electron basis, 288
One-electron potential, 291
One-particle CGF, 205
One-particle energies, 327
Operator form for density matrix, 168
Operator method (OM), 4, 5, 27
  for atom, 289
  convergence, 63
  convergence for CQAO, 200
  for density matrix, 167
  for 3-dimensional polaron, 349
  essential features, 59
  for exchange energy, 299
  features, 81
  for gamma function, 122
  for H atom in electric field, 264
  for H atom in fields, 251
  for H atom in magnetic field, 267
  for integrals, 117
  numerical solution, 224
  for quantum field model, 355
  for special functions, 123
  for strong coupling polaron, 332
  for sums, 117
  for two-dimensional exciton, 218
  zeroth-order approximation for TLS, 278
Operator of perturbation, 2
Operator representation of PF, 130
Optimal parameter choice, 30, 31, 41
Orthonormalization procedure, 324
Orthonormalized basis, 324
Oscillation period, 58
Oscillations of electron density, 316
Oscillator model of atom, 289

Pade approximant, 22
Parity operator, 73
Partial sum(s), 40
Partial summation, 24, 40
Particle-field Hamiltonian, 355
Particle number representation, 9
Partition function (PF), 129
  for asymmetric potential, 155
  for CQAO, 163
  for high order OM, 151
  for Morse oscillator, 158
  for polynomial potential, 136
  QAO, 133
  for quantum rotators, 136
Path integrals, 1
Periodic potential, 103
Periodic solutions, 51, 96
Permutation relation, 9
Perturbation theory, 2
Poisson equation, 291
Polaron, 331
  effective mass, 338
  excited states, 344, 345
  polarizability, 339
Potential of eighth order, 85
Power decrease of coefficients, 61
Principle of minimum sensitivity, 32
Projection operator, 73, 218
PT series, 7

Quantum anharmonic oscillator, 5
Quantum mechanics laws, 1
Quantum statistics, 5, 129
Quasi-classic approximation, 2, 15
Quasi-crossing, 348
Quasienergies, 103
Quasi-intersection, 352
Quasimomentum, 99
Quasi-stationary states, 65, 113, 261, 264
  of CQAO, 202

Rayleigh-Schrödinger, 7
Reconstruction of PT series, 11
Recurrence formulas, 40
Recurrence relations for TLS, 274
Recurrent equations, 201

Recurrent relations, 44
Regular perturbation theory, 203
Re-scaling the variable, 14
Ritz-Bubnov-Galerkin method, 17
Rotating wave approximation, 271
Rydberg states, 259

Schrödinger equation, 1, 6
Screened potential, 242
Second-order correction, 337
Second quantization representation, 28
Self-consistent field, 323
Self-consistent potential, 290, 291, 314
Self-similar approximation, 4
Self-similarity of potential, 313
Shell effects, 312
Shifted harmonic oscillator, 77
Single-mode quantum field, 252, 271
Singularities of energies, 20
Slater determinant, 326
Slater orbitals, 325
Small anharmonicity of CQAO, 195
Small parameter, 3
Specific representation, 39
Spin-orbital interaction, 328
Splitting of levels, 75
Spontaneous symmetry breaking, 72
Statistical operator, 166
Stirling's formula, 123
Strong anharmonicity, 35
Strong anharmonicity of CQAO, 196
Strong coupling, 2
  approximation, 13
  limit, 36
  polaron, 332
Strong magnetic field, 229
Successive approximations, 2
Sum over intermediate states, 205
Summation of asymptotic series, 21
Summation over transitions for He, 211
Supplementary parameters, 89

Thermal expansion, 179
Thermodynamical characteristics, 176
Thomas-Fermi equation, 115
Thomas-Fermi model, 287
Time evolution of TLS, 280
TLS in strong field, 280
Total atom energy in ECM, 326
Total energy of atom, 292, 298, 318
Transition length, 44
Transition matrix elements, 39
Trial density matrix, 169
Trial distribution function, 131
Trial function, 17
Trial state vector for PF, 130
Trial vector for PF, 130
Two-dimensional anharmonic oscillator, 216
Two-dimensional complex space, 253, 259
Two-dimensional donor state, 242
Two-dimensional exciton, 215
Two-electron atom, 203
Two-level system, 104, 252, 271
Two-particle CGF, 204, 207

Uniform convergence, 2
Uniformly available approximation, 4, 5
Unitary transformation, 96
  of frequency, 256
Universal algorithm, 5
Universal function, 297
  for Coulomb model, 318
  for potentials, 312
  for radii, 321
Universal integral equation, 295
Universality, 2
Unperturbed oscillator, 9
Upper border of the spectrum, 94

Vacuum of the excitations, 29
Variational method, 16
Variational parameters, 17
Virial relation, 322

Weak coupling limit, 34
Weak coupling polaron, 342
Weierstrass function, 125
Wentzel-Kramers-Brillouin approximation, 15

Zeroth approximation, 2, 32
  Hamiltonian, 82
  for PT, 131
Zeroth convergence radius, 11
Zone spectrum, 100

MIX
Papier aus verantwortungsvollen Quellen
Paper from responsible sources
FSC® C105338

If you have any concerns about our products,
you can contact us on
ProductSafety@springernature.com

In case Publisher is established outside the EU,
the EU authorized representative is:
Springer Nature Customer Service Center GmbH
Europaplatz 3, 69115 Heidelberg, Germany

Printed by Libri Plureos GmbH
in Hamburg, Germany